THE ROCK PHYSICS HANDBOOK

The science of rock physics addresses the relationship between the physical properties of rocks and geophysical observations. During the past three decades, scientists have discovered an increasing number of relations among properties that once appeared unrelated. *The Rock Physics Handbook* brings together much of the theory and data that form the foundation of rock physics with particular emphasis on the needs of interpreters of seismic data. In concise summaries, this book makes available a vast amount of laboratory and theoretical material in an accessible form.

Seventy-six articles cover a wide range of topics, including wave propagation, effective media, elasticity, pore fluid flow and diffusion, and AVO-AVOZ. The book also contains overviews of dispersion mechanisms, fluid substitution, and V_P–V_S relations. Useful empirical results derived from reservoir rocks, sediments, and granular media, as well as tables of mineral data and an atlas of reservoir rock properties complete the text.

This collection is presented in such a way that it can immediately be applied to solve real problems. Geophysics researchers and students as well as petroleum engineers and environmental geoscientists will value *The Rock Physics Handbook* for practical problem solving.

Gary Mavko received his Ph.D. in Geophysics from Stanford University in 1977. He is currently an Associate Professor (Research) at Stanford University. Tapan Mukerji received his Ph.D. in Geophysics from Stanford University in 1995. He is currently a Post-Doctoral Researcher at Stanford University. Jack Dvorkin received his Ph.D. in Mechanics and Mathematics from Moscow University in 1980. He is currently a Senior Research Scientist in the Department of Geophysics at Stanford University.

D1225730

STANFORD–CAMBRIDGE PROGRAM

The Stanford–Cambridge Program is an innovative publishing venture resulting from the collaboration between Cambridge University Press and Stanford University and its Press.

The Program provides a new international imprint for the teaching and communication of pure and applied sciences. Drawing on Stanford's eminent faculty and associated institutions, books within the Program reflect the high quality of teaching and research at Stanford University.

The Program includes textbooks at undergraduate and graduate levels and research monographs across a broad range of the sciences.

Cambridge University Press publishes and distributes books in the Stanford–Cambridge Program throughout the world.

THE ROCK PHYSICS HANDBOOK

TOOLS FOR SEISMIC ANALYSIS IN POROUS MEDIA

Gary Mavko
Stanford University

Tapan Mukerji
Stanford University

Jack Dvorkin
Stanford University

PUBLISHED BY THE PRESS SYNDICATE OF THE UNIVERSITY OF CAMBRIDGE
The Pitt Building, Trumpington Street, Cambridge, United Kingdom

CAMBRIDGE UNIVERSITY PRESS
The Edinburgh Building, Cambridge CB2 2RU, UK
40 West 20th Street, New York NY 10011–4211, USA
477 Williamstown Road, Port Melbourne, VIC 3207, Australia
Ruiz de Alarcón 13, 28014 Madrid, Spain
Dock House, The Waterfront, Cape Town 8001, South Africa

http://www.cambridge.org

First published 1998
First paperback edition 2003

Typeset in Franklin Gothic Demi and Times Roman

A catalogue record for this book is available from the British Library

Library of Congress Cataloguing-in-Publication Data

Mavko, Gary, 1949–
    The rock physics handbook : tools for seismic analysis in porous
media / Gary Mavko, Tapan Mukerji, Jack Dvorkin.
      p. cm.
    Includes index.
    ISBN 0 521 62068 6 hardback
    1. Rocks. 2. Geophysics. I. Mukerji, Tapan, 1965– .
II. Dvorkin, Jack. 1953– . III. Title.
QE431.6.P6M38 1998
552′.06–DC21   97-36653
CIP

ISBN 0 521 62068 6 hardback
ISBN 0 521 54344 4 paperback

CONTENTS

PART 10: APPENDIXES

PREFACE

I believe that the greatest practical impact of rock physics during the next few years will come not from new laboratory or theoretical discoveries but from making decades of existing results accessible to those who need them. This is not to say that rock physics is a finished subject. On the contrary, many aspects are still poorly understood and even controversial. Yet, particularly in applied fields, only a fraction of existing rock physics results are widely known. Our goal in preparing *The Rock Physics Handbook* was to help disseminate this information by distilling into a single volume part of the scattered and eclectic mass of knowledge that can already be useful for the rock physics interpretation of seismic data.

Our objective in preparing the handbook was to summarize in a convenient form many of the commonly needed theoretical and empirical relations of rock physics – those relations that we derive once every two years and then forget or find ourselves searching for in piles of articles, somewhere in that shelf of books, or on scraps of paper taped to the side of the filing cabinet.

Our approach was to present *results*, with a few of the key assumptions and limitations, and almost never any derivations. Our intention was to create a quick reference and not a textbook. Hence, we chose to encapsulate a broad range of topics rather than give in-depth coverage of a few. Even so, there are many topics that we have not addressed. We hope that the brevity of our discussions does not give the impression that application of any rock physics result to real rocks is free of pitfalls. We assume that the reader will be generally aware of the various topics, and, if not, we provide a few references to the more complete descriptions in books and journals.

The Rock Physics Handbook is presented in the form of seventy-six stand-alone articles. We wanted the user to be able to go directly to the topic of interest and to find all of the necessary information within a few pages without the need to refer to previous chapters, as in a conventional textbook. As a result, an occasional redundancy is evident in the explanatory text.

The handbook contains sections on wave propagation, AVO–AVOZ, effective media, elasticity and poroelasticity, and pore-fluid flow and diffusion, plus overviews of dispersion mechanisms, fluid substitution, and V_P–V_S relations. The book also presents empirical results derived from reservoir rocks, sediments, and granular media, as well as tables of mineral data and an atlas of reservoir rock properties. The emphasis throughout is primarily on seismic properties. We have also included commonly used models and relations for electrical and dielectric rock properties.

We believe that this book is complementary to other works. For in-depth discussions of specific rock physics topics, we recommend *Acoustics of Porous Media* by Bourbié, Coussy, and Zinszner; *Introduction to the Physics of Rocks* by Guéguen and Palciauskas; *Rock Physics and Phase Relations* edited by Ahrens; and *Offset Dependent Reflectivity – Theory and Practice of AVO Analysis* edited by Castagna and Backus. For excellent collections and discussions of classic rock physics papers we recommend *Seismic and Acoustic Velocities in Reservoir Rocks*, Volumes 1 and 2 edited by Wang and Nur, *Elastic Properties and Equations of State* edited by Shankland and Bass, and *Seismic Wave Attenuation* by Toksöz and Johnston.

We wish to thank the faculty, students, and industrial affiliates of the Stanford Rock Physics and Borehole Geophysics (SRB) project for many valuable comments and insights. We found discussions with Zhijing Wang, Thierry Cadoret, Ivar Brevik, Sue Raikes, Sverre Strandenes, Mike Batzle, and Jim Berryman particularly useful. Li Teng contributed the chapter on anisotropic AVOZ, and Ran Bachrach contributed to the chapter on dielectric properties. Ranie Lynds helped with the graphics and did a marvelous job of proofing and editing. Special thanks are extended to Barbara Mavko for many useful comments on content and style. And as always, we are indebted to Amos Nur, whose work, past and present, has helped to make the field of rock physics what it is today.

We hope you will find this handbook useful.

Gary Mavko

PART 1

BASIC TOOLS

1.1 THE FOURIER TRANSFORM

SYNOPSIS

The **Fourier transform** of $f(x)$ is defined as

$$F(s) = \int_{-\infty}^{\infty} f(x)e^{-i2\pi xs}\, dx$$

The inverse Fourier transform is given by

$$f(x) = \int_{-\infty}^{\infty} F(s)e^{+i2\pi xs}\, ds$$

EVENNESS AND ODDNESS

A function $E(x)$ is *even* if $E(x) = E(-x)$. A function $O(x)$ is *odd* if $O(x) = -O(-x)$.

The Fourier transform has the following properties for even and odd functions:

- *Even Functions*. The Fourier transform of an even function is even. A *real even* function transforms to a *real even* function. An *imaginary even* function transforms to an *imaginary even* function.

- *Odd Functions.* The Fourier transform of an odd function is odd. A *real odd* function transforms to an *imaginary odd* function. An *imaginary odd* function transforms to a *real odd* function (i.e., the "realness" flips when the Fourier transform of an odd function is taken).

$$
\begin{aligned}
\text{real even (RE)} &\rightarrow \text{real even (RE)} \\
\text{imaginary even (IE)} &\rightarrow \text{imaginary even (IE)} \\
\text{real odd (RO)} &\rightarrow \text{imaginary odd (IO)} \\
\text{imaginary odd (IO)} &\rightarrow \text{real odd (RO)}
\end{aligned}
$$

Any function can be expressed in terms of its even and odd parts:

$$f(x) = E(x) + O(x)$$

where

$$E(x) = \frac{1}{2}[f(x) + f(-x)]$$

$$O(x) = \frac{1}{2}[f(x) - f(-x)]$$

Then, for an arbitrary complex function we can summarize these relations as (Bracewell, 1965)

$$f(x) = \text{re}(x) + i\,\text{ie}(x) + \text{ro}(x) + i\,\text{io}(x)$$

$$F(s) = \text{RE}(s) + i\,\text{IE}(s) + \text{RO}(s) + i\,\text{IO}(s)$$

As a consequence, a real function $f(x)$ has a Fourier transform that is *hermitian*, $F(s) = F^*(-s)$, where * refers to the complex conjugate.

For a more general complex function, $f(x)$, we can tabulate some additional properties (Bracewell, 1965):

$$
\begin{aligned}
f(x) &\Leftrightarrow F(s) \\
f^*(x) &\Leftrightarrow F^*(-s) \\
f^*(-x) &\Leftrightarrow F^*(s) \\
f(-x) &\Leftrightarrow F(-s) \\
2\,\text{Re}\,f(x) &\Leftrightarrow F(s) + F^*(-s) \\
2\,\text{Im}\,f(x) &\Leftrightarrow F(s) - F^*(-s) \\
f(x) + f^*(-x) &\Leftrightarrow 2\,\text{Re}\,F(s) \\
f(x) - f^*(-x) &\Leftrightarrow 2\,\text{Im}\,F(s)
\end{aligned}
$$

The **convolution** of two functions $f(x)$ and $g(x)$ is

$$f(x) * g(x) = \int_{-\infty}^{+\infty} f(z)g(x-z)\,dz = \int_{-\infty}^{+\infty} f(x-z)g(z)\,dz$$

CONVOLUTION THEOREM

If $f(x)$ has Fourier transform $F(s)$, and $g(x)$ has Fourier transform $G(s)$, then the Fourier transform of the convolution $f(x) * g(x)$ is the product $F(s)G(s)$.

The **cross-correlation** of two functions $f(x)$ and $g(x)$ is

$$f^*(x) \star g(x) = \int_{-\infty}^{+\infty} f^*(z-x)g(z)\,dz = \int_{-\infty}^{+\infty} f^*(z)g(z+x)\,dz$$

where f^* refers to the complex conjugate of f. When the two functions are the same, $f^*(x) \star f(x)$ is called the **autocorrelation** of $f(x)$.

ENERGY SPECTRUM

The modulus squared of the Fourier transform $|F(s)|^2 = F(s)F^*(s)$ is sometimes called the **energy spectrum** or simply the **spectrum**.

If $f(x)$ has Fourier transform $F(s)$, then the autocorrelation of $f(x)$ has Fourier transform $|F(s)|^2$.

PHASE SPECTRUM

The Fourier transform $F(s)$ is most generally a complex function, which can be written as

$$F(s) = |F|e^{i\varphi} = \operatorname{Re} F(s) + i \operatorname{Im} F(s)$$

where $|F|$ is the modulus and φ is the **phase**, given by

$$\varphi = \tan^{-1}(\operatorname{Im} F(s)/\operatorname{Re} F(s))$$

$\varphi(s)$ is sometimes also called the **phase spectrum**.

Obviously, both the modulus and phase must be known to completely specify the Fourier transform $F(s)$ or its transform pair in the other domain $f(x)$. Consequently, an infinite number of functions $f(x) \Leftrightarrow F(s)$ are consistent with a given spectrum $|F(s)|^2$.

The **zero phase** equivalent function (or zero phase equivalent wavelet) corresponding to a given spectrum is

$$F(s) = |F(s)|$$

$$f(x) = \int_{-\infty}^{\infty} |F(s)|e^{+i2\pi xs}\,ds$$

which implies that $F(s)$ is real and $f(x)$ is hermitian. In the case of zero phase *real* wavelets, then, both $F(s)$ and $f(x)$ are real even functions.

The **minimum phase** equivalent function or wavelet corresponding to a spectrum is the unique one that is both *causal* and *invertible*. A simple way to compute the minimum phase equivalent of a spectrum $|F(s)|^2$ is to perform the following steps (Claerbout, 1992):

1) Take the logarithm, $B(s) = \ln|F(s)|$.
2) Take the Fourier transform, $B(s) \Rightarrow b(x)$.
3) Multiply $b(x)$ by zero for $x < 0$ and by 2 for $x > 0$. If done numerically, leave the values of b at zero and the Nyquist unchanged.
4) Transform back, giving $B(s) + i\varphi(s)$, where φ is the desired phase spectrum.
5) Take the complex exponential to yield the minimum phase function: $F_{mp}(s) = \exp[B(s) + i\varphi(s)] = |F(s)|e^{i\varphi(s)}$.
6) The causal minimum phase wavelet is the Fourier transform of $F_{mp}(s) \Rightarrow f_{mp}(x)$.

Another way of saying this is that the phase spectrum of the minimum phase equivalent function is the Hilbert transform (see Section 1.2 on the Hilbert transform) of the log of the energy spectrum.

SAMPLING THEOREM

A function $f(x)$ is said to be *band-limited* if its Fourier transform is nonzero only within a finite range of frequencies, $|s| < s_c$, where s_c is sometimes called the *cutoff frequency*. The function $f(x)$ is fully specified if sampled at equal spacing not to exceed $\Delta x = 1/(2s_c)$. Equivalently, a time series sampled at interval Δt adequately describes the frequency components out to the *Nyquist frequency* $f_N = 1/(2\Delta t)$. The numerical process of recovering the intermediate points between samples is to convolve with the *sinc function*:

$$2s_c \, \text{sinc}(2s_c x) = 2s_c \sin(\pi 2s_c x)/(\pi 2s_c x)$$

where

$$\text{sinc}(x) \equiv \frac{\sin(\pi x)}{\pi x}$$

which has the properties:

$$\text{sinc}(0) = 1$$

$$\text{sinc}(n) = 0 \quad n = \text{nonzero integer}$$

The Fourier transform of $\text{sinc}(x)$ is the boxcar function $\Pi(s)$:

$$\Pi(s) = \begin{cases} 0 & |s| > 1/2 \\ 1/2 & |s| = 1/2 \\ 1 & |s| < 1/2 \end{cases}$$

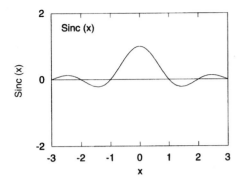 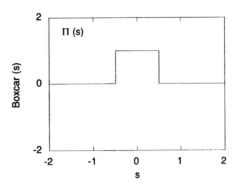

Figure 1.1.1

One can see from the convolution and similarity theorems below that convolving with $2s_c \, \mathrm{sinc}(2s_c x)$ is equivalent to multiplying by $\Pi(s/2s_c)$ in the frequency domain (i.e., zeroing out all frequencies $|s| > s_c$ and passing all frequencies $|s| < s_c$).

NUMERICAL DETAILS

Consider a band-limited function $g(t)$ sampled at N points at equal intervals: $g(0), g(\Delta t), g(2\Delta t), \ldots g((N-1)\Delta t)$. A typical fast Fourier transform (FFT) routine will yield N equally spaced values of the Fourier transform, $G(f)$, often arranged as

$$
1 \quad 2 \quad 3 \quad \cdots \quad \left(\frac{N}{2}+1\right) \quad \left(\frac{N}{2}+2\right) \quad \cdots \quad (N-1) \quad N
$$

$$
G(0) \;\; G(\Delta f) \;\; G(2\Delta f) \;\cdots\; G(\pm f_N) \;\; G(-f_N+\Delta f) \;\cdots\; G(-2\Delta f) \;\; G(-\Delta f)
$$

$$
\begin{aligned}
\text{time domain sample rate} \quad &\Delta t \\
\text{Nyquist frequency} \quad f_N &= 1/(2\Delta t) \\
\text{frequency domain sample rate} \quad \Delta f &= 1/(N\Delta t)
\end{aligned}
$$

Note that, because of "wraparound," the sample at $(N/2+1)$ represents both $\pm f_N$.

SPECTRAL ESTIMATION AND WINDOWING

It is often desirable in rock physics and seismic analysis to estimate the spectrum of a wavelet or seismic trace. The most common, easiest, and, in some ways, the worst way is simply to chop out a piece of the data, take the Fourier transform, and find its magnitude. The problem is related to sample length. If the true data function is $f(t)$, a small sample of the data can be thought of as

$$
f_{\text{sample}}(t) = \begin{cases} f(t), & a \le t \le b \\ 0, & \text{elsewhere} \end{cases}
$$

or

$$f_{\text{sample}}(t) = f(t)\Pi\left(\frac{t - \frac{a+b}{2}}{b - a}\right)$$

where $\Pi(t)$ is the boxcar function discussed above. Taking the Fourier transform of the data sample gives

$$F_{\text{sample}}(s) = F(s) * \left[|b - a| \operatorname{sinc}((b - a)s)e^{-i\pi(a+b)s}\right]$$

More generally, we can "window" the sample with some other function $w(t)$:

$$f_{\text{sample}}(t) = f(t)w(t)$$

yielding

$$F(s) * W(s)$$

Thus, the estimated spectrum can be highly contaminated by the Fourier transform of the window. This can be particularly severe in the analysis of ultrasonic waveforms in the laboratory, where often only the first 1–$1\frac{1}{2}$ cycles are windowed out. The solution to the problem is not easy, and there is an extensive literature (e.g., Jenkins and Watts, 1968; Marple, 1987) on spectral estimation. Our advice is to be aware of the artifacts of windowing and to experiment to determine how sensitive the results are, such as spectral ratio or phase velocity, to the choice of window size and shape.

FOURIER TRANSFORM THEOREMS

Table 1.1.1 summarizes some useful theorems (Bracewell, 1965). If $f(x)$ has Fourier transform $F(s)$, and $g(x)$ has Fourier transform $G(s)$, then the Fourier transform pairs in the x-domain and the s-domain are as follows:

TABLE 1.1.1

Theorem	x-domain		s-domain
Similarity	$f(ax)$	$\Leftrightarrow$	$\dfrac{1}{|a|}F\left(\dfrac{s}{a}\right)$
Addition	$f(x) + g(x)$	$\Leftrightarrow$	$F(s) + G(s)$
Shift	$f(x - a)$	$\Leftrightarrow$	$e^{-i2\pi as}F(s)$
Modulation	$f(x)\cos\omega x$	$\Leftrightarrow$	$\dfrac{1}{2}F\left(s - \dfrac{\omega}{2\pi}\right) + \dfrac{1}{2}F\left(s + \dfrac{\omega}{2\pi}\right)$
Convolution	$f(x) * g(x)$	$\Leftrightarrow$	$F(s)G(s)$
Autocorrelation	$f(x) * f^*(-x)$	$\Leftrightarrow$	$|F(s)|^2$
Derivative	$f'(x)$	$\Leftrightarrow$	$i2\pi s F(s)$

TABLE 1.1.2. Some additional theorems.

Derivative of convolution	$\dfrac{d}{dx}[f(x) * g(x)] = f'(x) * g(x) = f(x) * g'(x)$
Rayleigh	$\displaystyle\int_{-\infty}^{\infty} |f(x)|^2 \, dx = \int_{-\infty}^{\infty} |F(s)|^2 \, ds$
Power	$\displaystyle\int_{-\infty}^{\infty} f(x)g^*(x) \, dx = \int_{-\infty}^{\infty} F(s)G^*(s) \, ds$
(f and g real)	$\displaystyle\int_{-\infty}^{\infty} f(x)g(-x) \, dx = \int_{-\infty}^{\infty} F(s)G(s) \, ds$

TABLE 1.1.3. Some Fourier transform pairs.

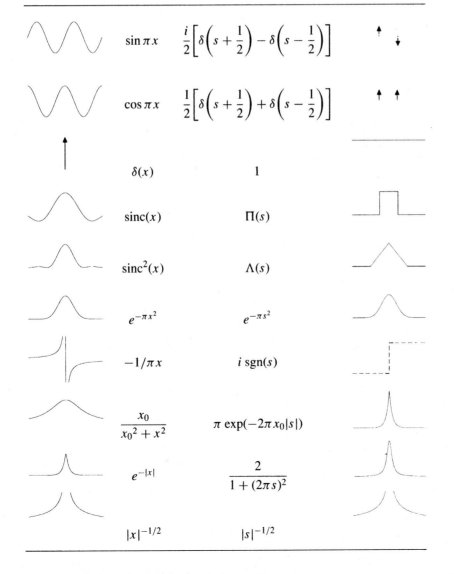

	$\sin \pi x$	$\dfrac{i}{2}\left[\delta\left(s + \dfrac{1}{2}\right) - \delta\left(s - \dfrac{1}{2}\right)\right]$
	$\cos \pi x$	$\dfrac{1}{2}\left[\delta\left(s + \dfrac{1}{2}\right) + \delta\left(s - \dfrac{1}{2}\right)\right]$
	$\delta(x)$	1
	$\mathrm{sinc}(x)$	$\Pi(s)$
	$\mathrm{sinc}^2(x)$	$\Lambda(s)$
	$e^{-\pi x^2}$	$e^{-\pi s^2}$
	$-1/\pi x$	$i \, \mathrm{sgn}(s)$
	$\dfrac{x_0}{x_0^2 + x^2}$	$\pi \exp(-2\pi x_0|s|)$
	$e^{-|x|}$	$\dfrac{2}{1 + (2\pi s)^2}$
	$|x|^{-1/2}$	$|s|^{-1/2}$

1.2 THE HILBERT TRANSFORM AND ANALYTIC SIGNAL

SYNOPSIS

The **Hilbert transform** of $f(x)$ is defined as

$$F_{Hi}(x) = \frac{1}{\pi} \int_{-\infty}^{\infty} \frac{f(x')\,dx'}{x' - x}$$

which can be expressed as a convolution of $f(x)$ with $(-1/\pi x)$ by

$$F_{Hi} = \frac{-1}{\pi x} * f(x)$$

The Fourier transform of $(-1/\pi x)$ is $(i\,\mathrm{sgn}(s))$, that is, $+i$ for positive s and $-i$ for negative s. Hence, applying the Hilbert transform keeps the Fourier amplitudes or spectrum the same but changes the phase. Under the Hilbert transform, $\sin(kx)$ gets converted to $\cos(kx)$, and $\cos(kx)$ gets converted to $-\sin(kx)$. Similarly, the Hilbert transforms of even functions are odd functions and vice versa.

The inverse of the Hilbert transform is itself the Hilbert transform with a change of sign:

$$f(x) = -\frac{1}{\pi} \int_{-\infty}^{\infty} \frac{F_{Hi}(x')\,dx'}{x' - x}$$

or

$$f(x) = -\left(\frac{-1}{\pi x}\right) * F_{Hi}$$

The **analytic signal** associated with a real function, $f(t)$, is the complex function

$$S(t) = f(t) - i\,F_{Hi}(t)$$

As discussed below, the Fourier transform of $S(t)$ is zero for negative frequencies.

The **instantaneous envelope** of the analytic signal is

$$E(t) = \sqrt{f^2(t) + F_{Hi}^2(t)}$$

The **instantaneous phase** of the analytic signal is

$$\varphi(t) = \arctan[-F_{Hi}(t)/f(t)]$$
$$= \mathrm{Im}[\ln(S(t))]$$

The **instantaneous frequency** of the analytic signal is

$$\omega = \frac{d\varphi}{dt} = \mathrm{Im}\left[\frac{d}{dt}\ln(S)\right] = \mathrm{Im}\left(\frac{1}{S}\frac{dS}{dt}\right)$$

Claerbout (1992) has suggested that ω can be numerically more stable if the denominator is rationalized and the functions are locally smoothed, as in the following equation:

$$\bar{\omega} = \mathrm{Im}\left[\frac{\langle S^*(t)\frac{dS(t)}{dt}\rangle}{\langle S^*(t)S(t)\rangle}\right]$$

where $\langle\cdot\rangle$ indicates some form of running average or smoothing.

CAUSALITY

The **impulse response**, $I(t)$, of a real physical system must be causal, that is,

$$I(t) = 0, \quad \text{for } t < 0$$

The Fourier transform $T(f)$ of the impulse response of a causal system is sometimes called the **Transfer Function**:

$$T(f) = \int_{-\infty}^{\infty} I(t)e^{-i2\pi ft}\,dt$$

$T(f)$ must have the property that the real and imaginary parts are Hilbert transform pairs. That is, $T(f)$ will have the form

$$T(f) = G(f) + iB(f)$$

where $B(f)$ is the Hilbert transform of $G(f)$.

$$B(f) = \frac{1}{\pi}\int_{-\infty}^{\infty}\frac{G(f')\,df'}{f'-f}$$

$$G(f) = -\frac{1}{\pi}\int_{-\infty}^{\infty}\frac{B(f')\,df'}{f'-f}$$

Similarly, if we reverse the domains, an analytic signal of the form

$$S(t) = f(t) - iF_{\mathrm{Hi}}(t)$$

must have a Fourier transform that is zero for negative frequencies. In fact, one convenient way to implement the Hilbert transform of a real function is by performing the following steps:

1) Take the Fourier transform.
2) Multiply the Fourier transform by zero for $f < 0$.
3) Multiply the Fourier transform by 2 for $f > 0$.
4) If done numerically, leave the samples at $f = 0$ and the Nyquist unchanged.
5) Take the inverse Fourier transform.

The imaginary part of the result will be the negative Hilbert transform of the real part.

1.3 STATISTICS AND LINEAR REGRESSION

SYNOPSIS

The **sample mean**, m, of a set of n data points, x_i, is the arithmetic average of the data values:

$$m = \frac{1}{n} \sum_{i=1}^{n} x_i$$

The **median** is the midpoint of the observed values if they are arranged in increasing order. The **sample variance**, σ^2, is the average squared difference of the observed values from the mean:

$$\sigma^2 = \frac{1}{n} \sum_{i=1}^{n} (x_i - m)^2$$

(An unbiased estimate of the **population variance** is often found by dividing the sum given above by $(n - 1)$ instead of by n.)

The **standard deviation**, σ, is the square root of the variance. The **mean deviation**, α, is

$$\alpha = \frac{1}{n} \sum_{i=1}^{n} |x_i - m|$$

REGRESSION

When trying to determine whether two different data variables, x and y, are related, we often estimate the **correlation coefficient**, ρ, given by (e.g., Young, 1962)

$$\rho = \frac{\frac{1}{n} \sum_{i=1}^{n} (x_i - m_x)(y_i - m_y)}{\sigma_x \sigma_y}, \quad \text{where } |\rho| \leq 1$$

where σ_x and σ_y are the standard deviations of the two distributions and m_x and m_y are their means. The correlation coefficient gives a measure of how close the points come to falling along a straight line. $|\rho| = 1$ if the points lie perfectly along a line, and $|\rho| < 1$ if there is scatter about the line. The numerator of this expression is the **covariance**, C_{xy}, which is defined as

$$C_{xy} = \frac{1}{n}\sum_{i=1}^{n}(x_i - m_x)(y_i - m_y)$$

It is important to remember that the correlation coefficient is a measure of the *linear* relation between x and y. If they are related in a nonlinear way, the correlation coefficient will be misleadingly small.

The simplest recipe for estimating the linear relation between two variables, x and y, is **linear regression**, in which we assume a relation of the form:

$$y = ax + b$$

The coefficients that provide the best fit to the measured values of y, in the least-squares sense, are

$$a = \rho\frac{\sigma_y}{\sigma_x} \qquad b = m_y - am_x$$

More explicitly,

$$a = \frac{n\sum x_i y_i - \left(\sum x_i\right)\left(\sum y_i\right)}{n\sum x_i^2 - \left(\sum x_i\right)^2} \qquad \text{slope}$$

$$b = \frac{\left(\sum y_i\right)\left(\sum x_i^2\right) - \left(\sum x_i y_i\right)\left(\sum x_i\right)}{n\sum x_i^2 - \left(\sum x_i\right)^2} \qquad \text{intercept}$$

The scatter or variation of y-values around the regression line can be described by the sum of the squared errors as

$$E^2 = \sum_{i=1}^{n}(y_i - \hat{y}_i)^2$$

where $\hat{y}_i$ is the value predicted from the regression line. This can be expressed as a variance around the regression line as

$$\hat{\sigma}_y^{\,2} = \frac{1}{n}\sum_{i=1}^{n}(y_i - \hat{y}_i)^2$$

The square of the correlation coefficient ρ is the **coefficient of determination**, often denoted by r^2, which is a measure of the regression variance relative to the total variance in the variable y expressed as

$$r^2 = \rho^2 = 1 - \frac{\text{variance of } y \text{ around the linear regression}}{\text{total variance of } y}$$

$$= 1 - \frac{\sum_{i=1}^{n}(y_i - \hat{y}_i)^2}{\sum_{i=1}^{n}(y_i - m_y)^2} = 1 - \frac{\hat{\sigma}_y^{\,2}}{\sigma_y^{\,2}}$$

The inverse relation is

$$\hat{\sigma}_y{}^2 = \sigma_y{}^2(1 - r^2)$$

Often, when doing a linear regression the choice of dependent and independent variables is arbitrary. The form above treats x as independent and exact and assigns errors to y. It often makes just as much sense to reverse their roles, and we can find a regression of the form

$$x = a'y + b'$$

Generally $a \neq 1/a'$ unless the data are perfectly correlated. In fact the correlation coefficient, ρ, can be written as $\rho = \sqrt{aa'}$.

The coefficients of the linear regression among three variables of the form

$$z = a + bx + cy$$

are given by

$$b = \frac{C_{xz}C_{yy} - C_{xy}C_{yz}}{C_{xx}C_{yy} - C_{xy}{}^2}$$

$$c = \frac{C_{xx}C_{yz} - C_{xy}C_{xz}}{C_{xx}C_{yy} - C_{xy}{}^2}$$

$$a = m_z - m_x b - m_y c$$

DISTRIBUTIONS

The **binomial distribution** gives the probability of n successes in N independent trials if p is the probability of success in any one trial. The binomial distribution is given by

$$f_{N,p}(n) = \binom{N}{n} p^n (1 - p)^{N-n}$$

The mean of the binomial distribution is given by

$$m_b = Np$$

and the variance of the binomial distribution is given by

$$\sigma_b{}^2 = Np(1 - p)$$

The **Poisson distribution** is the limit of the binomial distribution as $N \to \infty$ and $p \to 0$ so that $a = Np$ remains finite. The Poisson distribution is given by

$$f_a(n) = \frac{a^n e^{-a}}{n!}$$

The mean of the Poisson distribution is given by

$$m_p = a$$

and the variance of the Poisson distribution is given by

$$\sigma_p^2 = a$$

The **uniform** distribution is given by

$$f(x) = \begin{cases} \frac{1}{b-a}, & a \le x \le b \\ 0, & \text{elsewhere} \end{cases}$$

The mean of the uniform distribution is

$$m = \frac{(a+b)}{2}$$

and the standard deviation of the uniform distribution is

$$\sigma = \frac{|b-a|}{\sqrt{12}}$$

The **Gaussian** or **normal** distribution is given by

$$f(x) = \frac{1}{\sqrt{2\pi}\sigma} e^{-(x-m)^2/2\sigma^2}$$

where σ is the standard deviation and m is the mean. The mean deviation for the Gaussian distribution is

$$\alpha = \sigma\sqrt{\frac{2}{\pi}}$$

When n measurements are made of n quantities, the situation is described by the n-dimensional **multivariate Gaussian** pdf:

$$f_n(x) = \frac{1}{(2\pi)^{n/2}|\mathbf{C}|^{1/2}} \exp\left[-\frac{1}{2}(x-m)^T \mathbf{C}^{-1}(x-m)\right]$$

where $x^T = (x_1, x_2, \ldots, x_n)$ is the vector of observations, $m^T = (m_1, m_2, \ldots, m_n)$ is the vector of means of the individual distributions, and $\mathbf{C}$ is the covariance matrix:

$$\mathbf{C} = [C_{ij}]$$

where the individual covariances, C_{ij}, are as defined above. Notice that this reduces to the single variable normal distribution, when $n = 1$.

When the natural logarithm of a variable, $x = \ln(y)$, is normally distributed it belongs to a **lognormal distribution** expressed as

$$f(y) = \frac{1}{\sqrt{2\pi}\,\beta y} e^{-\frac{1}{2}\left(\frac{\ln(y)-\alpha}{\beta}\right)^2}$$

where α is the mean, and β^2 is the variance. The relations among the arithmetic and logarithmic parameters are

$$m = e^{\alpha + \beta^2/2} \qquad\qquad \alpha = \ln(m) - \beta^2/2$$

$$\sigma^2 = m^2\left(e^{\beta^2} - 1\right) \qquad \beta^2 = \ln\left(1 + \frac{\sigma^2}{m^2}\right)$$

1.4 COORDINATE TRANSFORMATIONS

SYNOPSIS

It is often necessary to transform vector and tensor quantities in one coordinate system to another more suited to a particular problem. Consider two right-hand rectangular Cartesian coordinates (x, y, z) and (x', y', z') with the same origin but with their axes rotated arbitrarily with respect to each other. The relative orientation of the two sets of axes is given by the direction cosines β_{ij} defined as the cosine of the angle between the i'-axis and the j-axis. The variables β_{ij} constitute the elements of a 3×3 rotation matrix $[\beta]$. Thus, β_{23} is the cosine of the angle between the 2-axis of the primed coordinate system and the 3-axis of the unprimed coordinate system.

The general transformation law for tensors is

$$M'_{ABCD\ldots} = \beta_{Aa}\beta_{Bb}\beta_{Cc}\beta_{Dd}\ldots M_{abcd\ldots}$$

where summation over repeated indices is implied. The left-hand subscripts $(A, B, C, D \ldots)$ on the β's match the subscripts of the transformed tensor $\mathbf{M}'$ on the left, and the right-hand subscripts $(a, b, c, d \ldots)$ match the subscripts of $\mathbf{M}$ on the right. Thus vectors, which are first-order tensors, transform as

$$v'_i = \beta_{ij} v_j$$

or, in matrix notation, as

$$\begin{pmatrix} v'_1 \\ v'_2 \\ v'_3 \end{pmatrix} = \begin{pmatrix} \beta_{11} & \beta_{12} & \beta_{13} \\ \beta_{21} & \beta_{22} & \beta_{23} \\ \beta_{31} & \beta_{32} & \beta_{33} \end{pmatrix} \begin{pmatrix} v_1 \\ v_2 \\ v_3 \end{pmatrix}$$

whereas second-order tensors, such as stresses and strains, obey

$$\sigma'_{ij} = \beta_{ik}\beta_{jl}\sigma_{kl}$$

or

$$[\sigma'] = [\beta][\sigma][\beta]^T$$

in matrix notation. Elastic stiffnesses and compliances are in general fourth-order tensors and hence transform according to

$$c'_{ijkl} = \beta_{ip}\beta_{jq}\beta_{kr}\beta_{ls}c_{pqrs}$$

Often c_{ijkl} and s_{ijkl} are expressed as 6×6 matrices C_{IJ} and S_{IJ} using the abbreviated 2-index notation, as defined in Section 2.2 on anisotropic elasticity. In this case, the usual tensor transformation law is no longer valid, and the change of coordinates is more efficiently performed with 6×6 **Bond transformation matrices M** and **N**, as explained below (Auld, 1990).

$$[C'] = [M][C][M]^T$$
$$[S'] = [N][S][N]^T$$

The elements of the 6×6 **M** and **N** matrices are given in terms of the direction cosines as follows:

$$\mathbf{M} = \begin{bmatrix} \beta_{11}^2 & \beta_{12}^2 & \beta_{13}^2 & 2\beta_{12}\beta_{13} & 2\beta_{13}\beta_{11} & 2\beta_{11}\beta_{12} \\ \beta_{21}^2 & \beta_{22}^2 & \beta_{23}^2 & 2\beta_{22}\beta_{23} & 2\beta_{23}\beta_{21} & 2\beta_{21}\beta_{22} \\ \beta_{31}^2 & \beta_{32}^2 & \beta_{33}^2 & 2\beta_{32}\beta_{33} & 2\beta_{33}\beta_{31} & 2\beta_{31}\beta_{32} \\ \beta_{21}\beta_{31} & \beta_{22}\beta_{32} & \beta_{23}\beta_{33} & \beta_{22}\beta_{33}+\beta_{23}\beta_{32} & \beta_{21}\beta_{33}+\beta_{23}\beta_{31} & \beta_{22}\beta_{31}+\beta_{21}\beta_{32} \\ \beta_{31}\beta_{11} & \beta_{32}\beta_{12} & \beta_{33}\beta_{13} & \beta_{12}\beta_{33}+\beta_{13}\beta_{32} & \beta_{11}\beta_{33}+\beta_{13}\beta_{31} & \beta_{11}\beta_{32}+\beta_{12}\beta_{31} \\ \beta_{11}\beta_{21} & \beta_{12}\beta_{22} & \beta_{13}\beta_{23} & \beta_{22}\beta_{13}+\beta_{12}\beta_{23} & \beta_{11}\beta_{23}+\beta_{13}\beta_{21} & \beta_{22}\beta_{11}+\beta_{12}\beta_{21} \end{bmatrix}$$

and

$$\mathbf{N} = \begin{bmatrix} \beta_{11}^2 & \beta_{12}^2 & \beta_{13}^2 & \beta_{12}\beta_{13} & \beta_{13}\beta_{11} & \beta_{11}\beta_{12} \\ \beta_{21}^2 & \beta_{22}^2 & \beta_{23}^2 & \beta_{22}\beta_{23} & \beta_{23}\beta_{21} & \beta_{21}\beta_{22} \\ \beta_{31}^2 & \beta_{32}^2 & \beta_{33}^2 & \beta_{32}\beta_{33} & \beta_{33}\beta_{31} & \beta_{31}\beta_{32} \\ 2\beta_{21}\beta_{31} & 2\beta_{22}\beta_{32} & 2\beta_{23}\beta_{33} & \beta_{22}\beta_{33}+\beta_{23}\beta_{32} & \beta_{21}\beta_{33}+\beta_{23}\beta_{31} & \beta_{22}\beta_{31}+\beta_{21}\beta_{32} \\ 2\beta_{31}\beta_{11} & 2\beta_{32}\beta_{12} & 2\beta_{33}\beta_{13} & \beta_{12}\beta_{33}+\beta_{13}\beta_{32} & \beta_{11}\beta_{33}+\beta_{13}\beta_{31} & \beta_{11}\beta_{32}+\beta_{12}\beta_{31} \\ 2\beta_{11}\beta_{21} & 2\beta_{12}\beta_{22} & 2\beta_{13}\beta_{23} & \beta_{22}\beta_{13}+\beta_{12}\beta_{23} & \beta_{11}\beta_{23}+\beta_{13}\beta_{21} & \beta_{22}\beta_{11}+\beta_{12}\beta_{21} \end{bmatrix}$$

The advantage of the Bond method for transforming stiffnesses and compliances is that it can be applied directly to the elastic constants given in 2-index notation, as they almost always are in handbooks and tables.

ASSUMPTIONS AND LIMITATIONS

Coordinate transformations presuppose right-hand rectangular coordinate systems.

PART 2

ELASTICITY AND HOOKE'S LAW

2.1 ELASTIC MODULI – ISOTROPIC FORM OF HOOKE'S LAW

SYNOPSIS

In an isotropic, linear elastic material, the stress and strain are related by **Hooke's law** as follows (e.g. Timoshenko and Goodier, 1934):

$$\sigma_{ij} = \lambda \delta_{ij} \varepsilon_{\alpha\alpha} + 2\mu \varepsilon_{ij}$$

or

$$\varepsilon_{ij} = \frac{1}{E}((1 + v)\sigma_{ij} - v\delta_{ij}\sigma_{\alpha\alpha})$$

where

ε_{ij} = elements of the strain tensor
σ_{ij} = elements of the stress tensor
$\varepsilon_{\alpha\alpha}$ = volumetric strain (sum over repeated index)
$\sigma_{\alpha\alpha}$ = mean stress times 3 (sum over repeated index)
δ_{ij} = 0 if $i \neq j$, 1 if $i = j$

In an isotropic, linear elastic medium, only two constants are needed to specify the stress–strain relation completely (for example, $[\lambda, \mu]$ in the first equation or $[E, \nu]$, which can be derived from $[\lambda, \mu]$, in the second equation). Other useful and convenient moduli can be defined but are always relatable to just two constants. For example, the

- **Bulk modulus**, K, defined as the ratio of hydrostatic stress, σ_0, to volumetric strain:

$$\sigma_0 = \frac{1}{3}\sigma_{\alpha\alpha} = K\varepsilon_{\alpha\alpha}$$

- **Shear modulus**, μ, defined as the ratio of shear stress to shear strain:

$$\sigma_{ij} = 2\mu\varepsilon_{ij}, \quad i \neq j$$

- **Young's modulus**, E, defined as the ratio of extensional stress to extensional strain in a *uniaxial stress* state:

$$\sigma_{zz} = E\varepsilon_{zz}, \qquad \sigma_{xx} = \sigma_{yy} = \sigma_{xy} = \sigma_{xz} = \sigma_{yz} = 0$$

TABLE 2.1.1. Relationships among elastic constants in an isotropic material (after Birch, 1961).

K	E	λ	ν	M	μ
$\lambda + 2\mu/3$	$\mu\dfrac{3\lambda + 2\mu}{\lambda + \mu}$	—	$\dfrac{\lambda}{2(\lambda + \mu)}$	$\lambda + 2\mu$	—
—	$9K\dfrac{K - \lambda}{3K - \lambda}$	—	$\dfrac{\lambda}{3K - \lambda}$	$3K - 2\lambda$	$3(K - \lambda)/2$
—	$\dfrac{9K\mu}{3K + \mu}$	$K - 2\mu/3$	$\dfrac{3K - 2\mu}{2(3K + \mu)}$	$K + 4\mu/3$	—
$\dfrac{E\mu}{3(3\mu - E)}$	—	$\mu\dfrac{E - 2\mu}{(3\mu - E)}$	$E/(2\mu) - 1$	$\mu\dfrac{4\mu - E}{3\mu - E}$	—
—	—	$3K\dfrac{3K - E}{9K - E}$	$\dfrac{3K - E}{6K}$	$3K\dfrac{3K + E}{9K - E}$	$\dfrac{3KE}{9K - E}$
$\lambda\dfrac{1 + \nu}{3\nu}$	$\lambda\dfrac{(1 + \nu)(1 - 2\nu)}{\nu}$	—	—	$\lambda\dfrac{1 - \nu}{\nu}$	$\lambda\dfrac{1 - 2\nu}{2\nu}$
$\mu\dfrac{2(1 + \nu)}{3(1 - 2\nu)}$	$2\mu(1 + \nu)$	$\mu\dfrac{2\nu}{1 - 2\nu}$	—	$\mu\dfrac{2 - 2\nu}{1 - 2\nu}$	—
—	$3K(1 - 2\nu)$	$3K\dfrac{\nu}{1 + \nu}$	—	$3K\dfrac{1 - \nu}{1 + \nu}$	$3K\dfrac{1 - 2\nu}{2 + 2\nu}$
$\dfrac{E}{3(1 - 2\nu)}$	—	$\dfrac{E\nu}{(1 + \nu)(1 - 2\nu)}$	—	$\dfrac{E(1 - \nu)}{(1 + \nu)(1 - 2\nu)}$	$\dfrac{E}{2 + 2\nu}$

- **Poisson's ratio**, which is defined as minus the ratio of lateral strain to axial strain in a *uniaxial stress* state:

$$v = -\frac{\varepsilon_{xx}}{\varepsilon_{zz}}, \qquad \sigma_{xx} = \sigma_{yy} = \sigma_{xy} = \sigma_{xz} = \sigma_{yz} = 0$$

- **P wave modulus**, $M = \rho V_P^2$, defined as the ratio of axial stress to axial strain in a *uniaxial strain* state:

$$\sigma_{zz} = M\varepsilon_{zz}, \qquad \varepsilon_{xx} = \varepsilon_{yy} = \varepsilon_{xy} = \varepsilon_{xz} = \varepsilon_{yz} = 0$$

Note that the moduli (λ, μ, K, E, M) all have the same units as stress (force/area), whereas Poisson's ratio is dimensionless.

Energy considerations require that the following relations always hold. If they do not, one should suspect experimental errors or that the material is not isotropic.

$$\lambda + \frac{2\mu}{3} \geq 0; \quad \mu \geq 0$$

or

$$-1 < v \leq 1/2; \quad E \geq 0$$

ASSUMPTIONS AND LIMITATIONS

The preceding equations assume isotropic, linear elastic media.

2.2 ANISOTROPIC FORM OF HOOKE'S LAW

SYNOPSIS

Hooke's law for a general anisotropic, linear elastic solid states that the stress σ_{ij} is linearly proportional to the strain ε_{ij} as expressed by

$$\sigma_{ij} = c_{ijkl}\varepsilon_{kl}$$

in which summation is implied over the repeated subscripts k and l. The **elastic stiffness tensor**, with elements c_{ijkl}, is a fourth-rank tensor obeying the laws of tensor transformation and has a total of eighty-one components. However, not all eighty-one components are independent. The symmetry of stresses and strains implies that

$$c_{ijkl} = c_{jikl} = c_{ijlk} = c_{jilk}$$

reducing the number of independent constants to thirty-six. Also the existence of a unique strain energy potential requires that

$$c_{ijkl} = c_{klij}$$

further reducing the number of independent constants to twenty-one. This is the maximum number of elastic constants that any medium can have. Additional restrictions imposed by symmetry considerations reduce the number much further. Isotropic, linear elastic materials, which have the maximum symmetry, are completely characterized by two independent constants, whereas materials with triclinic symmetry (lowest symmetry) require all twenty-one constants.

Alternatively, the strains may be expressed as a linear combination of the stresses by the expression

$$\varepsilon_{ij} = s_{ijkl}\sigma_{kl}$$

In this case s_{ijkl} are elements of the **elastic compliance tensor**, which has the same symmetry as the corresponding stiffness tensor. The compliance and stiffness are tensor inverses denoted by

$$c_{ijkl}s_{klmn} = I_{ijmn} = \frac{1}{2}(\delta_{im}\delta_{jn} + \delta_{in}\delta_{jm})$$

It is a standard practice in elasticity to use an abbreviated notation for the stresses, strains, and the stiffness and compliance tensors, for doing so simplifies some of the key equations (Auld, 1990). In this abbreviated notation the stresses and strains are written as six-element column vectors rather than as nine-element square matrices:

$$T = \begin{bmatrix} \sigma_1 = \sigma_{11} \\ \sigma_2 = \sigma_{22} \\ \sigma_3 = \sigma_{33} \\ \sigma_4 = \sigma_{23} \\ \sigma_5 = \sigma_{13} \\ \sigma_6 = \sigma_{12} \end{bmatrix} \qquad E = \begin{bmatrix} e_1 = \varepsilon_{11} \\ e_2 = \varepsilon_{22} \\ e_3 = \varepsilon_{33} \\ e_4 = 2\varepsilon_{23} \\ e_5 = 2\varepsilon_{13} \\ e_6 = 2\varepsilon_{12} \end{bmatrix}$$

Note the factor of 2 in the definitions of strains.

The four subscripts of the stiffness and compliance tensors are reduced to two. Each pair of indices $ij(kl)$ is replaced by one index $I(J)$ using the following convention:

$ij(kl)$	$I(J)$
11	1
22	2
33	3
23, 32	4
13, 31	5
12, 21	6

The relation, therefore, is $c_{IJ} = c_{ijkl}$ and $s_{IJ} = s_{ijkl}N$ where

$$N = \begin{cases} 1 & \text{for } I \text{ and } J = 1, 2, 3 \\ 2 & \text{for } I \text{ or } J = 4, 5, 6 \\ 4 & \text{for } I \text{ and } J = 4, 5, 6 \end{cases}$$

Note the difference in the definition of s_{IJ} from that of c_{IJ}. This results from the factors of two introduced in the definition of strains in the abbreviated notation.

CAUTION: Different definitions of strains are sometimes adopted, which move the factors of 2 and 4 from the compliances to the stiffnesses. However, the form given above is the more common convention. In the two-index notation, c_{IJ} and s_{IJ} can conveniently be represented as 6×6 matrices. However, they no longer follow the laws of tensor transformation. Care has to be taken when transforming from one coordinate system to another. One way is to go back to the four-index notation and then use the ordinary laws of coordinate transformation. A more efficient method is to use the **Bond transformation matrices**, which are explained in Section 1.4 on coordinate transformations.

The nonzero components of the more symmetric anisotropy classes commonly used in modeling rock properties are given below.

Isotropic – two independent constants:

$$\begin{bmatrix} c_{11} & c_{12} & c_{12} & 0 & 0 & 0 \\ c_{12} & c_{11} & c_{12} & 0 & 0 & 0 \\ c_{12} & c_{12} & c_{11} & 0 & 0 & 0 \\ 0 & 0 & 0 & c_{44} & 0 & 0 \\ 0 & 0 & 0 & 0 & c_{44} & 0 \\ 0 & 0 & 0 & 0 & 0 & c_{44} \end{bmatrix}, \quad c_{12} = c_{11} - 2c_{44}$$

The relations between the elements c and the Lamé's parameters λ and μ of isotropic linear elasticity are

$$c_{11} = \lambda + 2\mu \quad c_{12} = \lambda \quad c_{44} = \mu$$

Cubic – three independent constants:

$$\begin{bmatrix} c_{11} & c_{12} & c_{12} & 0 & 0 & 0 \\ c_{12} & c_{11} & c_{12} & 0 & 0 & 0 \\ c_{12} & c_{12} & c_{11} & 0 & 0 & 0 \\ 0 & 0 & 0 & c_{44} & 0 & 0 \\ 0 & 0 & 0 & 0 & c_{44} & 0 \\ 0 & 0 & 0 & 0 & 0 & c_{44} \end{bmatrix}$$

Hexagonal or Transversely isotropic – five independent constants:

$$
\begin{bmatrix}
c_{11} & c_{12} & c_{13} & 0 & 0 & 0 \\
c_{12} & c_{11} & c_{13} & 0 & 0 & 0 \\
c_{13} & c_{13} & c_{33} & 0 & 0 & 0 \\
0 & 0 & 0 & c_{44} & 0 & 0 \\
0 & 0 & 0 & 0 & c_{44} & 0 \\
0 & 0 & 0 & 0 & 0 & c_{66}
\end{bmatrix}, \quad c_{66} = \frac{1}{2}(c_{11} - c_{12})
$$

Orthorhombic – nine independent constants:

$$
\begin{bmatrix}
c_{11} & c_{12} & c_{13} & 0 & 0 & 0 \\
c_{12} & c_{22} & c_{23} & 0 & 0 & 0 \\
c_{13} & c_{23} & c_{33} & 0 & 0 & 0 \\
0 & 0 & 0 & c_{44} & 0 & 0 \\
0 & 0 & 0 & 0 & c_{55} & 0 \\
0 & 0 & 0 & 0 & 0 & c_{66}
\end{bmatrix}
$$

For *isotropic* symmetry the phase velocity of wave propagation is given by

$$
V_P = \sqrt{\frac{c_{11}}{\rho}}; \qquad V_S = \sqrt{\frac{c_{44}}{\rho}}
$$

where ρ is the density.

In *anisotropic* media there are, in general, three modes of propagation (quasi-longitudinal, quasi-shear and pure shear) with mutually orthogonal polarizations. For a medium with **transversely isotropic** (hexagonal) symmetry, the wave slowness surface is always rotationally symmetric about the axis of symmetry. The velocities of the three modes in any plane containing the symmetry axis are given as

quasi-longitudinal mode (transversely isotropic)

$$
V_P = (c_{11} \sin^2 \theta + c_{33} \cos^2 \theta + c_{44} + \sqrt{M})^{1/2}(2\rho)^{-1/2}
$$

quasi-shear mode (transversely isotropic)

$$
V_{SV} = (c_{11} \sin^2 \theta + c_{33} \cos^2 \theta + c_{44} - \sqrt{M})^{1/2}(2\rho)^{-1/2}
$$

pure shear mode (transversely isotropic)

$$
V_{SH} = \left(\frac{c_{66} \sin^2 \theta + c_{44} \cos^2 \theta}{\rho} \right)^{1/2}
$$

where

$$
M = [(c_{11} - c_{44}) \sin^2 \theta - (c_{33} - c_{44}) \cos^2 \theta]^2 + (c_{13} + c_{44})^2 \sin^2 2\theta
$$

and θ is the angle between the wave vector and the axis of symmetry. The five components of the stiffness tensor for a transversely isotropic material are obtained from five velocity measurements: $V_P(0°)$, $V_P(90°)$, $V_P(45°)$, $V_{SH}(90°)$, and $V_{SH}(0°) = V_{SV}(0°)$.

$$c_{11} = \rho V_P{}^2(90°)$$
$$c_{12} = c_{11} - 2\rho V_{SH}{}^2(90°)$$
$$c_{33} = \rho V_P{}^2(0°)$$
$$c_{44} = \rho V_{SH}{}^2(0°)$$

and

$$c_{13} = -c_{44}$$
$$+ \sqrt{4\rho^2 V_{P(45°)}{}^4 - 2\rho V_{P(45°)}{}^2(c_{11} + c_{33} + 2c_{44}) + (c_{11} + c_{44})(c_{33} + c_{44})}$$

For the more general **orthorhombic** symmetry the velocities of the three modes propagating in the three symmetry planes (XZ, YZ, and XY) are given as follows:

quasi-longitudinal mode (orthorhombic – XZ plane)

$$V_P = \left(c_{55} + c_{11} \sin^2 \theta + c_{33} \cos^2 \theta \right.$$
$$\left. + \sqrt{(c_{55} + c_{11} \sin^2 \theta + c_{33} \cos^2 \theta)^2 - 4A} \right)^{1/2} (2\rho)^{-1/2}$$

quasi-shear mode (orthorhombic – XZ plane)

$$V_{SV} = \left(c_{55} + c_{11} \sin^2 \theta + c_{33} \cos^2 \theta \right.$$
$$\left. - \sqrt{(c_{55} + c_{11} \sin^2 \theta + c_{33} \cos^2 \theta)^2 - 4A} \right)^{1/2} (2\rho)^{-1/2}$$

pure shear mode (orthorhombic – XZ plane)

$$V_{SH} = \left(\frac{c_{66} \sin^2 \theta + c_{44} \cos^2 \theta}{\rho} \right)^{1/2}$$

where

$$A = (c_{11} \sin^2 \theta + c_{55} \cos^2 \theta)(c_{55} \sin^2 \theta + c_{33} \cos^2 \theta)$$
$$- (c_{13} + c_{55})^2 \sin^2 \theta \cos^2 \theta$$

quasi-longitudinal mode (orthorhombic – YZ plane)

$$V_P = \left(c_{44} + c_{22} \sin^2 \theta + c_{33} \cos^2 \theta \right.$$
$$\left. + \sqrt{(c_{44} + c_{22} \sin^2 \theta + c_{33} \cos^2 \theta)^2 - 4B} \right)^{1/2} (2\rho)^{-1/2}$$

quasi-shear mode (orthorhombic – YZ plane)

$$V_{SV} = \left(c_{44} + c_{22} \sin^2 \theta + c_{33} \cos^2 \theta \right.$$
$$\left. - \sqrt{(c_{44} + c_{22} \sin^2 \theta + c_{33} \cos^2 \theta)^2 - 4B} \right)^{1/2} (2\rho)^{-1/2}$$

pure shear mode (orthorhombic – YZ plane)

$$V_{SH} = \left(\frac{c_{66} \sin^2 \theta + c_{55} \cos^2 \theta}{\rho} \right)^{1/2}$$

where

$$B = (c_{22} \sin^2 \theta + c_{44} \cos^2 \theta)(c_{44} \sin^2 \theta + c_{33} \cos^2 \theta) - (c_{23} + c_{44})^2 \sin^2 \theta \cos^2 \theta$$

quasi-longitudinal mode (orthorhombic – XY plane)

$$V_P = \left(c_{66} + c_{22} \sin^2 \varphi + c_{11} \cos^2 \varphi \right.$$
$$\left. + \sqrt{(c_{66} + c_{22} \sin^2 \varphi + c_{11} \cos^2 \varphi)^2 - 4C} \right)^{1/2} (2\rho)^{-1/2}$$

quasi-shear mode (orthorhombic – XY plane)

$$V_{SH} = \left(c_{66} + c_{22} \sin^2 \varphi + c_{11} \cos^2 \varphi \right.$$
$$\left. - \sqrt{(c_{66} + c_{22} \sin^2 \varphi + c_{11} \cos^2 \varphi)^2 - 4C} \right)^{1/2} (2\rho)^{-1/2}$$

pure shear mode (orthorhombic – XY plane)

$$V_{SV} = \left(\frac{c_{55} \cos^2 \varphi + c_{44} \sin^2 \varphi}{\rho} \right)^{1/2}$$

where

$$C = (c_{66} \sin^2 \varphi + c_{11} \cos^2 \varphi)(c_{22} \sin^2 \varphi + c_{66} \cos^2 \varphi)$$
$$- (c_{12} + c_{66})^2 \sin^2 \varphi \cos^2 \varphi$$

and the angles θ and φ are measured from the Z- and X-axes, respectively.

ASSUMPTIONS AND LIMITATIONS

The preceding equations assume anisotropic, linear elastic media.

2.3 THOMSEN'S NOTATION FOR WEAK ELASTIC ANISOTROPY

SYNOPSIS

A transversely isotropic elastic material is completely specified by five independent constants. In terms of the shortened notation (see Section 2.2 on elastic anisotropy) the elastic constants can be represented as

$$
\begin{bmatrix}
c_{11} & c_{12} & c_{13} & 0 & 0 & 0 \\
c_{12} & c_{11} & c_{13} & 0 & 0 & 0 \\
c_{13} & c_{13} & c_{33} & 0 & 0 & 0 \\
0 & 0 & 0 & c_{44} & 0 & 0 \\
0 & 0 & 0 & 0 & c_{44} & 0 \\
0 & 0 & 0 & 0 & 0 & c_{66}
\end{bmatrix}, \quad \text{where } c_{66} = \frac{1}{2}(c_{11} - c_{12})
$$

and where the 3-axis (z-axis) lies along the axis of symmetry.

Thomsen (1986) suggested the following convenient notation for this type of material, when only weakly anisotropic, in terms of the P-wave and S-wave velocities, α and β, propagating along the symmetry axis plus three additional constants:

$$
\alpha = \sqrt{c_{33}/\rho}
$$

$$
\beta = \sqrt{c_{44}/\rho}
$$

$$
\varepsilon = \frac{c_{11} - c_{33}}{2c_{33}}
$$

$$
\gamma = \frac{c_{66} - c_{44}}{2c_{44}}
$$

$$
\delta = \frac{(c_{13} + c_{44})^2 - (c_{33} - c_{44})^2}{2c_{33}(c_{33} - c_{44})}
$$

In terms of these constants, the three phase velocities can be written conveniently as

$$
V_{P}(\theta) \approx \alpha(1 + \delta \sin^2 \theta \cos^2 \theta + \varepsilon \sin^4 \theta)
$$

$$
V_{SV}(\theta) \approx \beta \left(1 + \frac{\alpha^2}{\beta^2}(\varepsilon - \delta) \sin^2 \theta \cos^2 \theta\right)
$$

$$
V_{SH}(\theta) \approx \beta(1 + \gamma \sin^2 \theta)
$$

where V_{SH} is the wavefront velocity of the pure shear wave, which has no component of polarization in the vertical (z) direction; V_{SV} is the pseudoshear wave polarized normal to the pure shear wave; and V_P is the pseudolongitudinal wave. The phase angle between the wavefront normal and the symmetry axis (z-axis) is denoted by θ.

The constant ε can be seen to describe the fractional difference of the P-wave velocities in the vertical and horizontal directions:

$$\varepsilon = \frac{V_P(90°) - V_P(0°)}{V_P(0°)}$$

and therefore best describes what is usually called the "P-wave anisotropy."

Similarly, the constant γ can be seen to describe the fractional difference of the SH-wave velocities between vertical and horizontal directions, which is equivalent to the difference between the vertical and horizontal polarizations of the horizontally propagating S-waves:

$$\gamma = \frac{V_{SH}(90°) - V_{SV}(90°)}{V_{SV}(90°)} = \frac{V_{SH}(90°) - V_{SH}(0°)}{V_{SH}(0°)}$$

USES

Thomsen's notation for weak elastic anisotropy is useful for conveniently characterizing the elastic constants of a transversely isotropic elastic medium.

ASSUMPTIONS AND LIMITATIONS

The preceding equations are based on the following assumptions:

• Material is linear, elastic, and transversely isotropic.
• Anisotropy is weak, so that $\varepsilon, \gamma, \delta \ll 1$.

2.4 STRESS-INDUCED ANISOTROPY IN ROCKS

SYNOPSIS

The closing of cracks under compressive stress (or, equivalently, the stiffening of compliant grain contacts) tends to increase the effective elastic moduli of rocks.

When the crack population is anisotropic, either in the original unstressed condition or as a result of the stress field, then this condition can impact the overall elastic anisotropy of the rock. Laboratory demonstrations of stress-induced anisotropy have been reported by numerous authors (Nur and Simmons, 1969; Lockner, Walsh, and Byerlee, 1977; Zamora and Poirier, 1990; Sayers, Van Munster, and King, 1990; Yin 1992; Cruts et al., 1995).

The simplest case to understand is a rock with a random (isotropic) distribution of cracks embedded in an isotropic mineral matrix. In the initial unstressed state the rock is elastically isotropic. If a *hydrostatic* compressive stress is applied, cracks in all directions respond similarly, and the rock remains isotropic but becomes stiffer. However, if a *uniaxial* compressive stress is applied, cracks with normals parallel or nearly parallel to the applied stress axis will tend to close preferentially, and the rock will take on an axial or transversely isotropic symmetry.

Figure 2.4.1 illustrates the effects of stress-induced crack alignment on seismic velocity anisotropy discovered in the laboratory by Nur and Simmons (1969). The crack porosity of the dry granite sample is essentially isotropic at low stress.

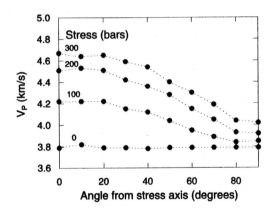

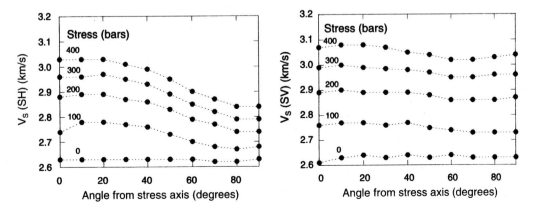

Figure 2.4.1

As uniaxial stress is applied, crack anisotropy is induced. The velocities (compressional and two polarizations of shear) clearly vary with direction relative to the stress-induced crack alignment. Table 2.4.1 below summarizes the elastic symmetries that result when various applied stress fields interact with various initial crack symmetries (Paterson and Weiss, 1961; Nur, 1971).

A rule of thumb is that a wave is most sensitive to cracks when its direction of propagation or direction of polarization is perpendicular (or nearly so) to the crack faces.

The most common approach to modeling the stress-induced anisotropy is to assume angular distributions of idealized penny-shaped cracks (Nur, 1971;

TABLE 2.4.1. Dependence of symmetry of induced velocity anisotropy on initial crack distribution and applied stress and its orientation.

Symmetry of initial crack distribution	Applied stress	Orientation of applied stress	Symmetry of induced velocity anisotropy	Number of elastic constants
Random	Hydrostatic		Isotropic	2
	Uniaxial		Axial	5
	Triaxial[a]		Orthorhombic	9
Axial	Hydrostatic		Axial	5
	Uniaxial	Parallel to axis of symmetry	Axial	5
	Uniaxial	Normal to axis of symmetry	Orthorhombic	9
	Uniaxial	Inclined	Monoclinic	13
	Triaxial[a]	Parallel to axis of symmetry	Orthorhombic	9
	Triaxial[a]	Inclined	Monoclinic	13
Orthorhombic	Hydrostatic		Orthorhombic	9
	Uniaxial	Parallel to axis of symmetry	Orthorhombic	9
	Uniaxial	Inclined in plane of symmetry	Monoclinic	13
	Uniaxial	Inclined	Triclinic	21
	Triaxial[a]	Parallel to axis of symmetry	Orthorhombic	9
	Triaxial[a]	Inclined in plane of symmetry	Monoclinic	13
	Triaxial[a]	Inclined	Triclinic	21

[a]Three generally unequal principal stresses.

Sayers, 1988a,b; Gibson and Toksöz, 1990). The stress dependence is introduced by assuming or inferring distributions or spectra of crack aspect ratios with various orientations.

The assumption is that a crack will close when the component of applied compressive stress normal to the crack faces causes a normal displacement of the crack faces equal to the original half-width of the crack. This allows us to estimate the crack *closing stress* as follows:

$$\sigma_{\text{close}} = \frac{3\pi(1 - 2v)}{4(1 - v^2)}\alpha K_0 = \frac{\pi}{2(1 - v)}\alpha\mu_0$$

where α is the aspect ratio of the crack, and v, μ_0, and K_0 are the Poisson's ratio, shear modulus, and bulk modulus of the mineral, respectively (see Section 2.6 on the deformation of inclusions and cavities in elastic solids). Hence, the thinnest cracks will close first, followed by thicker ones. This allows one to estimate, for a given aspect ratio distribution, how many cracks remain open in each direction for any applied stress field. These inferred crack distributions and their orientations can be put into one of the popular crack models (e.g., Hudson, 1981) to estimate the resulting effective elastic moduli of the rock. Although these penny-shaped crack models have been relatively successful and provide a useful physical interpretation, they are limited to low crack concentrations and may not effectively represent a broad range of crack geometries (see Section 4.10 on Hudson's model for cracked media).

As an alternative, Mavko, Mukerji, and Godfrey (1995) presented a simple recipe for estimating stress-induced velocity anisotropy directly from measured values of isotropic V_P and V_S versus hydrostatic pressure. This method differs from the inclusion models, because it is relatively independent of any assumed crack geometry and is not limited to small crack densities. To invert for a particular crack distribution, one needs to assume crack shapes and aspect ratio spectra. However, if we do not invert for a crack distribution but instead directly transform hydrostatic velocity-versus-pressure data to stress-induced velocity anisotropy, we can avoid the need for parameterization in terms of ellipsoidal cracks and the resulting limitations to low crack densities. In this sense, the method of Mavko et al. provides not only a simpler but also a more general solution to this problem, for ellipsoidal cracks are just one particular case of the general formulation.

The procedure is to estimate the generalized pore space compliance from the measurements of isotropic V_P and V_S. The physical assumption that the compliant part of the pore space is crack-like means that the pressure dependence of the generalized compliances is governed primarily by *normal* tractions resolved across cracks and defects. These defects can include grain boundaries and contact regions between clay platelets (Sayers, 1995). This assumption allows the measured pressure dependence to be mapped from the hydrostatic stress state to any applied nonhydrostatic stress.

The method applies to rocks that are approximately isotropic under hydrostatic stress and in which the anisotropy is caused by *crack closure* under stress. Sayers (1988b) found evidence for some stress-induced *opening* of cracks also, which is ignored in this method. The potentially important problem of stress–strain hysteresis is also ignored.

The anisotropic elastic compliance tensor $S_{ijkl}(\sigma)$ at any given stress state σ may be expressed as

$$\Delta S_{ijkl}(\sigma) = S_{ijkl}(\sigma) - S^0_{ijkl}$$

$$= \int_{\theta=0}^{\pi/2} \int_{\phi=0}^{2\pi} W'_{3333}(\hat{m}^T \sigma \hat{m}) m_i m_j m_k m_l \sin\theta \, d\theta \, d\phi$$

$$+ \int_{\theta=0}^{\pi/2} \int_{\phi=0}^{2\pi} \gamma(\hat{m}^T \sigma \hat{m}) W'_{3333}(\hat{m}^T \sigma \hat{m})[\delta_{ik} m_j m_l + \delta_{il} m_j m_k$$

$$+ \delta_{jk} m_i m_l + \delta_{jl} m_i m_k - 4 m_i m_j m_k m_l] \sin\theta \, d\theta \, d\phi$$

where

$$W'_{3333}(p) = \frac{1}{2\pi} \Delta S^{iso}_{jjkk}(p)$$

$$\gamma(p) = \frac{W'_{2323}(p)}{W'_{3333}(p)} = \frac{\Delta S^{iso}_{\alpha\beta\alpha\beta}(p) - \Delta S^{iso}_{\alpha\alpha\beta\beta}(p)}{4\Delta S^{iso}_{\alpha\alpha\beta\beta}(p)}$$

$$= \frac{\frac{3}{2}\Delta S^{iso}_{2323}(p) - 2\Delta S^{iso}_{1111}(p)}{12\Delta S^{iso}_{1111}(p) - 4\Delta S^{iso}_{2323}(p)}$$

The tensor S^0_{ijkl} denotes the reference compliance at some very large confining hydrostatic pressure when all of the compliant parts of the pore space are closed. The expression $\Delta S^{iso}_{ijkl}(p) = S^{iso}_{ijkl}(p) - S^0_{ijkl}$ describes the difference in the compliances under a hydrostatic effective pressure p and the reference compliance at high pressure. These are determined from measured P- and S-wave velocities versus hydrostatic pressure. The tensors elements W'_{3333} and W'_{2323} are the *measured* normal and shear crack compliances and include all interactions with neighboring cracks and pores. These could be approximated, for example, with the compliances of idealized ellipsoidal cracks interacting or not, but this would immediately reduce the generality. The expression $\hat{m} \equiv (\sin\theta \cos\phi, \sin\theta \sin\phi, \cos\theta)^T$ denotes the unit normal to the crack face, where θ and ϕ are the polar and azimuthal angles in a spherical coordinate system.

An important physical assumption in the preceding equations is that, for thin cracks, the *crack compliance tensor* W'_{ijkl} *is sparse*, and thus only W'_{3333}, W'_{1313}, and W'_{2323} are nonzero. This is a general property of planar crack formulations and reflects an approximate decoupling of normal and shear deformation of the crack and decoupling of the in-plane and out-of-plane deformation. This allows us to write $W'_{jjkk} \approx W'_{3333}$. Furthermore, it is assumed that the two unknown shear compliances are approximately equal, $W'_{1313} \approx W'_{2323}$. A second important physical assumption is that for a thin crack under any stress field, it is primarily the *normal* component of stress, $\sigma = \hat{m}^T \sigma \hat{m}$, resolved on the faces of a crack that causes it to close and to have a stress-dependent compliance. Any open crack will have both normal and shear deformation under normal and shear loading, but it is only the normal stress that determines crack closure.

Sometimes a symmetric second-rank compliance tensor B_{ij} is introduced to relate displacement discontinuity across cracks or contact regions to the applied traction across the faces (Kachanov, 1992; Sayers, 1995). This tensor can be related to W'_{ijkl} by

$$W'_{ijkl}\sigma_{kl} = \frac{1}{V} \int [B_{i\alpha}\sigma_{\alpha\beta}m_\beta m_j + B_{j\alpha}\sigma_{\alpha\beta}m_\beta m_i] \, dS_c$$

where the surface integral is taken over the surface of the crack S_c, and is normalized by the representative elementary volume V.

For the case of uniaxial stress σ_0 applied along the 3-axis to an initially isotropic rock, the normal stress in any direction is $\sigma_n = \sigma_0 \cos^2 \theta$. The rock takes on a tranversely isotropic symmetry, with five independent elastic constants. The five independent components of ΔS_{ijkl} become

$$\Delta S_{3333}^{\text{uni}} = 2\pi \int_0^{\pi/2} [1 - 4\gamma(\sigma_0 \cos^2 \theta)] W'_{3333}(\sigma_0 \cos^2 \theta) \cos^4 \theta \sin \theta \, d\theta$$

$$+ 2\pi \int_0^{\pi/2} 4\gamma(\sigma_0 \cos^2 \theta) W'_{3333}(\sigma_0 \cos^2 \theta) \cos^2 \theta \sin \theta \, d\theta$$

$$\Delta S_{1111}^{\text{uni}} = 2\pi \int_0^{\pi/2} \frac{3}{8}[1 - 4\gamma(\sigma_0 \cos^2 \theta)] W'_{3333}(\sigma_0 \cos^2 \theta) \sin^4 \theta \sin \theta \, d\theta$$

$$+ 2\pi \int_0^{\pi/2} 2\gamma(\sigma_0 \cos^2 \theta) W'_{3333}(\sigma_0 \cos^2 \theta) \sin^2 \theta \sin \theta \, d\theta$$

$$\Delta S_{1122}^{\text{uni}} = 2\pi \int_0^{\pi/2} \frac{1}{8}[1 - 4\gamma(\sigma_0 \cos^2 \theta)] W'_{3333}(\sigma_0 \cos^2 \theta) \sin^4 \theta \sin \theta \, d\theta$$

$$\Delta S_{1133}^{\text{uni}} = 2\pi \int_0^{\pi/2} \frac{1}{2}[1 - 4\gamma(\sigma_0 \cos^2 \theta)] W'_{3333}(\sigma_0 \cos^2 \theta) \sin^2 \theta \cos^2 \theta \sin \theta \, d\theta$$

$$\Delta S_{2323}^{\text{uni}} = 2\pi \int_0^{\pi/2} \frac{1}{2}[1 - 4\gamma(\sigma_0 \cos^2 \theta)] W'_{3333}(\sigma_0 \cos^2 \theta) \sin^2 \theta \cos^2 \theta \sin \theta \, d\theta$$

$$+ 2\pi \int_0^{\pi/2} \frac{1}{2}\gamma(\sigma_0 \cos^2 \theta) W'_{3333}(\sigma_0 \cos^2 \theta) \sin^2 \theta \sin \theta \, d\theta$$

$$+ 2\pi \int_0^{\pi/2} \gamma(\sigma_0 \cos^2 \theta) W'_{3333}(\sigma_0 \cos^2 \theta) \cos^2 \theta \sin \theta \, d\theta$$

Note that the terms in parentheses with $\gamma(\ )$ and $W'_{3333}(\ )$ are arguments to the γ and W'_{3333} pressure functions and not multiplicative factors.

ASSUMPTIONS AND LIMITATIONS

Most approaches assume an isotropic, linear, elastic solid mineral material. Methods based on ellipsoidal cracks or spherical contacts are limited to these idealized geometries and low crack densities.

2.5 STRAIN COMPONENTS AND EQUATIONS OF MOTION IN CYLINDRICAL AND SPHERICAL COORDINATE SYSTEMS

SYNOPSIS

The equations of motion and the expressions for small strain components in cylindrical and spherical coordinate systems differ from those in a rectangular coordinate system.

In the **cylindrical** coordinate system (r, ϕ, z), the coordinates are related to those in the rectangular coordinate system (x, y, z) as

$$r = \sqrt{x^2 + y^2}, \quad \tan(\phi) = \frac{y}{x}$$

$$x = r\cos(\phi), \quad y = r\sin(\phi), \quad z = z$$

The small strain components can be expressed through the displacements u_r, u_ϕ, and u_z (which are in the directions r, ϕ, and z, respectively) as

$$e_{rr} = \frac{\partial u_r}{\partial r}, \quad e_{\phi\phi} = \frac{1}{r}\frac{\partial u_\phi}{\partial \phi} + \frac{u_r}{r}, \quad e_{zz} = \frac{\partial u_z}{\partial z}$$

$$e_{r\phi} = \frac{1}{2}\left(\frac{1}{r}\frac{\partial u_r}{\partial \phi} + \frac{\partial u_\phi}{\partial r} - \frac{u_\phi}{r}\right)$$

$$e_{\phi z} = \frac{1}{2}\left(\frac{\partial u_\phi}{\partial z} + \frac{1}{r}\frac{\partial u_z}{\partial \phi}\right), \quad e_{zr} = \frac{1}{2}\left(\frac{\partial u_z}{\partial r} + \frac{\partial u_r}{\partial z}\right)$$

The equations of motion are

$$\frac{\partial \sigma_{rr}}{\partial r} + \frac{1}{r}\frac{\partial \sigma_{r\phi}}{\partial \phi} + \frac{\partial \sigma_{zr}}{\partial z} + \frac{\sigma_{rr} - \sigma_{\phi\phi}}{r} = \rho\frac{\partial^2 u_r}{\partial t^2}$$

$$\frac{\partial \sigma_{r\phi}}{\partial r} + \frac{1}{r}\frac{\partial \sigma_{\phi\phi}}{\partial \phi} + \frac{\partial \sigma_{\phi z}}{\partial z} + \frac{2\sigma_{r\phi}}{r} = \rho\frac{\partial^2 u_\phi}{\partial t^2}$$

$$\frac{\partial \sigma_{rz}}{\partial r} + \frac{1}{r}\frac{\partial \sigma_{\phi z}}{\partial \phi} + \frac{\partial \sigma_{zz}}{\partial z} + \frac{\sigma_{rz}}{r} = \rho\frac{\partial^2 u_z}{\partial t^2}$$

where ρ is density and t is time.

In the **spherical** coordinate system (r, ϕ, φ) the coordinates are related to those in the rectangular coordinate system (x, y, z) as

$$r = \sqrt{x^2 + y^2 + z^2}, \quad \tan(\phi) = \frac{y}{x}, \quad \cos(\varphi) = \frac{z}{\sqrt{x^2 + y^2 + z^2}}$$

$$x = r\sin(\varphi)\cos(\phi), \quad y = r\sin(\varphi)\sin(\phi), \quad z = r\cos(\varphi)$$

The small strain components can be expressed through the displacements u_r, u_ϕ, and u_φ (which are in the directions r, ϕ, and φ, respectively) as

$$e_{rr} = \frac{\partial u_r}{\partial r}, \qquad e_{\phi\phi} = \frac{1}{r\sin(\varphi)}\frac{\partial u_\phi}{\partial \phi} + \frac{u_r}{r} + \frac{u_\varphi}{r\tan(\varphi)}, \qquad e_{\varphi\varphi} = \frac{1}{r}\frac{\partial u_\varphi}{\partial \varphi} + \frac{u_r}{r}$$

$$e_{r\phi} = \frac{1}{2}\left(\frac{1}{r\sin(\varphi)}\frac{\partial u_r}{\partial \phi} + \frac{\partial u_\phi}{\partial r} - \frac{u_\phi}{r}\right)$$

$$e_{\phi\varphi} = \frac{1}{2}\left(\frac{1}{r}\frac{\partial u_\phi}{\partial \varphi} - \frac{u_\phi}{r\tan(\varphi)} + \frac{1}{r\sin(\varphi)}\frac{\partial u_\varphi}{\partial \phi}\right)$$

$$e_{r\varphi} = \frac{1}{2}\left(\frac{\partial u_\varphi}{\partial r} - \frac{u_\varphi}{r} + \frac{1}{r}\frac{\partial u_r}{\partial \varphi}\right)$$

The equations of motion are

$$\frac{\partial \sigma_{rr}}{\partial r} + \frac{1}{r\sin(\varphi)}\frac{\partial \sigma_{r\phi}}{\partial \phi} + \frac{1}{r}\frac{\partial \sigma_{r\varphi}}{\partial \varphi} + \frac{2\sigma_{rr} + \sigma_{r\varphi}\cot(\varphi) - \sigma_{\phi\phi} - \sigma_{\varphi\varphi}}{r} = \rho\frac{\partial^2 u_r}{\partial t^2}$$

$$\frac{\partial \sigma_{r\phi}}{\partial r} + \frac{1}{r\sin(\varphi)}\frac{\partial \sigma_{\phi\phi}}{\partial \phi} + \frac{1}{r}\frac{\partial \sigma_{\phi\varphi}}{\partial \varphi} + \frac{3\sigma_{r\phi} + \sigma_{\phi\varphi}\cot(\varphi)}{r} = \rho\frac{\partial^2 u_\phi}{\partial t^2}$$

$$\frac{\partial \sigma_{r\varphi}}{\partial r} + \frac{1}{r\sin(\varphi)}\frac{\partial \sigma_{\phi\phi}}{\partial \phi} + \frac{1}{r}\frac{\partial \sigma_{\phi\varphi}}{\partial \varphi} + \frac{3\sigma_{r\varphi} + (\sigma_{\varphi\varphi} - \sigma_{\phi\phi})\cot(\varphi)}{r} = \rho\frac{\partial^2 u_\varphi}{\partial t^2}$$

USES

The foregoing equations are used to solve elasticity problems where cylindrical or spherical geometries are most natural.

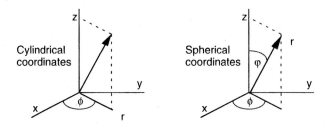

Figure 2.5.1

ASSUMPTIONS AND LIMITATIONS

The equations presented assume that the strains are small.

2.6 DEFORMATION OF INCLUSIONS AND CAVITIES IN ELASTIC SOLIDS

SYNOPSIS

Many problems in effective medium theory and poroelasticity can be solved or estimated in terms of the elastic behavior of cavities and inclusions. Some **static** and **quasistatic** results for cavities are presented here. It should be remembered that often these are also valid for certain limiting cases of dynamic problems. Excellent treatments of cavity deformation and pore compressibility are given by Jaeger and Cook (1969) and by Zimmerman (1991).

GENERAL PORE DEFORMATION

EFFECTIVE DRY COMPRESSIBILITY

Consider a homogeneous linear elastic solid that has an arbitrarily shaped pore space – either a single cavity or a collection of pores. The effective dry compressibility (reciprocal of the dry bulk modulus) of the porous solid can be written as

$$\frac{1}{K_{dry}} = \frac{1}{K_0} + \frac{\phi}{v_p} \frac{\partial v_p}{\partial \sigma}\bigg|_{dry}$$

where

$$K_{dry} = \text{effective bulk modulus of dry porous solid}$$
$$K_0 = \text{bulk modulus of the solid mineral material}$$
$$\phi = \text{porosity}$$
$$v_p = \text{pore volume}$$
$$\partial v_p / \partial \sigma |_{dry} = \text{derivative of pore volume with respect to externally applied hydrostatic stress}$$

and where we assume that no inelastic effects such as friction or viscosity are present. This is strictly true, regardless of pore geometry and pore concentration. The preceding equation can be slightly rewritten as

$$\frac{1}{K_{dry}} = \frac{1}{K_0} + \frac{\phi}{K_\phi}$$

where $K_\phi = v_p / \frac{\partial v_p}{\partial \sigma}|_{dry}$ is defined as the dry pore space stiffness. These equations state simply that the porous rock compressibility is equal to the intrinsic mineral compressibility plus an additional compressibility caused by the pore space.

CAUTION: **"Dry rock" is not the same as gas-saturated rock.** The dry frame modulus refers to the incremental bulk deformation resulting from an increment of applied confining pressure with pore pressure held constant. This corresponds to a "drained" experiment in which pore fluids can flow freely in or out of the sample to ensure constant pore pressure. Alternatively, the dry frame modulus can correspond to an undrained experiment in which the pore fluid has zero bulk modulus, and in which pore compressions therefore do not induce changes in pore pressure. This is approximately the case for an air-filled sample at standard temperature and pressure. However, at reservoir conditions, gas takes on a nonnegligible bulk modulus and should be treated as a saturating fluid.

An equivalent expression for the dry rock compressibility or bulk modulus is

$$K_{dry} = K_0(1 - \beta)$$

where β is sometimes called the **Biot coefficient**, which describes the ratio of pore volume change Δv_p to total bulk volume change ΔV under dry or drained conditions:

$$\beta = \frac{\Delta v_p}{\Delta V}\bigg|_{dry} = \frac{\phi K_{dry}}{K_\phi}$$

STRESS-INDUCED PORE PRESSURE

If this arbitrary pore space is filled with a pore fluid with bulk modulus, K_{fl}, the saturated solid is stiffer under compression than the dry solid because an increment d pressure is induced that resists the volumetric strain. The ratio of the re pressure, dP, to the applied compressive stress, $d\sigma$, is sometimes npton's coefficient and can be written as

$$\frac{dP}{d\sigma} = \frac{1}{1 + K_\phi\left(\frac{1}{K_\mathrm{fl}} - \frac{1}{K_0}\right)} = \frac{1}{1 + \phi\left(\frac{1}{K_\mathrm{fl}} - \frac{1}{K_0}\right)\left(\frac{1}{K_\mathrm{dry}} - \frac{1}{K_0}\right)^{-1}}$$

where K_ϕ is the dry pore space stiffness defined earlier in this section. For this to be true, the pore pressure must be uniform throughout the pore space, as will be the case if (1) there is only one pore, (2) all pores are well connected and the frequency and viscosity are low enough for any pressure differences to equilibrate, or (3) all pores have the same dry pore stiffness. Given these conditions, there is no additional limitation on pore geometry or concentration. All of the necessary information about pore stiffness and geometry is contained in the parameter K_ϕ.

SATURATED STRESS-INDUCED PORE VOLUME CHANGE

The corresponding change in fluid-saturated pore volume, v_p, caused by the remote stress is

$$\frac{1}{v_\mathrm{p}}\frac{dv_\mathrm{p}}{d\sigma}\bigg|_\mathrm{sat} = \frac{1}{K_\mathrm{fl}}\frac{dP}{d\sigma} = \frac{1/K_\mathrm{fl}}{1 + K_\phi\left(\frac{1}{K_\mathrm{fl}} - \frac{1}{K_0}\right)}$$

LOW-FREQUENCY SATURATED COMPRESSIBILITY

The low-frequency saturated bulk modulus, K_sat, can be derived from Gassmann's equation (see Section 6.3 on Gassmann). One equivalent form is

$$\frac{1}{K_\mathrm{sat}} = \frac{1}{K_0} + \frac{\phi}{K_\phi + \frac{K_0 K_\mathrm{fl}}{K_0 - K_\mathrm{fl}}} \approx \frac{1}{K_0} + \frac{\phi}{K_\phi + K_\mathrm{fl}}$$

where, again, all of the necessary information about pore stiffness and geometry is contained in the dry pore stiffness K_ϕ, and we must ensure that the stress-induced pore pressure is uniform throughout the pore space.

THREE-DIMENSIONAL ELLIPSOIDAL CAVITIES

Many effective media models are based on ellipsoidal inclusions or cavities. These are mathematically convenient shapes and allow quantitative estimates of, for example, K_ϕ, which was defined earlier in this section. Eshelby (1957) discovered that the strain, ε_{ij}, inside an ellipsoidal inclusion is homogeneous when a homogeneous strain, ε_{ij}^0, (or stress) is applied at infinity. Because the inclusion strain is homogeneous, operations such as determining the inclusion stress or integrating to get the displacement field are simple.

It is very important to remember that the following results assume a single isolated cavity in an infinite medium. Therefore, substituting them directly into the preceding formulas for dry and saturated moduli gives estimates that are strictly valid only for low concentrations of pores (see also Section 4.8 on self-consistent theories).

SPHERICAL CAVITY

For a single spherical cavity with volume $v_p = (4/3)\pi R^3$ and a hydrostatic stress, $d\sigma$, applied at infinity, the radial strain of the cavity is

$$\frac{dR}{R} = \frac{1}{K_0}\frac{(1-v)}{2(1-2v)}d\sigma$$

where v and K_0 are the Poisson's ratio and bulk modulus of the solid material. The change of pore volume is

$$dv_p = \frac{1}{K_0}\frac{3(1-v)}{2(1-2v)}v_p\,d\sigma$$

Then, the volumetric strain of the sphere is

$$\varepsilon_{ii} = \frac{dv_p}{v_p} = \frac{1}{K_0}\frac{3(1-v)}{2(1-2v)}d\sigma$$

and the single pore stiffness is

$$\frac{1}{K_\phi} = \frac{1}{v_p}\frac{dv_p}{d\sigma} = \frac{1}{K_0}\frac{3(1-v)}{2(1-2v)}$$

Remember that this estimate of K_ϕ assumes a single isolated spherical cavity in an infinite medium.

Now, under a remotely applied homogeneous *shear stress*, τ_0, corresponding to remote shear strain $\varepsilon_0 = \tau_0/2\mu_0$, the effective *shear strain* in the spherical cavity is

$$\varepsilon = \frac{15(1-v)}{2\mu_0(7-5v)}\tau_0$$

where μ_0 is the shear modulus of the solid. Note that this results in approximately twice the strain that would occur without the cavity.

PENNY-SHAPED CRACK – OBLATE SPHEROID

Consider a dry penny-shaped ellipsoidal cavity with semiaxes $a \ll b = c$. When a remote uniform tensional stress, $d\sigma$, is applied normal to the plane of the crack, each crack face undergoes an outward displacement, U, normal to the plane of the crack, given by the radially symmetric distribution

$$U(r) = \frac{4(1 - v^2)c\sqrt{1 - \left(\frac{r}{c}\right)^2}}{3\pi K_0(1 - 2v)} d\sigma$$

where r is the radial distance from the axis of the crack. For any arbitrary homogeneous remote stress, $d\sigma$ is the component of stress *normal* to the plane of the crack. Thus, $d\sigma$ can also be thought of as a remote hydrostatic stress field.

This displacement function is also an ellipsoid with semiminor axis, $U(r = 0)$, and semimajor axis, c. Therefore the volume change, dv, which is simply the integral of $U(r)$ over the faces of the crack, is just the volume of the displacement ellipsoid:

$$dv_p = \frac{4}{3}\pi U(0)c^2 = \frac{16c^3}{9K_0} \frac{(1 - v^2)}{(1 - 2v)} d\sigma$$

Then the volumetric strain of the cavity is

$$\varepsilon_{ii} = \frac{dv_p}{v_p} = \frac{4(c/a)}{3\pi K_0} \frac{(1 - v^2)}{(1 - 2v)} d\sigma$$

and the pore stiffness is

$$\frac{1}{K_\phi} = \frac{1}{v_p} \frac{dv_p}{d\sigma} = \frac{4(c/a)}{3\pi K_0} \frac{(1 - v^2)}{(1 - 2v)}$$

An interesting case is an applied compressive stress causing a displacement, $U(0)$, equal to the original half-width of the crack, thus closing the crack. Setting $U(0) = a$ allows one to compute the *closing stress*

$$\sigma_{close} = \frac{3\pi(1 - 2v)}{4(1 - v^2)} \alpha K_0 = \frac{\pi}{2(1 - v)} \alpha \mu_0$$

where $\alpha = (a/c)$ is the aspect ratio.

Now, under a remotely applied homogeneous *shear stress*, τ_0, corresponding to remote shear strain $\varepsilon_0 = \tau_0/2\mu_0$, the effective *shear strain* in the cavity is

$$\varepsilon = \tau_0 \frac{2(c/a)(1 - v)}{\pi \mu_0 (2 - v)}$$

NEEDLE-SHAPED PORE – PROLATE SPHEROID

Consider a dry needle-shaped ellipsoidal cavity with semiaxes $a \gg b = c$ and with pore volume $v = (4/3)\pi ac^2$. When a remote hydrostatic stress, $d\sigma$, is applied, the pore volume change is

$$dv_p = \frac{(5 - 4v)}{3K_0(1 - 2v)} v_p \, d\sigma$$

The volumetric strain of the cavity is then

$$\varepsilon_{ii} = \frac{dv_p}{v_p} = \frac{(5 - 4v)}{3K_0(1 - 2v)} \, d\sigma$$

and the pore stiffness is

$$\frac{1}{K_\phi} = \frac{1}{v_p} \frac{dv_p}{d\sigma} = \frac{(5 - 4v)}{3K_0(1 - 2v)}$$

Note that in the limit of very large a/c, these results are exactly the same as for a two-dimensional circular cylinder.

EXAMPLE

Estimate the increment of pore pressure induced in a water-saturated rock when a 1 bar increment of hydrostatic confining pressure is applied. Assume that the rock consists of stiff, spherical pores in a quartz matrix. Compare this with a rock with thin, penny-shaped cracks (aspect ratio $\alpha = 0.001$) in a quartz matrix. The moduli of the individual constituents are $K_{quartz} = 36$ GPa, $K_{water} = 2.2$ GPa, and $v_{quartz} = 0.07$.

The pore space stiffnesses are given by

$$K_{\phi\text{-sphere}} = K_{quartz} \frac{2(1 - 2v_{quartz})}{3(1 - v_{quartz})} = 22.2 \text{ GPa}$$

$$K_{\phi\text{-crack}} = \frac{3\pi\alpha K_{quartz}}{4} \frac{(1 - 2v_{quartz})}{(1 - v_{quartz}^2)} = 0.0733 \text{ GPa}$$

The pore pressure increment is computed from Skempton's coefficient.

$$\frac{\Delta P_{pore}}{\Delta P_{confining}} = B = \frac{1}{1 + K_\phi(K_{water}^{-1} - K_{quartz}^{-1})}$$

$$B_{sphere} = \frac{1}{1 + 22.2[(1/2.2) - (1/36)]} = 0.095$$

$$B_{crack} = \frac{1}{1 + 0.0733[(1/2.2) - (1/36)]} = 0.970$$

Therefore, the pore pressure induced in the spherical pores is 0.095 bar, and the pore pressure induced in the cracks is 0.98 bar.

TWO-DIMENSIONAL TUBES

A special two-dimensional case of long, tubular pores was treated by Mavko (1980) to describe melt or fluids arranged along the edges of grains. The cross-sectional shape is described by the equations

$$x = R\left(\cos\theta + \frac{1}{2+\gamma}\cos 2\theta\right)$$

$$y = R\left(-\sin\theta + \frac{1}{2+\gamma}\sin 2\theta\right)$$

where γ is a parameter describing the roundness, Figure 2.6.1.

Consider in particular the case on the left, $\gamma = 0$. The pore volume is $(1/2)\pi a R^2$, where $a \gg R$ is the length of the tube. When a remote hydrostatic stress, $d\sigma$, is applied, the pore volume change is

$$dv_{\mathrm{p}} = \frac{(13 - 4v - 8v^2)}{3K_0(1 - 2v)}v_{\mathrm{p}}\,d\sigma$$

The volumetric strain of the cavity is then

$$\varepsilon_{ii} = \frac{dv_{\mathrm{p}}}{v_{\mathrm{p}}} = \frac{(13 - 4v - 8v^2)}{3K_0(1 - 2v)}d\sigma$$

and the pore stiffness is

$$\frac{1}{K_\phi} = \frac{1}{v_p}\frac{dv_{\mathrm{p}}}{d\sigma} = \frac{(13 - 4v - 8v^2)}{3K_0(1 - 2v)}$$

In the extreme, $\gamma \to \infty$, the shape becomes a circular cylinder, and the expression

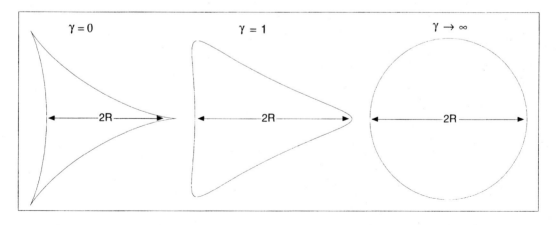

Figure 2.6.1

for pore stiffness, K_ϕ, is exactly the same as derived for the needle-shaped pores above. Note that the triangular cavity ($\gamma = 0$) has about half the pore stiffness of the circular one; that is, the triangular tube can give approximately the same effective modulus as the circular tube with about half the porosity.

CAUTION: These expressions for K_ϕ, dv_p, and ε_{ii} include an estimate of tube shortening as well as reduction in pore cross-sectional area under hydrostatic stress. Hence, the deformation is neither plane stress nor plane strain.

PLANE STRAIN

The *plane strain* compressibility in terms of the reduction in cross-sectional area A is given by

$$\frac{1}{K'_\phi} = \frac{1}{A}\frac{dA}{d\sigma} = \left\{ \begin{array}{ll} \frac{6(1-\nu)}{\mu_0} & \gamma \to 0 \\ \frac{2(1-\nu)}{\mu_0} & \gamma \to \infty \end{array} \right\}$$

The latter case ($\gamma \to \infty$) corresponds to a tube with a circular cross section and agrees (as it should) with the expression given below for the limiting case of a tube with an elliptical cross section with aspect ratio unity.

A general method of determining K'_ϕ for nearly arbitrarily shaped two-dimensional cavities under *plain strain* deformation was developed by Zimmerman (1986, 1991) and involves conformal mapping of the tube shape into circular pores. For example, pores with cross-sectional shapes that are n-sided hypotrochoids given by

$$x = \cos(\theta) + \frac{1}{(n-1)}\cos(n-1)\theta$$

$$y = -\sin(\theta) + \frac{1}{(n-1)}\sin(n-1)\theta$$

[examples labeled (1) in Table 2.6.1] have plane-strain compressibilities

$$\frac{1}{K'_\phi} = \frac{1}{A}\frac{dA}{d\sigma} = \frac{1}{K'^{cir}_\phi}\frac{1 + \frac{1}{(n-1)}}{1 - \frac{1}{(n-1)}}$$

where $1/K'^{cir}_\phi$ is the plane-strain compressibility of a circular tube given by

$$\frac{1}{K'^{cir}_\phi} = \frac{2(1-\nu)}{\mu_0}$$

Table 2.6.1 summarizes a few plane-strain pore compressibilities.

TABLE 2.6.1. Plane-strain compressibility normalized by the compressibility of a circular tube.

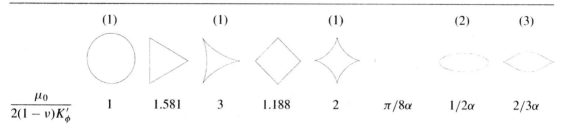

	(1)	(1)	(1)	(2)	(3)
$\dfrac{\mu_0}{2(1-v)K'_\phi}$	1	1.581	3	1.188	2	$\pi/8\alpha$	$1/2\alpha$	$2/3\alpha$

(1) $x = \cos(\theta) + \dfrac{1}{(n-1)}\cos(n-1)\theta$

$\quad 1 = -\sin(\theta) + \dfrac{1}{(n-1)}\sin(n-1)\theta$,   where n = number of sides

(2) $y = 2b\sqrt{1 - \left(\dfrac{x}{c}\right)^2}$ (ellipse)

(3) $y = 2b\left[1 - \left(\dfrac{x}{c}\right)^2\right]^{3/2}$ (nonelliptical, "tapered" crack)

TWO-DIMENSIONAL THIN CRACKS

A convenient description of very thin two-dimensional cracks is in terms of elastic line dislocations. Consider a crack lying along $-c < x < c$ in the $y = 0$ plane and very long in the z direction. The *total relative displacement* of the crack faces $u(x)$, defined as the displacement of the negative face ($y = 0^-$) relative to the positive face ($y = 0^+$), is related to the dislocation density function by

$$B(x) = -\frac{\partial u}{\partial x}$$

where $B(x)\,dx$ represents the total length of Burger's vectors of the dislocations lying between x and $x+dx$. The stress change in the plane of the crack that results from introduction of a dislocation line with unit Burger's vector at the origin is

$$\sigma = \frac{\mu_0}{2\pi Dx}$$

where $D = 1$ for screw dislocations and $D = (1 - v)$ for edge dislocations (v is Poisson's ratio, and μ_0 is the shear modulus). Edge dislocations can be used to describe mode I and mode II cracks; screw dislocations can be used to describe mode III cracks. The stress here is the component of traction in the crack plane parallel to the displacement: normal stress for mode I deformation, in-plane shear for mode II deformation, out-of-plane shear for mode III deformation.

Then the stress resulting from the distribution $B(x)$ is given by the convolution

$$\sigma(x) = \frac{\mu_0}{2\pi D} \int_{-c}^{c} \frac{B(x')\,dx'}{x - x'}$$

The special case of interest for nonfrictional cavities is the deformation for stress-free crack faces under a remote uniform tensional stress, $d\sigma$, acting normal to the plane of the crack. The outward displacement distribution of *each* crack face is given by (using edge dislocations)

$$U(x) = \frac{c(1 - v)\sqrt{1 - \left(\frac{x}{c}\right)^2}}{\mu_0} = \frac{2c(1 - v^2)\sqrt{1 - \left(\frac{x}{c}\right)^2}}{3K_0(1 - 2v)}$$

The volume change is then given by

$$\frac{dv_p}{d\sigma} = \pi U(0)ca = \frac{2\pi c^2 a}{3K_0} \frac{(1 - v^2)}{(1 - 2v)}$$

It is important to note that *these results for displacement and volume change apply to any two-dimensional crack of arbitrary cross section* as long as it is very thin and approximately planar. They are not limited to cracks of elliptical cross section.

For the special case in which the very thin crack is elliptical in cross section with half-width b in the thin direction, the volume is $v = \pi abc$, and the pore stiffness under plane-strain deformation is given by

$$\frac{1}{K'_\phi} = \frac{1}{A}\frac{dA}{d\sigma} = \frac{(c/b)(1 - v)}{\mu_0} = \frac{2(c/b)}{3K_0}\frac{(1 - v^2)}{(1 - 2v)}$$

Another special case is a crack of *nonelliptical* form (Mavko and Nur, 1978) with initial shape given by

$$U_0(x) = 2b\left[1 - \left(\frac{x}{c_0}\right)^2\right]^{3/2}$$

where c_0 is the crack half-length, and $2b$ is the maximum crack width. This crack is plotted in Figure 2.6.2. Note that unlike elliptical cracks that have rounded or blunted ends, this crack has tapered tips where faces make a smooth, tangent contact. If we apply a pressure, P, the crack shortens as well as thins, and the

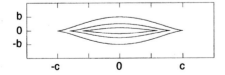 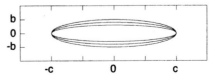

Figure 2.6.2. Nonelliptical crack shortens as well as narrows under compression. Elliptical crack only narrows.

pressure-dependent length is given by

$$c = c_0 \left[1 - \frac{2(1 - \nu)}{3\mu_0(b/c_0)} P \right]^{1/2}$$

Then the deformed shape is

$$U(x, P) = 2b \left(\frac{c}{c_0} \right)^3 \left[1 - \left(\frac{x}{c} \right)^2 \right]^{3/2} \qquad |x| \le c$$

An important consequence of the smoothly tapered crack tips and the gradual crack shortening is that there is no stress singularity at the crack tips. In this case, crack closure occurs (i.e., $U \to 0$) as the crack length goes to zero ($c \to 0$). The closing stress is

$$\sigma_{\text{close}} = \frac{3}{2(1 - \nu)} \alpha_0 \mu_0 = \frac{3}{4(1 - \nu^2)} \alpha_0 E_0$$

where $\alpha_0 = b/c_0$ is the original crack aspect ratio and μ_0 and E_0 are the shear and Young's moduli of the solid material, respectively. This expression is consistent with the usual rule of thumb that the crack-closing stress is numerically $\sim \alpha_0 E_0$. The exact factor depends on the details of the original crack shape. In comparison, the pressure required to close a two-dimensional elliptical crack of aspect ratio α_0 is

$$\sigma_{\text{close}} = \frac{1}{2(1 - \nu^2)} \alpha_0 E_0$$

ELLIPSOIDAL CRACKS OF FINITE THICKNESS

The pore compressibility under plane-strain deformation of a *two-dimensional* elliptical cavity of arbitrary aspect ratio α is given by (Zimmerman, 1991)

$$\frac{1}{K'_\phi} = \frac{1}{A} \frac{dA}{d\sigma} = \frac{1 - \nu}{\mu_0} \left[\alpha + \frac{1}{\alpha} \right] = \frac{2(1 - \nu^2)}{3K_0(1 - 2\nu)} \left[\alpha + \frac{1}{\alpha} \right]$$

where μ_0, K_0, and ν are the shear modulus, bulk modulus, and Poisson's ratio of the mineral material, respectively. Circular pores (tubes) correspond to aspect ratio $\alpha = 1$ and the pore compressibility is given as

$$\frac{1}{K'_\phi} = \frac{1}{A} \frac{dA}{d\sigma} = \frac{2(1 - \nu)}{\mu_0} = \frac{4(1 - \nu^2)}{3K_0(1 - 2\nu)}$$

USES

The equations presented in this section are useful for computing deformation of cavities in elastic solids and estimating effective moduli of porous solids.

ASSUMPTIONS AND LIMITATIONS

The equations presented in this section are based on the following assumptions:

- Solid material must be homogeneous, isotropic, linear, and elastic.
- Results for specific geometries, such as spheres and ellipsoids, are derived for single isolated cavities. Therefore, estimates of effective moduli based on these are limited to relatively low pore concentrations where pore elastic interaction is small.
- Pore pressure computations assume that the induced pore pressure is uniform throughout the pore space, which will be the case if (1) there is only one pore, (2) all pores are well connected and the frequency and viscosity are low enough for any pressure differences to equilibrate, or (3) all pores have the same dry pore stiffness.

2.7 DEFORMATION OF A CIRCULAR HOLE – BOREHOLE STRESSES

SYNOPSIS

Presented here are some solutions related to a circular hole in a stressed, linear, elastic, isotropic medium.

HOLLOW CYLINDER WITH INTERNAL AND EXTERNAL PRESSURES

The cylinder's internal radius is R_1, and the external radius is R_2. Hydrostatic stress p_1 is applied at the interior surface at R_1, and hydrostatic stress p_2 is applied at the exterior surface at R_2. The resulting (plane-strain) outward displacement U and radial and tangential stresses are

$$U = \frac{\left(p_2 R_2{}^2 - p_1 R_1{}^2\right)}{2(\lambda + \mu)\left(R_2{}^2 - R_1{}^2\right)} r + \frac{(p_2 - p_1)R_1{}^2 R_2{}^2}{2\mu\left(R_2{}^2 - R_1{}^2\right)} \frac{1}{r}$$

$$\sigma_{rr} = \frac{\left(p_2 R_2{}^2 - p_1 R_1{}^2\right)}{\left(R_2{}^2 - R_1{}^2\right)} - \frac{(p_2 - p_1)R_1{}^2 R_2{}^2}{\left(R_2{}^2 - R_1{}^2\right)} \frac{1}{r^2}$$

$$\sigma_{\theta\theta} = \frac{\left(p_2 R_2{}^2 - p_1 R_1{}^2\right)}{\left(R_2{}^2 - R_1{}^2\right)} + \frac{(p_2 - p_1)R_1{}^2 R_2{}^2}{\left(R_2{}^2 - R_1{}^2\right)} \frac{1}{r^2}$$

where λ and μ are the Lamé coefficient and shear modulus, respectively.

If $R_1 = 0$, we have the case of a *solid cylinder* under external pressure, with displacement and stress denoted by the following:

$$U = \frac{p_2 r}{2(\lambda + \mu)}$$

$$\sigma_{rr} = \sigma_{\theta\theta} = p_2$$

If, instead, $R_2 \to \infty$, then

$$U = \frac{p_2 r}{2(\lambda + \mu)} + \frac{(p_2 - p_1)R_1^2}{2\mu r}$$

$$\sigma_{rr} = p_2 \left(1 - \frac{R_1^2}{r^2}\right) + \frac{p_1 R_1^2}{r^2}$$

$$\sigma_{\theta\theta} = p_2 \left(1 + \frac{R_1^2}{r^2}\right) - \frac{p_1 R_1^2}{r^2}$$

These results for *plane strain* can be converted to *plane stress* by replacing v by $v/(1 + v)$, where v is the Poisson ratio.

CIRCULAR HOLE WITH PRINCIPAL STRESSES AT INFINITY

The circular hole with radius R lies along the z-axis. A principal stress, σ_{xx}, is applied at infinity. The stress solution is then

$$\sigma_{rr} = \frac{\sigma_{xx}}{2}\left(1 - \frac{R^2}{r^2}\right) + \frac{\sigma_{xx}}{2}\left(1 - \frac{4R^2}{r^2} + \frac{3R^4}{r^4}\right)\cos 2\theta$$

$$\sigma_{\theta\theta} = \frac{\sigma_{xx}}{2}\left(1 + \frac{R^2}{r^2}\right) - \frac{\sigma_{xx}}{2}\left(1 + \frac{3R^4}{r^4}\right)\cos 2\theta$$

$$\sigma_{r\theta} = -\frac{\sigma_{xx}}{2}\left(1 + \frac{2R^2}{r^2} - \frac{3R^4}{r^4}\right)\sin 2\theta$$

$$\frac{8\mu U_r}{R\sigma_{xx}} = (\chi - 1 + 2\cos 2\theta)\frac{r}{R} + \frac{2R}{r}\left[1 + \left(\chi + 1 - \frac{R^2}{r^2}\right)\cos 2\theta\right]$$

$$\frac{8\mu U_\theta}{R\sigma_{xx}} = \left[-\frac{2r}{R} + \frac{2R}{r}\left(1 - \chi - \frac{R^2}{r^2}\right)\right]\sin 2\theta$$

where θ is measured from the x-axis, and

$$\chi = 3 - 4v \quad \text{for plane strain}$$

$$\chi = \frac{3 - v}{1 + v} \quad \text{for plane stress}$$

At the cavity surface, $r = R$:

$$\sigma_{rr} = \sigma_{r\theta} = 0$$

$$\sigma_{\theta\theta} = \sigma_{xx}(1 - 2\cos 2\theta)$$

Thus, we see the well-known result that the borehole creates a stress concentration of $\sigma_{\theta\theta} = 3\sigma_{xx}$ at $\theta = 90°$.

STRESS CONCENTRATION AROUND AN ELLIPTICAL HOLE

If, instead, the borehole is elliptical in cross section with a shape denoted by (Lawn and Wilshaw, 1975)

$$\frac{x^2}{b^2} + \frac{y^2}{c^2} = 1$$

where b is the semiminor axis and c is the semimajor axis, and the principal stress σ_{xx} is applied at infinity, then the largest stress concentration occurs at the tip of the long axis ($y = c; x = 0$). This is the same location at $\theta = 90°$ as for the circular hole. The stress concentration is

$$\sigma_{\theta\theta} = \sigma_{xx}\left[1 + 2(c/\rho)^{1/2}\right]$$

where ρ is the radius of curvature at the tip given by

$$\rho = \frac{b^2}{c}$$

When $b \ll c$, the stress concentration is approximately

$$\frac{\sigma_{\theta\theta}}{\sigma_{xx}} \approx \frac{2c}{b} = 2\sqrt{\frac{c}{\rho}}$$

USES

The equations presented in this section can be used for the following:

- Estimating the stresses around a borehole resulting from tectonic stresses.
- Estimating the stresses and deformation of a borehole caused by changes in borehole fluid pressure.

ASSUMPTIONS AND LIMITATIONS

The equations presented in this section are based on the following assumptions:

- The material is linear, isotropic, and elastic.
- The borehole axis is along one of the principal axes of the remote stresses.

EXTENSIONS

More complicated remote stress fields can be constructed by superimposing the solutions for the corresponding principal stresses.

2.8 MOHR'S CIRCLES

SYNOPSIS

Mohr's circles provide a graphical representation of how the tractions on a plane depend on the angular orientation of the plane within a given stress field. Consider a stress state with principal stresses $\sigma_1 \geq \sigma_2 \geq \sigma_3$ and coordinate axes defined along the corresponding principal directions x_1, x_2, x_3. The traction vector, T, acting on a plane with outward unit normal vector, $n = (n_1, n_2, n_3)$, is given by Cauchy's formula as

$$T = \sigma n$$

where σ is the stress tensor. The components of n are the direction cosines of n relative to the coordinate axes and are denoted by

$$n_1 = \cos \phi$$

$$n_2 = \cos \gamma$$

$$n_3 = \cos \theta$$

and

$$n_1{}^2 + n_2{}^2 + n_3{}^2 = 1$$

where ϕ, γ, and θ are the angles between n and the axes x_1, x_2, x_3 (Figure 2.8.1).

The normal component of traction, σ, and the *magnitude* of the shear component, τ, acting on the plane are given by

$$\sigma = n_1{}^2\sigma_1 + n_2{}^2\sigma_2 + n_3{}^2\sigma_3$$

$$\tau^2 = n_1{}^2\sigma_1{}^2 + n_2{}^2\sigma_2{}^2 + n_3{}^2\sigma_3{}^2 - \sigma^2$$

THREE-DIMENSIONAL MOHR'S CIRCLE

The numerical values of σ and τ can be read graphically from the three-dimensional Mohr's circle shown in Figure 2.8.2. All permissible values of σ and τ must lie in the shaded area.

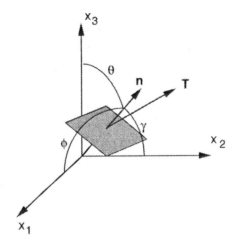

Figure 2.8.1

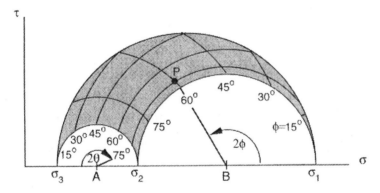

Figure 2.8.2. Three-dimensional Mohr's circle.

To determine σ and τ from the orientation of the plane (ϕ, γ, and θ), perform the following procedures:

1) Plot $\sigma_1 \geq \sigma_2 \geq \sigma_3$ on the horizontal axis and construct the three circles as shown. The outer circle is centered at $(\sigma_1 + \sigma_3)/2$ and has radius $(\sigma_1 - \sigma_3)/2$. The left inner circle is centered at $(\sigma_2 + \sigma_3)/2$ and has radius $(\sigma_2 - \sigma_3)/2$. The right inner circle is centered at $(\sigma_1 + \sigma_2)/2$ and has radius $(\sigma_1 - \sigma_2)/2$.

2) Mark angles 2θ and 2ϕ on the small circles centered at **A** and **B**. For example, $\phi = 60°$ plots at $2\phi = 120°$ from the horizontal, and $\theta = 75°$ plots at $2\theta = 150°$ from the horizontal, as shown. Be certain to include the factor of 2, and note the different directions defined for the positive angles.

3) Draw a circle centered at point **A** that intersects the right small circle at the mark for ϕ.

4) Draw another circle centered at point **B** that intersects the left small circle at the point for θ.

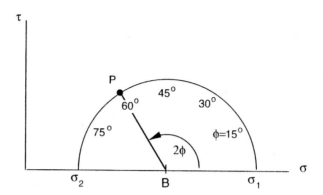

Figure 2.8.3. Two-dimensional Mohr's circle.

5) The intersection of the two constructed circles at point **P** gives the values of σ and τ.

Reverse the procedure to determine the orientation of the plane having particular values of σ and τ.

TWO-DIMENSIONAL MOHR'S CIRCLE

When the plane of interest contains one of the principal axes, the tractions on the plane depend only on the two remaining principal stresses, and using Mohr's circle is therefore simplified. For example, when $\theta = 90°$, i.e., the (x_3 axis lies in the plane of interest), all stress states lie on the circle centered at **B** in Figure 2.8.2. The stresses then depend only on σ_1 and σ_2 and on the angle ϕ, and we need only draw the single circle, as shown in Figure 2.8.3.

USES

The Mohr's circle is used for graphical determination of normal and shear tractions acting on a plane of arbitrary orientation relative to the principal stresses.

PART 3

SEISMIC WAVE PROPAGATION

3.1 SEISMIC VELOCITIES

SYNOPSIS

The velocities of various types of seismic waves in homogeneous, isotropic, elastic media are given by

$$V_{\mathrm{P}} = \sqrt{\frac{K + (4/3)\mu}{\rho}} = \sqrt{\frac{\lambda + 2\mu}{\rho}}$$

$$V_{\mathrm{S}} = \sqrt{\frac{\mu}{\rho}}$$

$$V_{\mathrm{E}} = \sqrt{\frac{E}{\rho}}$$

where

V_{P} = P-wave velocity
V_{S} = S-wave velocity
V_{E} = extensional wave velocity in a narrow bar

and

$$\rho = \text{density}$$
$$K = \text{bulk modulus}$$
$$\mu = \text{shear modulus}$$
$$\lambda = \text{Lamé's coefficient}$$
$$E = \text{Young's modulus}$$
$$\nu = \text{Poisson's ratio}$$

In terms of Poisson's ratio one can also write

$$\frac{V_P^2}{V_S^2} = \frac{2(1-\nu)}{(1-2\nu)}$$

$$\frac{V_E^2}{V_P^2} = \frac{(1+\nu)(1-2\nu)}{(1-\nu)}$$

$$\frac{V_E^2}{V_S^2} = 2(1+\nu)$$

$$\nu = \frac{V_P^2 - 2V_S^2}{2\left(V_P^2 - V_S^2\right)} = \frac{V_E^2 - 2V_S^2}{2V_S^2}$$

The various wave velocities are related by

$$\frac{V_P^2}{V_S^2} = \frac{4 - \frac{V_E^2}{V_S^2}}{3 - \frac{V_E^2}{V_S^2}}$$

$$\frac{V_E^2}{V_S^2} = \frac{3\frac{V_P^2}{V_S^2} - 4}{\frac{V_P^2}{V_S^2} - 1}$$

The elastic moduli can be extracted from measurements of density and any two wave velocities. For example,

$$\mu = \rho V_S^2$$

$$K = \rho\left(V_P^2 - \left(\frac{4}{3}\right)V_S^2\right)$$

$$E = \rho V_E^2$$

$$\nu = \frac{\left(V_P^2 - 2V_S^2\right)}{2\left(V_P^2 - V_S^2\right)}$$

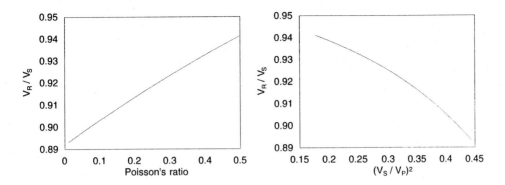

Figure 3.1.1. Rayleigh wave phase velocity normalized by shear velocity.

The **Rayleigh wave** phase velocity V_R at the surface of an isotropic homogeneous elastic half-space is given by the solution to the equation (White, 1983)

$$\left(2 - \frac{V_R^2}{V_S^2}\right)^2 - 4\left(1 - \frac{V_R^2}{V_P^2}\right)^{1/2}\left(1 - \frac{V_R^2}{V_S^2}\right)^{1/2} = 0$$

(Note that this equation in Bourbié, Coussy, and Zinszner, 1987 is in error.) The wave speed is plotted in Figure 3.1.1 and is given by

$$\frac{V_R^2}{V_S^2} = \left[-\frac{q}{2} + \left(\frac{q^2}{4} + \frac{p^3}{27}\right)^{1/2}\right]^{1/3} + \left[-\frac{q}{2} - \left(\frac{q^2}{4} + \frac{p^3}{27}\right)^{1/2}\right]^{1/3} + \frac{8}{3}$$

$$\text{for } \left(\frac{q^2}{4} + \frac{p^3}{27}\right) > 0$$

$$\frac{V_R^2}{V_S^2} = -2\left(\frac{-p}{3}\right)^{1/2}\cos\left[\frac{\pi - \cos^{-1}(-27q^2/4p^3)^{1/2}}{3}\right] + \frac{8}{3}$$

$$\text{for } \left(\frac{q^2}{4} + \frac{p^3}{27}\right) < 0$$

$$p = \frac{8}{3} - \frac{16V_S^2}{V_P^2} \qquad q = \frac{272}{27} - \frac{80V_S^2}{3V_P^2}$$

The Rayleigh velocity at the surface of a homogeneous elastic half-space is nondispersive (independent of frequency).

ASSUMPTIONS AND LIMITATIONS

These equations assume isotropic, linear, elastic media.

3.2 PHASE, GROUP, AND ENERGY VELOCITIES

SYNOPSIS

In the physics of wave propagation we often talk about different velocities (the phase, group, and energy velocities: V_p, V_g, and V_e, respectively) associated with the wave phenomenon. In laboratory measurements of core sample velocities using finite bandwidth signals and finite-sized transducers, the velocity given by the first arrival does not always correspond to an easily identified velocity.

A general time harmonic wave may be defined as

$$U(x, t) = U_0(x) \cos[\omega t - p(x)]$$

where ω is the angular frequency and U_0 and p are functions of position x; U can be any field of interest such as pressure, stress, or electromagnetic fields. The surfaces given by $p(x) = $ constant are called cophasal or wave surfaces. In particular, for plane waves, $p(x) = k \cdot x$ where k is the wave vector, or the propagation vector, and is in the direction of propagation. For the phase to be the same at (x, t) and $(x + dx, t + dt)$ we must have

$$\omega \, dt - (\text{grad } p) \cdot dx = 0$$

from which the **phase velocity** is defined as

$$V_p = \frac{\omega}{|\text{grad } p|}$$

For *plane waves* grad $p = k$, and hence $V_p = \omega/k$. The reciprocal of the phase velocity is often called the **slowness**, and a polar plot of slowness versus direction of propagation is termed the slowness surface. Phase velocity is the speed of advance of the cophasal surfaces. Born and Wolf (1980) consider the phase velocity to be devoid of any physical significance because it does not correspond to the velocity of propagation of any signal and cannot be directly determined experimentally.

Waves encountered in rock physics are rarely perfectly monochromatic but instead have a finite bandwidth, $\Delta\omega$, centered around some mean frequency $\bar{\omega}$. The wave may be regarded as a superposition of monochromatic waves of different frequencies, which then gives rise to the concept of wave packets or wave groups. Wave packets, or modulation on a wave containing a finite band of frequencies, propagate with the **group velocity** defined as

$$V_g = \frac{1}{\left|\text{grad}\left(\frac{\partial p}{\partial \omega}\right)_{\bar{\omega}}\right|}$$

which for plane waves becomes

$$V_g = \left(\frac{\partial \omega}{\partial k} \right)_{\tilde{\omega}}$$

The group velocity may be considered to be the velocity of propagation of the envelope of a modulated carrier wave. The group velocity can also be expressed in various equivalent ways as

$$V_g = V_p - \lambda \frac{d V_p}{d \lambda}$$

$$V_g = V_p + k \frac{d V_p}{d k}$$

or

$$\frac{1}{V_g} = \frac{1}{V_p} - \frac{\omega}{V_p^2} \frac{d V_p}{d \omega}$$

These equations show that the group velocity is different from the phase velocity when the phase velocity is *frequency dependent, direction dependent,* or both. When the phase velocity is frequency dependent (and hence different from the group velocity), the medium is said to be dispersive. Dispersion is termed *normal* if group velocity decreases with frequency and *anomalous* or *inverse* if it increases with frequency (Bourbié et al., 1987; Elmore and Heald, 1985). In elastic, isotropic media, dispersion can arise as a result of geometric effects such as propagation along waveguides. As a rule such geometric dispersion (Rayleigh waves, waveguides) is *normal* (i.e., the group velocity decreases with frequency). In a homogeneous viscoelastic medium, on the other hand, dispersion is *anomalous* or *inverse* and arises owing to intrinsic dissipation.

The **energy velocity** V_e represents the velocity at which energy propagates and may be defined as

$$V_e = \frac{P_{av}}{E_{av}}$$

where

$$P_{av} = \text{average power flow density}$$
$$E_{av} = \text{average total energy density}$$

In *isotropic, homogeneous, elastic media* all three velocities are the same. In a *lossless homogeneous* medium (of arbitrary symmetry), V_g and V_e are identical, and energy propagates with the group velocity. In this case the energy velocity may be obtained from the group velocity, which is usually somewhat easier to compute. If the medium is not strongly dispersive and a wave group can travel a measurable distance without appreciable "smearing" out, the group velocity *may* be considered to represent the velocity at which the energy is propagated (though this is not strictly true in general).

In *anisotropic, homogeneous, elastic* media, the phase velocity, in general, differs from the group velocity (which is equal to the energy velocity because the medium is elastic) except along certain symmetry directions, where they coincide. The direction in which V_e is deflected away from k (which is also the direction of V_p) is obtained from the slowness surface, for V_e ($=V_g$ in elastic media) must always be normal to the slowness surface (Auld, 1990).

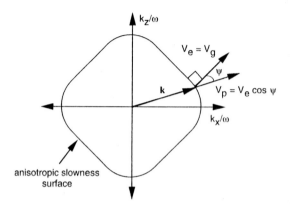

Figure 3.2.1. In anisotropic media energy propagates along V_e, which is always normal to the slowness surface and in general is deflected away from V_p and the wave vector k.

The group velocity in anisotropic media may be calculated by differentiation of the dispersion relation obtained in an implicit form from the Christoffel equation given by

$$|k^2 c_{ijkl} n_j n_l - \rho \omega^2 \delta_{ik}| = \Phi(\omega, k_x, k_y, k_z) = 0$$

where c_{ijkl} is the stiffness tensor, n_i are the direction cosines of k, ρ is the density, and δ_{ij} is the Kronecker delta. The group velocity is then evaluated as

$$V_g = -\frac{\nabla_k \Phi}{\partial \Phi / \partial \omega}$$

where the gradient is with respect to k_x, k_y, k_z.

The concept of group velocity is not strictly applicable to attenuating viscoelastic media, but the energy velocity is still well defined (White, 1983). The energy propagation velocity in dissipative medium is neither the group velocity nor the phase velocity *except* when

1) the medium is infinite, homogeneous, linear, and viscoelastic, and
2) the wave is monochromatic and homogeneous (i.e., planes of equal phase are parallel to planes of equal amplitude, or, in other words, the real and imaginary parts of the complex wave vector point in the same direction [in general they do not], in which case the energy velocity is equal to the phase velocity [Bourbié et al., 1987; Ben-Menahem and Singh, 1981]).

For the special case of a Voigt solid (see Section 3.6 on viscoelasticity) the energy transport velocity is equal to the phase velocity at all frequencies. For wave propagation in dispersive, viscoelastic media, one sometimes defines the limit

$$V_\infty = \lim_{\omega \to \infty} V_p(\omega)$$

which describes the propagation of a well-defined wavefront and is referred to as the **signal velocity** (Beltzer, 1988).

Sometimes it is not clear which velocities are represented by the recorded travel times in laboratory ultrasonic core sample measurements, especially when the sample is anisotropic. For elastic materials, there is no ambiguity for propagation along symmetry directions because the phase and group velocities are identical. For nonsymmetry directions, the energy does not necessarily propagate straight up the axis of the core from the transducer to the receiver. Numerical modeling of laboratory experiments (Dellinger and Vernik, 1992) indicates that, for typical transducer widths (≈ 10 mm), the recorded travel times correspond closely to the *phase velocity*. Accurate measurement of group velocity along nonsymmetry directions would essentially require point transducers less than 2 mm wide.

According to Bourbié et al. (1987), the velocity measured by a resonant-bar standing-wave technique corresponds to the phase velocity.

ASSUMPTIONS AND LIMITATIONS

In general, phase, group, and energy velocities may differ from each other in both magnitude and direction. Under certain conditions two or more of them may become identical. For homogeneous, linear, isotropic, elastic media all three are the same.

3.3 IMPEDANCE, REFLECTIVITY, AND TRANSMISSIVITY

SYNOPSIS

The impedance, I, of an elastic medium is the ratio of stress to particle velocity (Aki and Richards, 1980) and is given by ρV, where ρ is the density and V is the wave propagation velocity. At a plane interface between two thick, homogeneous, isotropic, elastic layers, the normal incidence reflectivity for waves traveling from medium 1 to medium 2 is the ratio of the displacement amplitude, A_r, of the reflected wave to that of the incident wave, A_i, and is given by

$$R_{12} = \frac{A_r}{A_i} = \frac{I_2 - I_1}{I_2 + I_1} = \frac{\rho_2 V_2 - \rho_1 V_1}{\rho_2 V_2 + \rho_1 V_1}$$

This expression for the reflection coefficient is obtained when the *particle displacements are measured with respect to the direction of the wave vector* (equivalent to the slowness vector or direction of propagation). A displacement is taken to be positive when its component along the interface has the same phase (or same direction) as the component of the wave vector along the interface. For P-waves, this means that positive displacement is along the direction of propagation. Thus, a positive reflection coefficient implies that a compression is reflected as a compression, whereas a negative reflection coefficient implies a phase inversion (Sheriff, 1991). When the *displacements are measured with respect to a space-fixed coordinate system*, and not with respect to the wave vector, the reflection coefficient is given by

$$R_{12} = \frac{A_r}{A_i} = \frac{I_1 - I_2}{I_2 + I_1} = \frac{\rho_1 V_1 - \rho_2 V_2}{\rho_2 V_2 + \rho_1 V_1}$$

The normal incidence transmissivity in both coordinate systems is

$$T_{12} = \frac{A_t}{A_i} = \frac{2I_1}{I_2 + I_1} = \frac{2\rho_1 V_1}{\rho_2 V_2 + \rho_1 V_1}$$

where A_t is the displacement amplitude of the transmitted wave. Continuity at the interface requires

$$A_i + A_r = A_t$$

$$1 + R = T$$

This choice of signs for A_i and A_r is for a space-fixed coordinate system. Note that the reflection and transmission coefficients for wave amplitudes can be greater than 1. Sometimes the reflection and transmission coefficients are defined in terms of scaled displacements A', which are proportional to the square root of energy flux (Aki and Richards, 1980; Kennett, 1983). The scaled displacements are given by

$$A' = A\sqrt{\rho V \cos \theta}$$

where θ is the angle between the wave vector and the normal to the interface. The normal incidence reflection and transmission coefficients in terms of these scaled displacements are

$$R'_{12} = \frac{A'_r}{A'_i} = R_{12}$$

$$T'_{12} = \frac{A'_t}{A'_i} = T_{12} \frac{\sqrt{\rho_2 V_2}}{\sqrt{\rho_1 V_1}} = \frac{2\sqrt{\rho_1 V_1 \rho_2 V_2}}{\rho_2 V_2 + \rho_1 V_1}$$

Reflectivity and transmissivity for energy fluxes, R^e and T^e, respectively, are given by the squares of the reflection and transmission coefficients for scaled displacements. For normal incidence they are

$$R_{12}^e = \frac{E_r}{E_i} = \left(R_{12}'\right)^2 = \frac{(\rho_1 V_1 - \rho_2 V_2)^2}{(\rho_2 V_2 + \rho_1 V_1)^2}$$

$$T_{12}^e = \frac{E_t}{E_i} = \left(T_{12}'\right)^2 = \frac{4\rho_1 V_1 \rho_2 V_2}{(\rho_2 V_2 + \rho_1 V_1)^2}$$

where E_i, E_r, and E_t are the incident, reflected, and transmitted energy fluxes, respectively. Conservation of energy at an interface where no trapping of energy occurs requires that

$$E_i = E_r + E_t$$

$$1 = R^e + T^e$$

The reflection and transmission coefficients for energy fluxes can never be greater than 1.

ROUGH SURFACES

Random interface roughness at scales smaller than the wavelength causes incoherent scattering and a decrease in amplitude of the coherent reflected and transmitted waves. This could be one of the explanations for the observation that amplitudes of multiples in synthetic seismograms are often larger than the amplitudes of corresponding multiples in the data (Frazer, 1994). Kuperman (1975) gives results that modify the reflectivity and transmissivity to include scattering losses at the interface. With the mean squared departure from planarity of the rough interface denoted by σ^2, the modified coefficients are

$$\tilde{R}_{12} = R_{12}\left(1 - 2k_1^2\sigma^2\right) = R_{12}[1 - 8\pi^2(\sigma/\lambda_1)^2]$$

$$\tilde{T}_{12} = T_{12}\left(1 - \frac{1}{2}(k_1 - k_2)^2\sigma^2\right) = T_{12}\left[1 - 2\pi^2(\lambda_2 - \lambda_1)^2\left(\frac{\sigma}{\lambda_1\lambda_2}\right)^2\right]$$

where $k_1 = \omega/V_1$, $k_2 = \omega/V_2$ are the wavenumbers, and λ_1 and λ_2 are the wavelengths in media 1 and 2, respectively.

USES

The equations presented in this section can be used for the following purposes:

- To calculate amplitudes and energy fluxes of reflected and transmitted waves at interfaces in elastic media.
- To estimate the decrease in wave amplitude caused by scattering losses during reflection and transmission at rough interfaces.

ASSUMPTIONS AND LIMITATIONS

The equations presented in this section apply only under the following conditions:

- Normal incidence, plane-wave, time harmonic propagation in isotropic, linear, elastic media with a single interface.
- No energy losses or trapping at the interface.
- Rough surface results are valid for small deviations from planarity (small σ).

3.4 REFLECTIVITY AND AVO

SYNOPSIS

The **seismic impedance** is the product of velocity and density (see Section 3.3) as expressed by

$$I_P = \rho V_P$$

$$I_S = \rho V_S$$

where

$$
\begin{aligned}
I_P, I_S &= \text{P- and S-wave impedances} \\
V_P, V_S &= \text{P- and S-wave velocities} \\
\rho &= \text{density}
\end{aligned}
$$

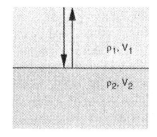

Figure 3.4.1

At an interface between two thick homogeneous, isotropic, elastic layers, the **normal incidence reflectivity**, defined as the ratio of reflected wave amplitude to incident wave amplitude, is

$$R_{PP} = \frac{\rho_2 V_{P2} - \rho_1 V_{P1}}{\rho_2 V_{P2} + \rho_1 V_{P1}} = \frac{I_{P2} - I_{P1}}{I_{P2} + I_{P1}}$$

$$\approx \frac{1}{2} \ln (I_{P2}/I_{P1})$$

$$R_{SS} = \frac{\rho_2 V_{S2} - \rho_1 V_{S1}}{\rho_2 V_{S2} + \rho_1 V_{S1}} = \frac{I_{S2} - I_{S1}}{I_{S2} + I_{S1}}$$

$$\approx \frac{1}{2} \ln (I_{S2}/I_{S1})$$

where R_{PP} is the normal incidence P-to-P reflectivity, R_{SS} is the S-to-S reflectivity, and the subscripts 1 and 2 refer to the first and second media, respectively.

The logarithmic approximation is reasonable for $|R| < 0.5$ (Castagna, 1993). A normally incident P-wave generates only reflected and transmitted P-waves. A normally incident S-wave generates only reflected and transmitted S-waves. There is no mode conversion.

AVO: AMPLITUDE VARIATIONS WITH OFFSET

For nonnormal incidence, the situation is more complicated. An incident P-wave generates reflected P- and S-waves and transmitted P- and S-waves. The reflection and transmission coefficients depend on the angle of incidence as well as on the material properties of the two layers. An excellent review is given by Castagna (1993).

The angles of the incident, reflected, and transmitted rays (Figure 3.4.2) are related by Snell's law as follows

$$p = \frac{\sin \theta_1}{V_{P1}} = \frac{\sin \theta_2}{V_{P2}} = \frac{\sin \phi_1}{V_{S1}} = \frac{\sin \phi_2}{V_{S2}}$$

where p is the **ray parameter**.

The complete solution for the amplitudes of transmitted and reflected P- and S-waves for both incident P- and S-waves is given by the Zoeppritz (1919) equations (Aki and Richards, 1980; Castagna, 1993).

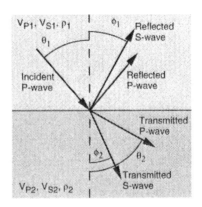

Figure 3.4.2

Aki and Richards (1980) give the results in the following convenient matrix form:

$$\begin{pmatrix} \overset{\downarrow\,\uparrow}{PP} & \overset{\downarrow\,\uparrow}{SP} & \overset{\uparrow\,\uparrow}{PP} & \overset{\uparrow\,\uparrow}{SP} \\ \overset{\downarrow\,\uparrow}{PS} & \overset{\downarrow\,\uparrow}{SS} & \overset{\uparrow\,\uparrow}{PS} & \overset{\uparrow\,\uparrow}{SS} \\ \overset{\downarrow\,\downarrow}{PP} & \overset{\downarrow\,\downarrow}{SP} & \overset{\uparrow\,\downarrow}{PP} & \overset{\uparrow\,\downarrow}{SP} \\ \overset{\downarrow\,\downarrow}{PS} & \overset{\downarrow\,\downarrow}{SS} & \overset{\uparrow\,\downarrow}{PS} & \overset{\uparrow\,\downarrow}{SS} \end{pmatrix} = \mathbf{M}^{-1}\mathbf{N}$$

where each matrix element is a reflection coefficient. The first letter designates the type of incident wave, and the second letter designater the type of reflected wave. The arrows indicate downward $\downarrow$ and upward $\uparrow$ propagation. The matrices **M** and **N** are given by

$$\mathbf{M} = \begin{pmatrix} -\sin \theta_1 & -\cos \phi_1 & \sin \theta_2 & \cos \phi_2 \\ \cos \theta_1 & -\sin \phi_1 & \cos \theta_2 & -\sin \phi_2 \\ 2\rho_1 V_{S1} \sin \phi_1 \cos \theta_1 & \rho_1 V_{S1}(1 - 2\sin^2 \phi_1) & 2\rho_2 V_{S2} \sin \phi_2 \cos \theta_2 & \rho_2 V_{S2}(1 - 2\sin^2 \phi_2) \\ -\rho_1 V_{P1}(1 - 2\sin^2 \phi_1) & \rho_1 V_{S1} \sin 2\phi_1 & \rho_2 V_{P2}(1 - 2\sin^2 \phi_2) & -\rho_2 V_{S2} \sin 2\phi_2 \end{pmatrix}$$

$$\mathbf{N} = \begin{pmatrix} \sin \theta_1 & \cos \phi_1 & -\sin \theta_2 & -\cos \phi_2 \\ \cos \theta_1 & -\sin \phi_1 & \cos \theta_2 & -\sin \phi_2 \\ 2\rho_1 V_{S1} \sin \phi_1 \cos \theta_1 & \rho_1 V_{S1}(1 - 2\sin^2 \phi_1) & 2\rho_2 V_{S2} \sin \phi_2 \cos \theta_2 & \rho_2 V_{S2}(1 - 2\sin^2 \phi_2) \\ \rho_1 V_{P1}(1 - 2\sin^2 \phi_1) & -\rho_1 V_{S1} \sin 2\phi_1 & -\rho_2 V_{P2}(1 - 2\sin^2 \phi_2) & \rho_2 V_{S2} \sin 2\phi_2 \end{pmatrix}$$

The most useful results are the P-to-P reflectivity, $R_{PP} = \overset{\downarrow\uparrow}{PP}$, and P-to-S reflectivity, $R_{PS} = \overset{\downarrow\uparrow}{PS}$, given explicitly by Aki and Richards (1980) as follows:

$$R_{PP} = \left[\left(b\frac{\cos\theta_1}{V_{P1}} - c\frac{\cos\theta_2}{V_{P2}} \right) F - \left(a + d\frac{\cos\theta_1}{V_{P1}}\frac{\cos\phi_2}{V_{S2}} \right) Hp^2 \right] \Big/ D$$

$$R_{PS} = \left[-2\frac{\cos\theta_1}{V_{P1}} \left(ab + cd\frac{\cos\theta_2}{V_{P2}}\frac{\cos\phi_2}{V_{S2}} \right) pV_{P1} \right] \Big/ (V_{S1} D)$$

where

$$a = \rho_2(1 - 2\sin^2\phi_2) - \rho_1(1 - 2\sin^2\phi_1)$$

$$b = \rho_2(1 - 2\sin^2\phi_2) + 2\rho_1 \sin^2\phi_1$$

$$c = \rho_1(1 - 2\sin^2\phi_1) + 2\rho_2 \sin^2\phi_2$$

$$d = 2\left(\rho_2 V_{S2}^2 - \rho_1 V_{S1}^2\right)$$

$$D = EF + GHp^2 = (\det \mathbf{M})/(V_{P1} V_{P2} V_{S1} V_{S2})$$

$$E = b\frac{\cos\theta_1}{V_{P1}} + c\frac{\cos\theta_2}{V_{P2}}$$

$$F = b\frac{\cos\phi_1}{V_{S1}} + c\frac{\cos\phi_2}{V_{S2}}$$

$$G = a - d\frac{\cos\theta_1}{V_{P1}}\frac{\cos\phi_2}{V_{S2}}$$

$$H = a - d\frac{\cos\theta_2}{V_{P2}}\frac{\cos\phi_1}{V_{S1}}$$

APPROXIMATE FORMS

Although the complete Zoeppritz equations can be evaluated numerically, it is often useful and more insightful to use one of the simpler approximations.

Bortfeld (1961) linearized the Zoeppritz equations by assuming small contrasts between layer properties as follows:

$$R_{PP}(\theta_1) \approx \frac{1}{2}\ln\left(\frac{V_{P2}\rho_2 \cos\theta_1}{V_{P1}\rho_1 \cos\theta_2}\right) + \left(\frac{\sin\theta_1}{V_{P1}}\right)^2 \left(V_{S1}^2 - V_{S2}^2\right)\left[2 + \frac{\ln\left(\frac{\rho_2}{\rho_1}\right)}{\ln\left(\frac{V_{S2}}{V_{S1}}\right)}\right]$$

Aki and Richards (1980) also derived a simplified form by assuming small layer contrasts. The results are conveniently expressed in terms of contrasts in V_P, V_S, and ρ as follows:

$$R_{PP}(\theta) \approx \frac{1}{2}(1 - 4p^2 V_S^2)\frac{\Delta\rho}{\rho} + \frac{1}{2\cos^2\theta}\frac{\Delta V_P}{V_P} - 4p^2 V_S^2\frac{\Delta V_S}{V_S}$$

$$R_{PS}(\theta) \approx \frac{-pV_p}{2\cos\phi}\left[\left(1 - 2V_S^2 p^2 + 2V_S^2\frac{\cos\theta\cos\phi}{V_P\ V_S}\right)\frac{\Delta\rho}{\rho}\right.$$
$$\left. - \left(4V_S^2 p^2 - 4V_S^2\frac{\cos\theta\cos\phi}{V_P\ V_S}\right)\frac{\Delta V_S}{V_S}\right]$$

where

$$p = \frac{\sin\theta_1}{V_{P1}} \qquad\qquad \theta = (\theta_2 + \theta_1)/2$$
$$\Delta\rho = \rho_2 - \rho_1 \qquad\qquad \rho = (\rho_2 + \rho_1)/2$$
$$\Delta V_P = V_{P2} - V_{P1} \qquad\qquad V_P = (V_{P2} + V_{P1})/2$$
$$\Delta V_S = V_{S2} - V_{S1} \qquad\qquad V_S = (V_{S2} + V_{S1})/2$$

Often, θ is approximated as θ_1.

This can be rewritten in the familiar form:

$$R_{PP}(\theta) \approx R_{P0} + B\sin^2\theta + C[\tan^2\theta - \sin^2\theta]$$

or

$$R_{PP}(\theta) \approx \frac{1}{2}\left(\frac{\Delta V_P}{V_P} + \frac{\Delta\rho}{\rho}\right) + \left[\frac{1}{2}\frac{\Delta V_P}{V_P} - 2\frac{V_S^2}{V_P^2}\left(\frac{\Delta\rho}{\rho} + 2\frac{\Delta V_S}{V_S}\right)\right]\sin^2\theta$$
$$+ \frac{1}{2}\frac{\Delta V_P}{V_P}[\tan^2\theta - \sin^2\theta]$$

This form can be interpreted in terms of different angular ranges (Castagna, 1993). In the above equations R_{P0} is the normal incidence reflection coefficient as expressed by

$$R_{P0} = \frac{I_{P2} - I_{P1}}{I_{P2} + I_{P1}} \approx \frac{\Delta I_P}{2I_P} \approx \frac{1}{2}\left(\frac{\Delta V_P}{V_P} + \frac{\Delta\rho}{\rho}\right)$$

The parameter B describes the variation at intermediate offsets and is often called the **AVO gradient**, and C dominates at far offsets near the critical angle.

Shuey (1985) presented a similar approximation where the AVO gradient is expressed in terms of the Poisson's ratio v as follows:

$$R_{PP}(\theta_1) \approx R_{P0} + \left[E R_{P0} + \frac{\Delta v}{(1-v)^2}\right]\sin^2\theta_1 + \frac{1}{2}\frac{\Delta V_P}{V_P}[\tan^2\theta_1 - \sin^2\theta_1]$$

where

$$R_{P0} = \frac{1}{2}\left(\frac{\Delta V_P}{V_P} + \frac{\Delta\rho}{\rho}\right)$$

$$E = F - 2(1 + F)\left(\frac{1 - 2v}{1 - v}\right)$$

$$F = \frac{\Delta V_P/V_P}{\Delta V_P/V_P + \Delta\rho/\rho}$$

and

$$\Delta v = v_2 - v_1$$
$$v = (v_2 + v_1)/2$$

Smith and Gidlow (1987) offered a further simplification to the Aki–Richards equation by removing the dependence on density using Gardner's equation (see Section 7.9) as follows:

$$\rho \propto V^{1/4}$$

giving

$$R_{PP}(\theta) \approx c \frac{\Delta V_P}{V_P} + d \frac{\Delta V_S}{V_S}$$

where

$$c = \frac{5}{8} - \frac{1}{2} \frac{V_S^2}{V_P^2} \sin^2 \theta + \frac{1}{2} \tan^2 \theta$$

$$d = -4 \frac{V_S^2}{V_P^2} \sin^2 \theta$$

Wiggins, Kenny, and McClure (1983) showed that when $V_P \approx 2 V_S$, the AVO gradient is approximately (Spratt, Goins, and Fitch, 1993)

$$B \approx R_{P0} - 2 R_{S0}$$

given that the P and S normal incident reflection coefficients are

$$R_{P0} \approx \frac{1}{2} \left(\frac{\Delta V_P}{V_P} + \frac{\Delta \rho}{\rho} \right)$$

$$R_{S0} \approx \frac{1}{2} \left(\frac{\Delta V_S}{V_S} + \frac{\Delta \rho}{\rho} \right)$$

Hilterman (1989) suggested the following slightly modified form:

$$R_{PP}(\theta) \approx R_{P0} \cos^2 \theta + PR \sin^2 \theta$$

where R_{P0} is the normal incidence reflection coefficient and

$$PR = \frac{v_2 - v_1}{(1 - v)^2}$$

This modified form has the interpretation that the near-offset traces reveal the P-wave impedance, and the intermediate-offset traces image contrasts in Poisson's ratio (Castagna, 1993).

ASSUMPTIONS AND LIMITATIONS

The equations presented in this section apply in the following cases:

- The rock is linear, isotropic, and elastic.

- Plane-wave propagation is assumed.
- Most of the simplified forms assume small contrasts in material properties across the boundary and angles of incidence less than about $30°$.

3.5 AVOZ IN ANISOTROPIC ENVIRONMENTS

SYNOPSIS

An incident wave at a boundary between two anisotropic media can generate reflected quasi-P-waves and quasi-S-waves as well as transmitted quasi-P-waves and quasi-S-waves (Auld, 1990). In general, the reflection and transmission coefficients vary with offset and azimuth. The **AVOZ** (amplitude variation with offset and azimuth) can be detected by three-dimensional seismic surveys and is a useful seismic attribute for reservoir characterization.

Brute-force modeling of AVOZ by solving the Zoeppritz (1919) equations can be complicated and unintuitive for several reasons: For anisotropic media in general, the two shear waves are separate (shear wave birefringence); the slowness surfaces are nonspherical and are not necessarily convex; and the polarization vectors are neither parallel nor perpendicular to the propagation vectors.

Schoenberg and Protázio (1992) give explicit solutions for the plane-wave reflection and transmission problem in terms of submatrices of the coefficient matrix of the Zoeppritz equations. The most general case of the explicit solutions is applicable to monoclinic media with a mirror plane of symmetry parallel to the reflecting plane. Let **R** and **T** represent the **reflection and transmission matrices**, respectively,

$$\mathbf{R} = \begin{bmatrix} R_{PP} & R_{SP} & R_{TP} \\ R_{PS} & R_{SS} & R_{TS} \\ R_{PT} & R_{ST} & R_{TT} \end{bmatrix}$$

$$\mathbf{T} = \begin{bmatrix} T_{PP} & T_{SP} & T_{TP} \\ T_{PS} & T_{SS} & T_{TS} \\ T_{PT} & T_{ST} & T_{TT} \end{bmatrix}$$

where the first subscript denotes the type of incident wave and the second subscript denotes the type of reflected or transmitted wave. For "weakly" anisotropic media, the subscript P denotes the P-wave, S denotes one quasi-S-wave, and T denotes the other quasi-S-wave (i.e., the tertiary or third wave). As a convention for real $s_{3P}{}^2$, $s_{3S}{}^2$, and $s_{3T}{}^2$,

$$s_{3P}{}^2 < s_{3S}{}^2 < s_{3T}{}^2$$

where s_{3i} is the vertical component of the phase slowness of the ith wave type when the reflecting plane is horizontal. An imaginary value for any of the vertical slownesses implies that the corresponding wave is inhomogeneous or evanescent. The impedance matrices are defined as

$$
\mathbf{X} = \begin{bmatrix}
e_{P1} & e_{S1} & e_{T1} \\
e_{P2} & e_{S2} & e_{T2} \\
-(C_{13}e_{P1} + C_{36}e_{P2})s_1 & -(C_{13}e_{S1} + C_{36}e_{S2})s_1 & -(C_{13}e_{T1} + C_{36}e_{T2})s_1 \\
-(C_{23}e_{P2} + C_{36}e_{P1})s_2 & -(C_{23}e_{S2} + C_{36}e_{S1})s_2 & -(C_{23}e_{T2} + C_{36}e_{T1})s_2 \\
-C_{33}e_{P3}s_{3P} & -C_{33}e_{S3}s_{3S} & -C_{33}e_{T3}s_{3T}
\end{bmatrix}
$$

$$
\mathbf{Y} = \begin{bmatrix}
-(C_{55}s_1 + C_{45}s_2)e_{P3} & -(C_{55}s_1 + C_{45}s_2)e_{S3} & -(C_{55}s_1 + C_{45}s_2)e_{T3} \\
-(C_{55}e_{P1} + C_{45}e_{P2})s_{3P} & -(C_{55}e_{S1} + C_{45}e_{S2})s_{3S} & -(C_{55}e_{T1} + C_{45}e_{T2})s_{3T} \\
-(C_{45}s_1 + C_{44}s_2)e_{P3} & -(C_{45}s_1 + C_{44}s_2)e_{S3} & -(C_{45}s_1 + C_{44}s_2)e_{T3} \\
-(C_{45}e_{P1} + C_{44}e_{P2})s_{3P} & -(C_{45}e_{S1} + C_{44}e_{S2})s_{3S} & -(C_{45}e_{T1} + C_{44}e_{T2})s_{3T} \\
e_{P3} & e_{S3} & e_{T3}
\end{bmatrix}
$$

where s_1 and s_2 are the horizontal components of the phase slowness vector; e_P, e_S, and e_T are the associated eigenvectors evaluated from the Christoffel equations (see Section 3.2), and C_{IJ} denotes elements of the stiffness matrix of the incident medium. The $\mathbf{X}'$ and $\mathbf{Y}'$ are the same as above except that primed parameters (transmission medium) replace unprimed parameters (incidence medium). When neither $\mathbf{X}$ nor $\mathbf{Y}$ is singular and $(\mathbf{X}^{-1}\mathbf{X}' + \mathbf{Y}^{-1}\mathbf{Y}')$ is invertible, the reflection and transmission coefficients can be written as

$$\mathbf{T} = 2(\mathbf{X}^{-1}\mathbf{X}' + \mathbf{Y}^{-1}\mathbf{Y}')^{-1}$$

$$\mathbf{R} = (\mathbf{X}^{-1}\mathbf{X}' - \mathbf{Y}^{-1}\mathbf{Y}')(\mathbf{X}^{-1}\mathbf{X}' + \mathbf{Y}^{-1}\mathbf{Y}')^{-1}$$

Schoenberg and Protázio (1992) point out that a singularity occurs at a horizontal slowness for which an interface wave (e.g., a Stoneley wave) exists. When $\mathbf{Y}$ is singular, straightforward matrix manipulations yield

$$\mathbf{T} = 2\mathbf{Y}'^{-1}\mathbf{Y}(\mathbf{X}^{-1}\mathbf{X}'\mathbf{Y}'^{-1}\mathbf{Y} + \mathbf{I})^{-1}$$

$$\mathbf{R} = (\mathbf{X}^{-1}\mathbf{X}'\mathbf{Y}'^{-1}\mathbf{Y} + \mathbf{I})^{-1}(\mathbf{X}^{-1}\mathbf{X}'\mathbf{Y}'^{-1}\mathbf{Y} - \mathbf{I})$$

Similarly, $\mathbf{T}$ and $\mathbf{R}$ can also be written without $\mathbf{X}^{-1}$ when $\mathbf{X}$ is singular as

$$\mathbf{T} = 2\mathbf{X}'^{-1}\mathbf{X}(\mathbf{I} + \mathbf{Y}^{-1}\mathbf{Y}'\mathbf{X}'^{-1}\mathbf{X})^{-1}$$

$$\mathbf{R} = (\mathbf{I} + \mathbf{Y}^{-1}\mathbf{Y}'\mathbf{X}'^{-1}\mathbf{X})^{-1}(\mathbf{I} - \mathbf{Y}^{-1}\mathbf{Y}'\mathbf{X}'^{-1}\mathbf{X})$$

Alternative solutions can be found by assuming that $\mathbf{X}'$ and $\mathbf{Y}'$ are invertible

$$\mathbf{R} = (\mathbf{Y}'^{-1}\mathbf{Y} + \mathbf{X}'^{-1}\mathbf{X})^{-1}(\mathbf{Y}'^{-1}\mathbf{Y} - \mathbf{X}'^{-1}\mathbf{X})$$

$$\mathbf{T} = 2\mathbf{X}'^{-1}\mathbf{X}(\mathbf{Y}'^{-1}\mathbf{Y} + \mathbf{X}'^{-1}\mathbf{X})^{-1}\mathbf{Y}'^{-1}\mathbf{Y}$$

$$= 2\mathbf{Y}'^{-1}\mathbf{Y}(\mathbf{Y}'^{-1}\mathbf{Y} + \mathbf{X}'^{-1}\mathbf{X})^{-1}\mathbf{X}'^{-1}\mathbf{X}$$

These formulas allow more straightforward calculations when the media have at least monoclinic symmetry with a horizontal symmetry plane.

APPROXIMATE FORMS

For a wave traveling in anisotropic media, there will generally be out-of-plane motion unless the wave path is in a symmetry plane. These symmetry planes include all vertical planes in **TIV** (transversely isotropic with vertical symmetry axis) media, the symmetry planes in **TIH** (transversely isotropic with horizontal symmetry axis), and orthorhombic media. In this case, the quasi-P- and the quasi-S-waves in the symmetry plane uncouple from the quasi-S-wave polarized transversely to the symmetry plane. For weakly anisotropic media, we can use simple analytical formulas (Banik, 1987; Thomsen, 1993; Rüger, 1995, 1996; Chen, 1995) to compute AVOZ responses at the interface of anisotropic media that can be either TIV, TIH, or orthorhombic. The analytical formulas give more insight into the dependence of AVOZ on anisotropy.

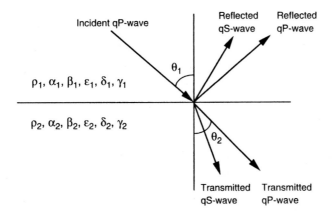

Figure 3.5.1. Reflected and transmitted rays caused by a P-wave incident at a boundary between two anisotropic media.

Thomsen (1986) introduced the following notation for weak transversely isotropic media with density ρ:

$$\alpha = \sqrt{\frac{C_{33}}{\rho}} \qquad \varepsilon = \frac{C_{11} - C_{33}}{2C_{33}}$$

$$\beta = \sqrt{\frac{C_{44}}{\rho}} \qquad \gamma = \frac{C_{66} - C_{44}}{2C_{44}}$$

$$\delta = \frac{(C_{13} + C_{44})^2 - (C_{33} - C_{44})^2}{2C_{33}(C_{33} - C_{44})}$$

The P-wave reflection coefficient for weakly anisotropic TIV media in the limit of small impedance contrast is given by (Thomsen, 1993)

$$R_{PP}(\theta) = R_{PP\text{-iso}}(\theta) + R_{PP\text{-aniso}}(\theta)$$

$$R_{PP\text{-iso}}(\theta) \approx \frac{1}{2}\left(\frac{\Delta Z}{\bar{Z}}\right) + \frac{1}{2}\left[\frac{\Delta\alpha}{\bar{\alpha}} - \left(\frac{2\bar{\beta}}{\bar{\alpha}}\right)^2\frac{\Delta G}{\bar{G}}\right]\sin^2(\theta)$$

$$+ \frac{1}{2}\left(\frac{\Delta\alpha}{\bar{\alpha}}\right)\sin^2(\theta)\tan^2(\theta)$$

$$R_{PP\text{-aniso}}(\theta) \approx \frac{\Delta\delta}{2}\sin^2(\theta) + \frac{\Delta\varepsilon}{2}\sin^2(\theta)\tan^2(\theta)$$

where

$\theta = (\theta_2 + \theta_1)/2$	$\Delta\varepsilon = \varepsilon_2 - \varepsilon_1$	$\Delta\gamma = \gamma_2 - \gamma_1$
$\bar{p} = (\rho_1 + \rho_2)/2$	$\Delta\rho = \rho_2 - \rho_1$	$\Delta\delta = \delta_2 - \delta_1$
$\bar{\alpha} = (\alpha_1 + \alpha_2)/2$	$\Delta\alpha = \alpha_2 - \alpha_1$	
$\bar{\beta} = (\beta_1 + \beta_2)/2$	$\Delta\beta = \beta_2 - \beta_1$	
$\bar{G} = (G_1 + G_2)/2$	$\Delta G = G_2 - G_1$	$G = \rho\beta^2$
$\bar{Z} = (Z_1 + Z_2)/2$	$\Delta Z = Z_2 - Z_1$	$Z = \rho\alpha$

In the preceding and following equations, Δ indicates a difference and an overbar indicates an average of the corresponding quantity. In TIH media, reflectivity will vary with azimuth, ϕ, as well as offset or incident angle θ. Rüger (1995, 1996) and Chen (1995) derived the P-wave reflection coefficient in the symmetry planes for reflections at the boundary of two TIH media sharing the same symmetry axis. At a horizontal interface between two TIH media with horizontal symmetry axis x_1 and vertical axis x_3, the P-wave reflectivity in the vertical symmetry plane parallel to the x_1 symmetry axis can be written as

$$R_{PP}(\phi = 0°, \theta) \approx R_{PP\text{-iso}}(\theta) + \left[\frac{\Delta\delta^{(V)}}{2} + \left(\frac{2\bar{\beta}}{\bar{\alpha}}\right)^2\Delta\gamma\right]\sin^2(\theta)$$

$$+ \frac{\Delta\varepsilon^{(V)}}{2}\sin^2(\theta)\tan^2(\theta)$$

where azimuth ϕ is measured from the x_1-axis and incident angle θ is defined with respect to x_3. The isotropic part $R_{PP\text{-iso}}(\theta)$ is the same as before. In the preceding expression

$$\alpha = \sqrt{\frac{C_{33}}{\rho}} \qquad \varepsilon^{(V)} = \frac{C_{11} - C_{33}}{2C_{33}}$$

$$\beta = \sqrt{\frac{C_{44}}{\rho}} \qquad \delta^{(V)} = \frac{(C_{13} + C_{55})^2 - (C_{33} - C_{55})^2}{2C_{33}(C_{33} - C_{55})}$$

$$\beta^\perp = \sqrt{\frac{C_{55}}{\rho}} \qquad \gamma^{(V)} = \frac{C_{66} - C_{44}}{2C_{44}} = -\frac{\gamma}{1 + 2\gamma}$$

In the vertical symmetry plane perpendicular to the symmetry axis, the P-wave reflectivity is the same as the isotropic solution:

$$R_{PP}(\phi = 90°, \theta) = R_{PP\text{-iso}}(\theta)$$

In nonsymmetry planes, Rüger (1996) derived the P-wave reflectivity $R_{PP}(\phi, \theta)$ using a perturbation technique as follows:

$$R_{PP}(\phi, \theta) \approx R_{PP\text{-iso}}(\theta) + \left\{ \left[\frac{\Delta\delta^{(V)}}{2} + \left(\frac{2\bar{\beta}}{\bar{\alpha}}\right)^2 \Delta\gamma \right] \cos^2(\phi) \right\} \sin^2(\theta)$$

$$+ \left\{ \left[\frac{\Delta\varepsilon^{(V)}}{2} \right] \cos^4(\phi) + \left[\frac{\Delta\delta^{(V)}}{2} \right] \sin^2(\phi) \cos^2(\phi) \right\} \sin^2(\theta) \tan^2(\theta)$$

Similarly, the anisotropic parameters in orthorhombic media are given by Chen (1995) and Tsvankin (1997) as follows

$$\varepsilon^{(1)} = \frac{C_{22} - C_{33}}{2C_{33}}$$

$$\varepsilon^{(2)} = \frac{C_{11} - C_{33}}{2C_{33}}$$

$$\delta^{(1)} = \frac{(C_{23} + C_{44})^2 - (C_{33} - C_{44})^2}{2C_{33}(C_{33} - C_{44})}$$

$$\delta^{(2)} = \frac{(C_{13} + C_{55})^2 - (C_{33} - C_{55})^2}{2C_{33}(C_{33} - C_{55})}$$

$$\gamma^{(2)} = \frac{C_{44} - C_{55}}{2C_{55}}$$

The parameters $\varepsilon^{(1)}$ and $\delta^{(1)}$ are Thomsen's parameters for the equivalent TIV media in the 2–3 plane. Similarly, $\varepsilon^{(2)}$ and $\delta^{(2)}$ are Thomsen's parameters for the equivalent TIV media in the 1–3 plane; γ_2 represents the velocity anisotropy between two shear-wave modes traveling along the z-axis. The difference in the approximate P-wave reflection coefficient in the two vertical symmetry planes (with x_3 as the vertical axis) of orthorhombic media is given by Rüger (1995, 1996) in the following form

$$R_{PP}^{[x_1, x_3]} - R_{PP}^{[x_2, x_3]} \approx \left[\frac{\Delta\delta^{(2)} - \Delta\delta^{(1)}}{2} + \left(\frac{2\bar{\beta}}{\bar{\alpha}}\right)^2 \Delta\gamma_2 \right] \sin^2(\theta)$$

$$+ \left[\frac{\Delta\varepsilon^{(2)} - \Delta\varepsilon^{(1)}}{2} \right] \sin^2(\theta) \tan^2(\theta)$$

The equations are good approximations up to 30–40° angle of incidence.

ASSUMPTIONS AND LIMITATIONS

The equations presented in this section apply under the following conditions:

- The rock is linear elastic.
- Approximate forms apply to the P–P reflection at near offset for slightly contrasting, weakly anisotropic media.

3.6 VISCOELASTICITY AND Q

SYNOPSIS

Materials are **linear elastic** when the stress is proportional to strain:

$$\frac{\sigma_{11} + \sigma_{22} + \sigma_{33}}{3} = K(\varepsilon_{11} + \varepsilon_{22} + \varepsilon_{33}) \qquad \text{volumetric}$$

$$\sigma_{ij} = 2\mu\varepsilon_{ij}, \quad i \neq j \qquad \text{shear}$$

$$\sigma_{ij} = \lambda\delta_{ij}\varepsilon_{kk} + 2\mu\varepsilon_{ij} \qquad \text{general isotropic}$$

where σ_{ij} and ε_{ij} are the stress and strain, K is the bulk modulus, μ is the shear modulus, and λ is Lamé's coefficient.

In contrast, **linear, viscoelastic** materials also depend on rate or history, which can be incorporated by using time derivatives. For example, shear stress and shear strain may be related by using one of the following simple models:

$$\dot{\varepsilon}_{ij} = \frac{\dot{\sigma}_{ij}}{2\mu} + \frac{\sigma_{ij}}{2\eta} \qquad \text{Maxwell solid} \qquad (1)$$

$$\sigma_{ij} = 2\eta\dot{\varepsilon}_{ij} + 2\mu\varepsilon \qquad \text{Voigt solid} \qquad (2)$$

$$\eta\dot{\sigma}_{ij} + (E_1 + E_2)\sigma_{ij} = E_2(\eta\dot{\varepsilon}_{ij} + E_1\varepsilon_{ij}) \qquad \text{Standard linear solid} \qquad (3)$$

where E_1 and E_2 are additional elastic moduli and η is a material constant resembling viscosity.

More generally, one can incorporate higher-order derivatives, as follows:

$$\sum_{i=0}^{N} a_i \frac{\partial^i \varepsilon}{\partial t^i} = \sum_{j=0}^{M} b_j \frac{\partial^j \sigma}{\partial t^j}$$

Similar equations would be necessary to describe the generalizations of other elastic constants such as K.

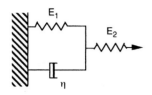

Figure 3.6.1. Schematic of a spring and dashpot system whose force displacement relation is described by the same equation as the standard linear solid.

It is customary to represent these equations with mechanical spring and dashpot models such as that for the standard linear solid shown in Figure 3.6.1

Consider a wave propagating in a viscoelastic solid so that the displacement, for example, is given by

$$u(x, t) = u_0 \exp[-\alpha(\omega)x] \exp[i(\omega t - kx)] \quad (4)$$

Then at any point in the solid the stress and strain are out of phase

$$\sigma = \sigma_0 \exp[i(\omega t - kx)] \quad (5)$$

$$\varepsilon = \varepsilon_0 \exp[i(\omega t - kx - \varphi)] \quad (6)$$

The ratio of stress to strain at the point is the complex modulus, $M(\omega)$.

The **quality factor**, Q, is a measure of how dissipative the material is. The lower the Q, the larger is the dissipation. There are several ways to express Q. One precise way is as the ratio of the imaginary and real parts of the complex modulus:

$$\frac{1}{Q} = \frac{M_I}{M_R}$$

In terms of energies, Q can be expressed as

$$\frac{1}{Q} = \frac{\Delta W}{2\pi W}$$

where ΔW is the energy dissipated per cycle of oscillation and W is the peak strain energy during the cycle. In terms of the spatial attenuation factor, α, in equation (4),

$$\frac{1}{Q} = \frac{\alpha V}{\pi f}$$

where V is the velocity and f is the frequency. In terms of the wave amplitudes of an oscillatory signal with period τ,

$$\frac{1}{Q} \approx \frac{1}{\pi} \ln \left[\frac{u(t)}{u(t + \tau)} \right]$$

which measures the amplitude loss per cycle. This is sometimes called the **logarithmic decrement**. Finally, in terms of the phase delay φ between the stress and strain, as in equations (5) and (6)

$$\frac{1}{Q} \approx \tan(\varphi)$$

Winkler and Nur (1979) showed that if we define $Q_E = E_R/E_I$, $Q_K = K_R/K_I$, and $Q_\mu = \mu_R/\mu_I$, where the subscripts R and I denote real and imaginary parts of the Young's, bulk, and shear moduli, respectively, and if the attenuation is small, then the various Q factors can be related through the following equations:

$$\frac{(1 - v)(1 - 2v)}{Q_P} = \frac{(1 + v)}{Q_E} - \frac{2v(2 - v)}{Q_S} \qquad \frac{Q_P}{Q_S} = \frac{\mu_I}{K_I + \frac{4}{3}\mu_I} \frac{V_P{}^2}{V_S{}^2}$$

$$\frac{(1 + v)}{Q_K} = \frac{3(1 - v)}{Q_P} - \frac{2(1 - 2v)}{Q_S} \qquad \frac{3}{Q_E} = \frac{(1 - 2v)}{Q_K} + \frac{2(1 + v)}{Q_S}$$

One of the following relations always occurs (Bourbié et al., 1987):

$$Q_K > Q_P > Q_E > Q_S \quad \text{for high } V_P/V_S \text{ ratios}$$
$$Q_K < Q_P < Q_E < Q_S \quad \text{for low } V_P/V_S \text{ ratios}$$
$$Q_K = Q_P = Q_E = Q_S$$

The **spectral ratio** method is a popular way to estimate Q in both the laboratory and the field. Because $1/Q$ is a measure of the fractional loss of energy per cycle of oscillation, after a fixed distance of propagation there is a tendency for shorter wavelengths to be attenuated more than longer wavelengths. If the amplitude of the propagating wave is

$$u(x, t) = u_0 \exp[-\alpha(\omega)x] = u_0 \exp\left[-\frac{\pi f}{VQ}x\right]$$

we can compare the spectral amplitudes at two different distances and determine Q from the slope of the logarithmic decrement:

$$\ln\left[\frac{S(f, x_2)}{S(f, x_1)}\right] = -\frac{\pi f}{QV}(x_2 - x_1)$$

A useful illustrative example is the **standard linear solid** (Zener, 1948) in equation (3). If we assume sinusoidal motion

$$\varepsilon = \varepsilon_0 e^{i\omega t}$$

$$\sigma = \sigma_0 e^{i\omega t}$$

and substitute into equation (3), we can write

$$\sigma_0 = M(\omega)\varepsilon_0$$

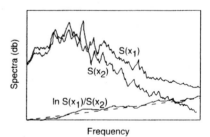

Figure 3.6.2. The slope of the log of the spectral ratio (difference of the spectra in db) can be interpreted in terms of Q.

with the complex frequency-dependent modulus

$$M(\omega) = \frac{E_2(E_1 + i\omega\eta)}{E_1 + E_2 + i\omega\eta}$$

In the limits of low frequency and high frequency, we get the limiting moduli

$$M_0 = \frac{E_2 E_1}{E_1 + E_2}, \qquad \omega \to 0$$

$$M_\infty = E_2, \qquad \omega \to \infty$$

Note that at very low frequencies and very high frequencies the moduli are real and independent of frequency, and thus in these limits the material behaves elastically. It is useful to rewrite the frequency-dependent complex modulus in terms of these limits:

$$M(\omega) = \frac{M_\infty\left[M_0 + i\frac{\omega}{\omega_r}(M_\infty M_0)^{1/2}\right]}{M_\infty + i\frac{\omega}{\omega_r}(M_\infty M_0)^{1/2}}$$

and

$$\mathrm{Re}[M(\omega)] = \frac{M_0 M_\infty\left[1 + \left(\frac{\omega}{\omega_r}\right)^2\right]}{M_\infty + \left(\frac{\omega}{\omega_r}\right)^2 M_0}$$

$$\mathrm{Im}[M(\omega)] = \frac{\frac{\omega}{\omega_r}\sqrt{M_0 M_\infty}(M_\infty - M_0)}{M_\infty + \left(\frac{\omega}{\omega_r}\right)^2 M_0}$$

where

$$\omega_r = \frac{\sqrt{E_1(E_1 + E_2)}}{\eta}$$

Similarly we can write Q as a function of frequency:

$$\frac{1}{Q} = \frac{M_\mathrm{I}(\omega)}{M_\mathrm{R}(\omega)}$$

$$\frac{1}{Q} = \frac{E_2}{\sqrt{E_1(E_1 + E_2)}}\frac{\left(\frac{\omega}{\omega_r}\right)}{1 + \left(\frac{\omega}{\omega_r}\right)^2}$$

The maximum attenuation

$$\left(\frac{1}{Q}\right)_{\max} = \frac{1}{2}\frac{E_2}{\sqrt{E_1(E_1 + E_2)}}$$

$$\left(\frac{1}{Q}\right)_{\max} = \frac{1}{2}\frac{M_\infty - M_0}{\sqrt{M_\infty M_0}}$$

occurs at $\omega = \omega_r$. This is sometimes written as

$$\left(\frac{1}{Q}\right)_{\max} = \frac{1}{2}\frac{\Delta M}{\bar{M}} \tag{7}$$

where $\Delta M/\bar{M} = (M_\infty - M_0)/\bar{M}$ is the **Modulus defect** and $\bar{M} = \sqrt{M_\infty M_0}$.

Liu, Anderson, and Kanamori (1976) considered the **nearly constant Q model** in which simple attenuation mechanisms are combined so that the attenuation is nearly a constant over a finite range of frequencies. One can then write

$$\frac{V(\omega)}{V(\omega_0)} = 1 + \frac{1}{\pi Q}\ln\left(\frac{\omega}{\omega_0}\right)$$

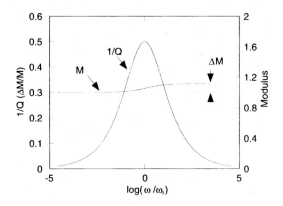

**Figure 3.6.3. Schematic of the standard linear
solid in the frequency domain.**

which relates the velocity dispersion within the band of constant Q to the value
of Q and the frequency. For large Q, this can be approximated as

$$\left(\frac{1}{Q}\right)_{max} \approx \frac{\pi}{\log\left(\frac{\omega}{\omega_0}\right)}\left(\frac{1}{2}\frac{M-M_0}{M_0}\right) \tag{8}$$

where M and M_0 are the moduli at two different frequencies ω and ω_0 within
the band where Q is nearly constant. Note the resemblance of this expression to
equation (7) for the standard linear solid.

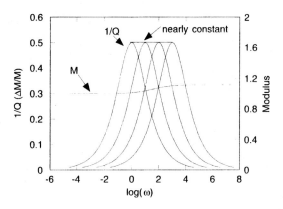

**Figure 3.6.4. Schematic of the nearly constant Q
model in the frequency domain.**

Kjartansson (1979) considered the **constant Q model** in which Q is strictly
constant. In this case the complex modulus and Q are related by

$$M(\omega) = M_0\left(\frac{i\omega}{\omega_0}\right)^{2\gamma}$$

where

$$\gamma = \frac{1}{\pi} \tan^{-1}\left(\frac{1}{Q}\right)$$

For large Q, this can be approximated as

$$\left(\frac{1}{Q}\right)_{max} \approx \frac{\pi}{\log\left(\frac{\omega}{\omega_0}\right)}\left(\frac{1}{2}\frac{M - M_0}{M_0}\right)$$

where M and M_0 are the moduli at two different frequencies ω and ω_0. Note the resemblance of this expression to equation (7) for the standard linear solid and equation (8) for the nearly constant Q model.

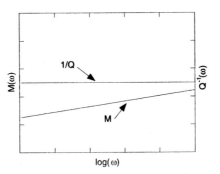

Figure 3.6.5. Schematic of the constant Q model in the frequency domain.

USES

The equations presented in this section are used for phenomenological modeling of attenuation and velocity dispersion of seismic waves.

ASSUMPTIONS AND LIMITATIONS

The equations presented in this section assume that the material is linear, dissipative, and causal.

3.7 KRAMERS–KRONIG RELATIONS BETWEEN VELOCITY DISPERSION AND Q

SYNOPSIS

For linear viscoelastic systems, causality requires that there be a very specific relation between velocity or modulus dispersion and Q; that is, if the dispersion is completely characterized for all frequencies then Q is known for all frequencies and vice versa.

We can write a viscoelastic constitutive law between stress and strain components as

$$\sigma(t) = \frac{dr}{dt} * \varepsilon(t)$$

where $r(t)$ is the relaxation function and $*$ denotes convolution. Then in the Fourier domain we can write

$$\tilde{\sigma}(\omega) = M(\omega)\tilde{\varepsilon}(\omega)$$

where $M(\omega)$ is the complex modulus. For $r(t)$ to be causal, in the frequency domain the real and imaginary parts of $M(\omega)/(i\omega)$ must be Hilbert transform pairs (Bourbié et al., 1987):

$$M_I(\omega) = \frac{\omega}{\pi} \int_{-\infty}^{+\infty} \frac{M_R(\alpha) - M_R(0)}{\alpha} \frac{d\alpha}{\alpha - \omega}$$

$$M_R(\omega) - M_R(0) = -\frac{\omega}{\pi} \int_{-\infty}^{+\infty} \frac{M_I(\alpha)}{\alpha} \frac{d\alpha}{\alpha - \omega}$$

where $M_R(0)$ is the real part of the modulus at zero frequency, which results because there is an instantaneous elastic response of a viscoelastic material. If we express this in terms of

$$Q^{-1} = \frac{M_I(\omega)\,\text{sgn}\,(\omega)}{M_R(\omega)}$$

then we get

$$Q^{-1}(\omega) = \frac{|\omega|}{\pi M_R(\omega)} \int_{-\infty}^{+\infty} \frac{M_R(\alpha) - M_R(0)}{\alpha} \frac{d\alpha}{\alpha - \omega} \tag{9}$$

and its inverse:

$$M_R(\omega) - M_R(0) = \frac{-\omega}{\pi} \int_{-\infty}^{+\infty} \frac{Q^{-1}(\alpha)M_R(\alpha)}{|\alpha|} \frac{d\alpha}{(\alpha - \omega)} \tag{10}$$

From these we see the expected result that a larger attenuation generally is associated with larger dispersion. Zero attenuation requires zero velocity dispersion.

One never has more than partial information about the frequency dependence of velocity and Q, but the Kramers–Kronig relation allows us to put some constraints on the material behavior. For example, Lucet (1989) measured velocity and attenuation at two frequencies (≈ 1 kHz and 1 MHz) and used the Kramers–Kronig relations to compare the differences with various viscoelastic models, as shown schematically in Figure 3.7.1. Using equation (10), we can determine the expected ratio of low-frequency modulus or velocity, V_{lf}, and high-frequency modulus or velocity, V_{hf}, for various functional forms of Q (for example, constant Q or nearly constant Q). In all cases, linear viscoelastic behavior should lead to an intercept

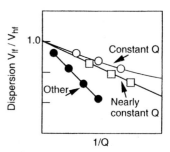

Figure 3.7.1. Lucet's (1989) use of the Kramers–Kronig relations to compare high- and low-frequency measured velocities and Q with various viscoelastic models.

of $V_{lf}/V_{hf} = 1$ at $1/Q = 0$. Mechanisms with peaked attenuation curves between the measurement points will generally cause a larger dispersion, which appears as a steeper negative slope.

USES

The Kramers–Kronig equations can be used to relate velocity dispersion and Q in linear viscoelastic materials.

ASSUMPTIONS AND LIMITATIONS

The Kramers–Kronig equations apply when the material is linear and causal.

3.8 WAVES IN LAYERED MEDIA: FULL-WAVEFORM SYNTHETIC SEISMOGRAMS

SYNOPSIS

One of the approaches for computing wave propagation in layered media is the use of propagator matrices (Aki and Richards, 1980; Claerbout, 1985). The wave variables of interest (usually stresses and particle velocity or displacements) at the top and bottom of the stack of layers are related by a product of propagator matrices, one for each layer. The calculations are done in the frequency domain and include the effects of all multiples. For waves traveling perpendicularly to n layers with layer velocities, densities, and thicknesses V_k, ρ_k, and d_k, respectively

$$\begin{bmatrix} S \\ W \end{bmatrix}_n = \prod_{k=1}^{n} \mathbf{A}_k \begin{bmatrix} S \\ W \end{bmatrix}_1$$

where S and W are the Fourier transforms of the wave variables σ and w, respectively. For normal-incidence P-waves, σ is interpreted as the normal stress across each interface, and w is the normal component of particle velocity. For normal-incidence S-waves, σ is the shear traction across each interface, and w is the tangential component of the particle velocity. Each layer matrix $\mathbf{A}_k$ has the form

$$
\mathbf{A}_k = \begin{bmatrix} \cos\left(\dfrac{\omega d_k}{V_k}\right) & i\rho_k V_k \sin\left(\dfrac{\omega d_k}{V_k}\right) \\[3ex] \dfrac{i}{\rho_k V_k}\sin\left(\dfrac{\omega d_k}{V_k}\right) & \cos\left(\dfrac{\omega d_k}{V_k}\right) \end{bmatrix}
$$

where ω is the angular frequency.

Kennett (1974, 1983) used the invariant imbedding method to generate the response of a layered medium recursively by adding one layer at a time. The overall reflection and transmission matrices, $\hat{\mathbf{R}}_D$ and $\hat{\mathbf{T}}_D$, respectively, for downgoing waves through a stack of layers are given by the following recursion relations:

$$
\hat{\mathbf{R}}_D^{(k)} = \mathbf{R}_D^{(k)} + \mathbf{T}_U^{(k)} \mathbf{E}_D^{(k)} \hat{\mathbf{R}}_D^{(k+1)} \mathbf{E}_D^{(k)} \left[\mathbf{I} - \mathbf{R}_U^{(k)} \mathbf{E}_D^{(k)} \hat{\mathbf{R}}_D^{(k+1)} \mathbf{E}_D^{(k)}\right]^{-1} \mathbf{T}_D^{(k)}
$$

$$
\hat{\mathbf{T}}_D^{(k)} = \hat{\mathbf{T}}_D^{(k+1)} \mathbf{E}_D^{(k)} \left[\mathbf{I} - \mathbf{R}_U^{(k)} \mathbf{E}_D^{(k)} \hat{\mathbf{R}}_D^{(k+1)} \mathbf{E}_D^{(k)}\right]^{-1} \mathbf{T}_D^{(k)}
$$

where $\mathbf{R}_D^{(k)}$, $\mathbf{T}_D^{(k)}$, $\mathbf{R}_U^{(k)}$, and $\mathbf{T}_U^{(k)}$ are just the single-interface downward and upward reflection and transmission matrices for the kth interface:

$$
\mathbf{R}_D^{(k)} = \begin{bmatrix} \overset{\downarrow\,\uparrow}{PP} & \overset{\downarrow\,\uparrow}{SP}\left(\dfrac{V_{P(k-1)}\cos\theta_{k-1}}{V_{S(k-1)}\cos\phi_{k-1}}\right)^{1/2} \\[3ex] \overset{\downarrow\,\uparrow}{PS}\left(\dfrac{V_{S(k-1)}\cos\phi_{k-1}}{V_{P(k-1)}\cos\theta_{k-1}}\right)^{1/2} & \overset{\downarrow\,\uparrow}{SS} \end{bmatrix}
$$

$$
\mathbf{T}_D^{(k)} = \begin{bmatrix} \overset{\downarrow\,\downarrow}{PP}\left(\dfrac{\rho_k V_{P(k)}\cos\theta_k}{\rho_{k-1} V_{P(k-1)}\cos\theta_{k-1}}\right)^{1/2} & \overset{\downarrow\,\downarrow}{SP}\left(\dfrac{\rho_k V_{P(k)}\cos\theta_k}{\rho_{k-1} V_{S(k-1)}\cos\phi_{k-1}}\right)^{1/2} \\[3ex] \overset{\downarrow\,\downarrow}{PS}\left(\dfrac{\rho_k V_{S(k)}\cos\phi_k}{\rho_{k-1} V_{P(k-1)}\cos\theta_{k-1}}\right)^{1/2} & \overset{\downarrow\,\downarrow}{SS}\left(\dfrac{\rho_k V_{S(k)}\cos\phi_k}{\rho_{k-1} V_{S(k-1)}\cos\phi_{k-1}}\right)^{1/2} \end{bmatrix}
$$

$$
\mathbf{R}_U^{(k)} = \begin{bmatrix} \overset{\uparrow\,\downarrow}{PP} & \overset{\uparrow\,\downarrow}{SP}\left(\dfrac{V_{P(k)}\cos\theta_k}{V_{S(k)}\cos\phi_k}\right)^{1/2} \\[3ex] \overset{\uparrow\,\downarrow}{PS}\left(\dfrac{V_{S(k)}\cos\phi_k}{V_{P(k)}\cos\theta_k}\right)^{1/2} & \overset{\uparrow\,\downarrow}{SS} \end{bmatrix}
$$

$$
\mathbf{T}_U^{(k)} = \begin{bmatrix} \overset{\uparrow\,\uparrow}{PP}\left(\dfrac{\rho_{k-1} V_{P(k-1)}\cos\theta_{k-1}}{\rho_k V_{P(k)}\cos\theta_k}\right)^{1/2} & \overset{\uparrow\,\uparrow}{SP}\left(\dfrac{\rho_{k-1} V_{P(k-1)}\cos\theta_{k-1}}{\rho_k V_{S(k)}\cos\phi_k}\right)^{1/2} \\[3ex] \overset{\uparrow\,\uparrow}{PS}\left(\dfrac{\rho_{k-1} V_{S(k-1)}\cos\phi_{k-1}}{\rho_k V_{P(k)}\cos\theta_k}\right)^{1/2} & \overset{\uparrow\,\uparrow}{SS}\left(\dfrac{\rho_{k-1} V_{S(k-1)}\cos\phi_{k-1}}{\rho_k V_{S(k)}\cos\phi_k}\right)^{1/2} \end{bmatrix}
$$

with

$$
\begin{pmatrix}
\overset{\downarrow\uparrow}{PP} & \overset{\downarrow\uparrow}{SP} & \overset{\uparrow\uparrow}{PP} & \overset{\uparrow\uparrow}{SP} \\
\overset{\downarrow\uparrow}{PS} & \overset{\downarrow\uparrow}{SS} & \overset{\uparrow\uparrow}{PS} & \overset{\uparrow\uparrow}{SS} \\
\overset{\downarrow\downarrow}{PP} & \overset{\downarrow\downarrow}{SP} & \overset{\uparrow\downarrow}{PP} & \overset{\uparrow\downarrow}{SP} \\
\overset{\downarrow\downarrow}{PS} & \overset{\downarrow\downarrow}{SS} & \overset{\uparrow\downarrow}{PS} & \overset{\uparrow\downarrow}{SS}
\end{pmatrix}
= \mathbf{M}^{-1}\mathbf{N}
$$

$$
\mathbf{M} =
\begin{bmatrix}
-\sin\theta_{k-1} & -\cos\phi_{k-1} & \sin\theta_k & \cos\phi_k \\
\cos\theta_{k-1} & -\sin\phi_{k-1} & \cos\theta_k & -\sin\phi_k \\
2I_{S(k-1)}\sin\phi_{k-1}\cos\theta_{k-1} & I_{S(k-1)}(1-2\sin^2\phi_{k-1}) & 2I_{S(k)}\sin\phi_k\cos\theta_k & I_{S(k)}(1-2\sin^2\phi_k) \\
-I_{P(k-1)}(1-2\sin^2\phi_{k-1}) & I_{S(k-1)}\sin 2\phi_{k-1} & I_{P(k)}(1-2\sin^2\phi_k) & -I_{S(k)}\sin 2\phi_k
\end{bmatrix}
$$

$$
\mathbf{N} =
\begin{bmatrix}
\sin\theta_{k-1} & \cos\phi_{k-1} & -\sin\theta_k & -\cos\phi_k \\
\cos\theta_{k-1} & -\sin\phi_{k-1} & \cos\theta_k & -\sin\phi_k \\
2I_{S(k-1)}\sin\phi_{k-1}\cos\theta_{k-1} & I_{S(k-1)}(1-2\sin^2\phi_{k-1}) & 2I_{S(k)}\sin\phi_k\cos\theta_k & I_{S(k)}(1-2\sin^2\phi_k) \\
I_{P(k-1)}(1-2\sin^2\phi_{k-1}) & -I_{S(k-1)}\sin 2\phi_{k-1} & -I_{P(k)}(1-2\sin^2\phi_k) & I_{S(k)}\sin 2\phi_k
\end{bmatrix}
$$

where $I_{P(k)} = \rho_k V_{P(k)}$, and $I_{S(k)} = \rho_k V_{S(k)}$ are the P and S impedances respectively of the kth layer, and θ_k and ϕ_k are the angles made by the P- and S-wave vectors with the normal to the kth interface. The elements of the reflection and transmission matrices $\mathbf{R}_D^{(k)}$, $\mathbf{T}_D^{(k)}$, $\mathbf{R}_U^{(k)}$, and $\mathbf{T}_U^{(k)}$ are the reflection and transmission coefficients for scaled displacements, which are proportional to the square root of the energy flux. The scaled displacement u' is related to the displacement u by $u' = u\sqrt{\rho V \cos\theta}$.

For normal-incidence wave propagation with no mode conversions, the reflection and transmission matrices reduce to the scalar coefficients:

$$
R_D^{(k)} = \frac{\rho_{k-1}V_{k-1} - \rho_k V_k}{\rho_{k-1}V_{k-1} + \rho_k V_k}, \qquad R_U^{(k)} = -R_D^{(k)}
$$

$$
T_D^{(k)} = \frac{2\sqrt{\rho_{k-1}V_{k-1}\rho_k V_k}}{\rho_{k-1}V_{k-1} + \rho_k V_k}, \qquad T_U^{(k)} = T_D^{(k)}
$$

The phase shift operator for propagation across each new layer is given by $\mathbf{E}_D^{(k)}$

$$
\mathbf{E}_D^{(k)} =
\begin{bmatrix}
\exp\left(i\omega d_k \cos\theta_k / V_P^{(k)}\right) & 0 \\
0 & \exp\left(i\omega d_k \cos\phi_k / V_S^{(k)}\right)
\end{bmatrix}
$$

where θ_k and ϕ_k are the angles between the normal to the layers and the directions of propagation of P- and S-waves, respectively. The terms $\hat{\mathbf{R}}_D$ and $\hat{\mathbf{T}}_D$ are functions of ω and represent the overall transfer functions of the layered medium in the frequency domain. Time-domain seismograms are obtained by multiplying the overall transfer function by the Fourier transform of the source wavelet and then doing an inverse transform.

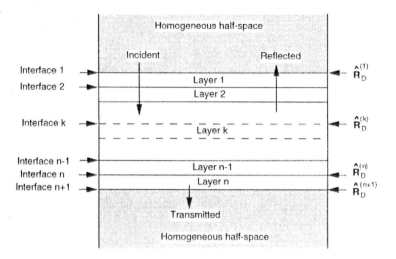

Figure 3.8.1

The recursion starts at the base of the layering at interface $n+1$ (Figure 3.8.1). Setting $\mathbf{R}_D^{(n+1)} = \hat{\mathbf{R}}_D^{(n+1)} = 0$ and $\mathbf{T}_D^{(n+1)} = \hat{\mathbf{T}}_D^{(n+1)} = \mathbf{I}$ simulates a stack of layers overlying a semiinfinite homogeneous half-space with properties equal to those of the last layer, layer n. The recursion relations are stepped up through the stack of layers one at a time to finally give $\hat{\mathbf{R}}_D^{(1)}$ and $\hat{\mathbf{T}}_D^{(1)}$, the overall reflection and transmission response for the whole stack.

EXAMPLE

Calculate the P-wave normal-incidence overall reflection and transmission functions $\hat{R}_D^{(1)}$ and $\hat{T}_D^{(1)}$ recursively for a three-layered medium with layer properties as follows:

$$V_{P(1)} = 4000\,\text{m/s}, \qquad \rho_1 = 2300\,\text{kg/m}^3, \qquad d_1 = 100\,\text{m}$$

$$V_{P(2)} = 3000\,\text{m/s}, \qquad \rho_2 = 2100\,\text{kg/m}^3, \qquad d_2 = 50\,\text{m}$$

$$V_{P(3)} = 5000\,\text{m/s}, \qquad \rho_3 = 2500\,\text{kg/m}^3, \qquad d_3 = 200\,\text{m}$$

The recursion starts with $\hat{R}_D^{(4)} = 0$ and $\hat{T}_D^{(4)} = 1$. The normal-incidence reflection and transmission coefficients at interface 3 are

$$R_D^{(3)} = \frac{\rho_2 V_{P(2)} - \rho_3 V_{P(3)}}{\rho_2 V_{P(2)} + \rho_3 V_{P(3)}} = -0.33, \qquad R_U^{(3)} = -R_D^{(3)}$$

$$T_D^{(3)} = \frac{2\sqrt{\rho_2 V_{P(2)} \rho_3 V_{P(3)}}}{\rho_2 V_{P(2)} + \rho_3 V_{P(3)}} = 0.94, \qquad T_U^{(3)} = T_D^{(3)}$$

and the phase factor for propagation across layer 3 is

$$E_D^{(3)} = \exp(i2\pi f d_3 / V_{P(3)}) = \exp(i2\pi f 200/5000)$$

where f is the frequency. The recursion relations give

$$\hat{R}_D^{(3)} = R_D^{(3)} + \frac{T_U^{(3)} E_D^{(3)} \hat{R}_D^{(4)} E_D^{(3)}}{1 - R_U^{(3)} E_D^{(3)} \hat{R}_D^{(4)} E_D^{(3)}} T_D^{(3)}$$

$$\hat{T}_D^{(3)} = \frac{\hat{T}_D^{(4)} E_D^{(3)}}{1 - R_U^{(3)} E_D^{(3)} \hat{R}_D^{(4)} E_D^{(3)}} T_D^{(3)}$$

The recursion is continued in a similar manner until finally we get $\hat{R}_D^{(1)}$ and $\hat{T}_D^{(1)}$.

The matrix inverse

$$\left[I - R_U^{(k)} E_D^{(k)} \hat{R}_D^{(k+1)} E_D^{(k)} \right]^{-1}$$

is referred to as the reverberation operator and includes the response caused by all internal reverberations. In the series expansion of the matrix inverse

$$\left[I - R_U^{(k)} E_D^{(k)} \hat{R}_D^{(k+1)} E_D^{(k)} \right]^{-1} = I + R_U^{(k)} E_D^{(k)} \hat{R}_D^{(k+1)} E_D^{(k)}$$
$$+ R_U^{(k)} E_D^{(k)} \hat{R}_D^{(k+1)} E_D^{(k)} R_U^{(k)} E_D^{(k)} \hat{R}_D^{(k+1)} E_D^{(k)} + \cdots$$

the first term represents the primaries, and each successive term corresponds to higher-order multiples. Truncating the expansion to $m + 1$ terms includes m internal multiples in the approximation. The full multiple sequence is included with the exact matrix inverse.

USES

The methods described in this section can be used to compute full-wave seismograms, which include the effects of multiples for wave propagation in layered media.

ASSUMPTIONS AND LIMITATIONS

The algorithms described in this section assume the following:

- Layered medium with no lateral heterogeneities.
- Layers are isotropic linear elastic.
- Plane-wave, time-harmonic propagation.

3.9 WAVES IN LAYERED MEDIA: STRATIGRAPHIC FILTERING AND VELOCITY DISPERSION

SYNOPSIS

Waves in layered media undergo attenuation and velocity dispersion caused by multiple scattering at the layer interfaces. The effective phase slowness of normally incident waves through layered media depends on the relative scales of the wavelength and layer thicknesses and may be written as $S_{\text{eff}} = S_{\text{rt}} + S_{\text{st}}$. The term S_{rt} is the ray theory slowness of the direct ray that does not undergo any reflections and is just the thickness-weighted average of the individual layer slownesses. The individual slownesses may be complex to account for intrinsic attenuation. The excess slowness S_{st} (sometimes called the stratigraphic slowness) arises because of multiple scattering within the layers. A flexible approach for calculating the effective slowness and travel time follows from Kennett's (1974) invariant imbedding formulation for the transfer function of a layered medium. The layered medium, of total thickness L, consists of layers with velocities (1/slownesses), densities, and thicknesses, V_j, ρ_j, and l_j, respectively.

The complex stratigraphic slowness is frequency dependent and can be calculated recursively (Frazer, 1994) by

$$S_{\text{st}} = \frac{1}{i\omega L} \sum_{j=1}^{n} \ln\left(\frac{t_j}{1 - R_j\theta_j^2 r_j}\right)$$

As each new layer $j + 1$ is added to the stack of j layers, R is updated according to

$$R_{j+1} = -r_{j+1} + \frac{R_j\theta_{j+1}^2 t_{j+1}^2}{1 - R_j\theta_{j+1}^2 r_{j+1}}$$

(with $R_0 = 0$) and the term

$$\ln\left[t_{j+1}\left(1 - R_{j+1}\theta_{j+1}^2 r_{j+1}\right)^{-1}\right]$$

is accumulated in the sum. In the above expressions, t_j and r_j are the transmission and reflection coefficients defined as

$$t_j = \frac{2\sqrt{\rho_j V_j \rho_{j+1} V_{j+1}}}{\rho_j V_j + \rho_{j+1} V_{j+1}}$$

$$r_j = \frac{\rho_{j+1} V_{j+1} - \rho_j V_j}{\rho_j V_j + \rho_{j+1} V_{j+1}}$$

whereas $\theta_j = \exp(i\omega l_j/V_j)$ is the phase shift for propagation across layer j, and ω is the angular frequency. The total travel time is $T = T_{rt} + T_{st}$, where T_{rt} is the ray theory travel time given by

$$T_{rt} = \sum_{j=1}^{n} \frac{l_j}{V_j}$$

and T_{st} is given by

$$T_{st} = \mathrm{Re}\left[\frac{1}{i\omega}\sum_{j=1}^{n}\ln\left(\frac{t_j}{1 - R_j\theta_j^2 r_j}\right)\right]$$

The deterministic results given above are not restricted to small perturbations in the material properties or statistically stationary geology.

EXAMPLE

Calculate the excess stratigraphic travel time caused by multiple scattering for a normally incident P-wave traveling through a three-layered medium with layer properties as follows:

$$V_{P(1)} = 4,000 \,\mathrm{m/s}, \qquad \rho_1 = 2,300 \,\mathrm{kg/m^3}, \qquad l_1 = 100 \,\mathrm{m}$$
$$V_{P(2)} = 3,000 \,\mathrm{m/s}, \qquad \rho_2 = 2,100 \,\mathrm{kg/m^3}, \qquad l_2 = 50 \,\mathrm{m}$$
$$V_{P(3)} = 5,000 \,\mathrm{m/s}, \qquad \rho_3 = 2,500 \,\mathrm{kg/m^3}, \qquad l_3 = 200 \,\mathrm{m}$$

The excess travel time is given by

$$T_{st} = \mathrm{Re}\left[\frac{1}{i\omega}\sum_{j=1}^{n}\ln\left(\frac{t_j}{1 - R_j\theta_j^2 r_j}\right)\right]$$

The recusrion begins with $R_0 = 0$.

$$R_1 = -r_1 + \frac{R_0\theta_1^2 t_1^2}{1 - R_0\theta_1^2 r_1}$$

where

$$t_1 = \frac{2\sqrt{\rho_1 V_1 \rho_2 V_2}}{\rho_1 V_1 + \rho_2 V_2} = 0.98$$

$$r_1 = \frac{\rho_2 V_2 - \rho_1 V_1}{\rho_1 V_1 + \rho_2 V_2} = -0.19$$

$\theta_1 = \exp(i2\pi f l_1/V_1) = \exp(i2\pi f\, 100/4000)$ with f as the frequency.

The recursion is continued to obtain R_2 and R_3. Setting $t_3 = 1$ and $r_3 = 0$ simulates an impedance-matching homogeneous infinite half-space beneath layer 3. Finally, the excess travel time, which is a function of the frequency, is obtained by taking the real part of the sum as follows:

$$T_{st} = \mathrm{Re}\left\{ \frac{1}{i2\pi f}\left[\ln\left(\frac{t_1}{1 - R_1\theta_1^2 r_1} \right) + \ln\left(\frac{t_2}{1 - R_2\theta_2^2 r_2} \right) + \ln\left(\frac{t_3}{1 - R_3\theta_3^2 r_3} \right) \right] \right\}$$

The effect of the layering can be thought of as a filter that attenuates the input wavelet and introduces a delay. The function

$$A(\omega) = \exp(i\omega x S_{st}) = \exp(i\omega T_{rt} S_{st}/S_{rt})$$

(where S_{rt} is assumed real in the absence of any intrinsic attenuation) is sometimes called the stratigraphic filter.

The **O'Doherty–Anstey formula** (O'Doherty and Anstey, 1971; Banik, Lerche, and Shuey, 1985)

$$|A(\omega)| \approx \exp(-\hat{R}(\omega)T_{rt})$$

approximately relates the amplitude of the stratigraphic filter to the power spectrum $\hat{R}(\omega)$ of the reflection coefficient time series $r(\tau)$ where

$$\tau(x) = \int_0^x dx'/V(x')$$

is the one-way travel time. Initially the O'Doherty–Anstey formula was obtained by a heuristic approach (O'Doherty and Anstey, 1971). Later, various authors substantiated the result by using statistical ensemble averages of wavefields (Banik et al., 1985), deterministic formulations (Resnick, Lerche, and Shuey, 1986), and the concepts of self-averaged values and wave localization (Shapiro and Zien, 1993). Resnick et al. (1986) showed that the O'Doherty–Anstey formula is obtained as an approximation from the exact frequency-domain theory of Resnick et al. by neglecting quadratic terms in the Riccatti equation of Resnick et al. Another equivalent way of expressing the O'Doherty–Anstey relation is

$$\frac{\mathrm{Im}(S_{st})}{S_{rt}} \approx \frac{1}{2Q} \approx \frac{\hat{R}(\omega)}{\omega} = \frac{1}{2}\omega\hat{M}(2\omega)$$

Here $1/Q$ is the scattering attenuation caused by the multiples, and $\hat{M}(\omega)$ is the power spectrum of the logarithmic impedance fluctuations of the medium, $\ln[\rho(\tau)V(\tau)] - \langle \ln[\rho(\tau)V(\tau)] \rangle$, where $\langle \cdot \rangle$ denotes a stochastic ensemble average. Because the filter is minimum phase, $\omega\,\mathrm{Re}(S_{st})$ and $\omega\,\mathrm{Im}(S_{st})$ are a Hilbert transform pair

$$\frac{\mathrm{Re}(S_{st})}{S_{rt}} \approx \frac{\delta t}{T_{rt}} \approx \frac{H\{\hat{R}(\omega)\}}{\omega}$$

where $H\{\cdot\}$ denotes the Hilbert transform and δt is the excess travel caused by multiple reverberations.

Shapiro and Zien (1993) generalized the O'Doherty–Anstey formula for non-normal incidence. The derivation is based on a small perturbation analysis and requires the fluctuations of material parameters to be small ($<30\%$). The generalized formula for plane pressure (scalar) waves in an acoustic medium incident at an angle θ with respect to the layer normal is

$$|A(\omega)| \approx \exp\left[-\frac{\hat{R}(\omega \cos\theta)}{\cos^4\theta} T_{\text{rt}}\right]$$

whereas

$$|A(\omega)| \approx \exp\left[-\frac{(2\cos^2\theta - 1)^2 \hat{R}(\omega \cos\theta)}{\cos^4\theta} T_{\text{rt}}\right]$$

for SH-waves in an elastic medium (Shapiro, Hubral, and Zien, 1994).

For a perfectly periodic stratified medium made up of two constituents with phase velocities V_1, V_2; densities ρ_1, ρ_2; and thicknesses l_1, l_2, the velocity dispersion relation may be obtained from the **Floquet solution** (Christensen, 1991) for periodic media:

$$\cos\left[\frac{\omega(l_1 + l_2)}{V}\right] = \cos\left(\frac{\omega l_1}{V_1}\right)\cos\left(\frac{\omega l_2}{V_2}\right) - \chi \sin\left(\frac{\omega l_1}{V_1}\right)\sin\left(\frac{\omega l_2}{V_2}\right)$$

$$\chi = \frac{(\rho_1 V_1)^2 + (\rho_2 V_2)^2}{2\rho_1\rho_2 V_1 V_2}$$

The Floquet solution is valid for arbitrary contrasts in the layer properties. If the spatial period ($l_1 + l_2$) is an integer multiple of one-half wavelength, multiple reflections are in phase and add constructively, resulting in a large total accumulated reflection. The frequency at which this **Bragg scattering** condition is satisfied is called the Bragg frequency. Waves cannot propagate within a stopband around the Bragg frequency.

USES

The results described in this section can be used to estimate velocity dispersion and attenuation caused by scattering for normal-incidence wave propagation in layered media.

ASSUMPTIONS AND LIMITATIONS

The methods described in this section apply under the following conditions:

- Layers are isotropic, linear elastic with no lateral variation.

- Propagation is normal to the layers except for the generalized O'Doherty–Anstey formula.
- Plane-wave, time-harmonic propagation is assumed.

3.10 WAVES IN LAYERED MEDIA: FREQUENCY-DEPENDENT ANISOTROPY AND DISPERSION

SYNOPSIS

Waves in layered media undergo attenuation and velocity dispersion caused by multiple scattering at the layer interfaces. Thinly layered media also give rise to velocity anisotropy. At low frequencies this phenomenon is usually described by the Backus average. Velocity anisotropy and dispersion in a multilayered medium are two aspects of the same phenomenon and are related to the frequency- and angle-dependent transmissivity resulting from multiple scattering in the medium. Shapiro et al. (1994) and Shapiro and Hubral (1995) have presented a whole-frequency-range statistical theory for the angle-dependent transmissivity of layered media for scalar waves (pressure waves in fluids) and elastic waves. The theory encompasses the Backus average in the low-frequency limit and ray theory in the high-frequency limit. The formulation avoids the problem of ensemble averaging versus measurements for a single realization by working with parameters that are averaged by the wave-propagation process itself for sufficiently long propagation paths. The results are obtained in the limit when the path length tends to infinity. Practically, this means the results are applicable when path lengths are very much longer than the characteristic correlation lengths of the medium.

The slowness (s) and density (ρ) distributions of the stack of layers (or a continuous inhomogeneous one-dimensional medium) are assumed to be realizations of random stationary processes. The fluctuations of the physical parameters are small ($<30\%$) compared with their constant mean values (denoted by subscripts 0):

$$s^2(z) = \left(1/c_0^2\right)[1 + \varepsilon_s(z)]$$

$$\rho(z) = \rho_0[1 + \varepsilon_\rho(z)]$$

where the fluctuating parts $\varepsilon_s(z)$ (the squared slowness fluctuation) and $\varepsilon_\rho(z)$ (the density fluctuation) have zero means by definition. The depth coordinate is denoted by z, and the x- and y-axes lie in the plane of the layers. The velocity

$$c_0 = \langle s^2 \rangle^{-1/2}$$

corresponds to the average squared slowness of the medium. Instead of the squared slowness fluctuations, the random medium may also be characterized by the P- and S-velocity fluctuations, α and β, respectively, as follows:

$$\alpha(z) = \langle \alpha \rangle [1 + \varepsilon_\alpha(z)] = \alpha_0 [1 + \varepsilon_\alpha(z)]$$

$$\beta(z) = \langle \beta \rangle [1 + \varepsilon_\beta(z)] = \beta_0 [1 + \varepsilon_\beta(z)]$$

In the case of small fluctuations $\varepsilon_\alpha \approx -\varepsilon_s/2$, and $\langle \alpha \rangle \approx c_0 (1 + 3\sigma_{\alpha\alpha}{}^2/2)$, where $\sigma_{\alpha\alpha}{}^2$ is the normalized variance (variance divided by the square of the mean) of the velocity fluctuations. The horizontal wavenumber k_x is related to the incidence angle θ by

$$k_x = k_0 \sin\theta = \omega p$$

where $k_0 = \omega/c_0$, $p = \sin\theta/c_0$ is the horizontal component of the slowness (also called the ray parameter), and ω is the angular frequency. For elastic media, depending on the type of the incident wave, $p = \sin\theta/\alpha_0$ or $p = \sin\theta/\beta_0$. The various autocorrelation and cross correlation functions of the density and velocity fluctuations are denoted by

$$B_{\rho\rho}(\xi) = \langle \varepsilon_\rho(z)\varepsilon_\rho(z+\xi) \rangle$$

$$B_{\alpha\alpha}(\xi) = \langle \varepsilon_\alpha(z)\varepsilon_\alpha(z+\xi) \rangle$$

$$B_{\beta\beta}(\xi) = \langle \varepsilon_\beta(z)\varepsilon_\beta(z+\xi) \rangle$$

$$B_{\alpha\beta}(\xi) = \langle \varepsilon_\alpha(z)\varepsilon_\beta(z+\xi) \rangle$$

$$B_{\alpha\rho}(\xi) = \langle \varepsilon_\alpha(z)\varepsilon_\rho(z+\xi) \rangle$$

$$B_{\beta\rho}(\xi) = \langle \varepsilon_\beta(z)\varepsilon_\rho(z+\xi) \rangle$$

These correlation functions can often be obtained from sonic and density logs. The corresponding normalized variances and cross-variances are given by

$$\sigma_{\rho\rho}{}^2 = B_{\rho\rho}(0), \qquad \sigma_{\alpha\alpha}{}^2 = B_{\alpha\alpha}(0), \qquad \sigma_{\beta\beta}{}^2 = B_{\beta\beta}(0),$$

$$\sigma_{\alpha\beta}{}^2 = B_{\alpha\beta}(0), \qquad \sigma_{\alpha\rho}{}^2 = B_{\alpha\rho}(0), \qquad \sigma_{\beta\rho}{}^2 = B_{\beta\rho}(0)$$

The real part of the effective vertical wavenumber for pressure waves in acoustic media is (neglecting higher than second order powers of the fluctuations)

$$k_z = k_z^{\text{stat}} - k_0{}^2 \cos^2\theta \int_0^\infty d\xi\, B(\xi) \sin(2k_0\xi\cos\theta)$$

$$k_z^{\text{stat}} = k_0 \cos\theta \left(1 + \sigma_{\rho\rho}{}^2/2 + \sigma_{\rho\alpha}{}^2/\cos^2\theta\right)$$

$$B(\xi) = B_{\rho\rho}(\xi) + 2B_{\rho\alpha}(\xi)/\cos^2\theta + B_{\alpha\alpha}(\xi)/\cos^4\theta$$

For waves in an elastic layered medium, the real part of the vertical wavenumber for P, SV, and SH waves is given by

$$k_z^{\text{P}} = \lambda_a + \omega \hat{A}_{\text{P}} - \omega^2 \int_0^\infty d\xi \, [B_{\text{P}}(\xi) \sin(2\xi\lambda_a)$$

$$+ B_{\text{BB}}(\xi) \sin(\xi\lambda_-) + B_{\text{DD}}(\xi) \sin(\xi\lambda_+)]$$

$$k_z^{\text{SV}} = \lambda_b + \omega \hat{A}_{\text{SV}} - \omega^2 \int_0^\infty d\xi \, [B_{\text{SV}}(\xi) \sin(2\xi\lambda_b)$$

$$- B_{\text{BB}}(\xi) \sin(\xi\lambda_-) + B_{\text{DD}}(\xi) \sin(\xi\lambda_+)]$$

$$k_z^{\text{SH}} = \lambda_b + \omega \hat{A}_{\text{SH}} - \omega^2 \int_0^\infty d\xi \, [B_{\text{SH}}(\xi) \sin(2\xi\lambda_b)]$$

and the imaginary part of the vertical wavenumber (which is related to the attenuation coefficient due to scattering) is

$$\gamma_{\text{P}} = \omega^2 \int_0^\infty d\xi \, [B_{\text{P}}(\xi) \cos(2\xi\lambda_a) + B_{\text{BB}}(\xi) \cos(\xi\lambda_-) + B_{\text{DD}}(\xi) \cos(\xi\lambda_+)]$$

$$\gamma_{\text{SV}} = \omega^2 \int_0^\infty d\xi \, [B_{\text{SV}}(\xi) \cos(2\xi\lambda_b) + B_{\text{BB}}(\xi) \cos(\xi\lambda_-) + B_{\text{DD}}(\xi) \cos(\xi\lambda_+)]$$

$$\gamma_{\text{SH}} = \omega^2 \int_0^\infty d\xi \, [B_{\text{SH}}(\xi) \cos(2\xi\lambda_b)]$$

where $\lambda_a = \omega\sqrt{1/\alpha_0^2 - p^2}$, $\lambda_b = \omega\sqrt{1/\beta_0^2 - p^2}$, $\lambda_+ = \lambda_a + \lambda_b$, and $\lambda_- = \lambda_b - \lambda_a$ (only real valued $\lambda_{a,b}$ are considered). The other quantities in the preceding expressions are

$$B_{\text{P}}(\xi) = (X^2\alpha_0^2)^{-1}[B_{\rho\rho}(\xi)C_1^2 + 2B_{\alpha\rho}(\xi)C_1 + 2B_{\beta\rho}(\xi)C_1C_2$$

$$+ 2B_{\alpha\beta}(\xi)C_2 + B_{\alpha\alpha}(\xi) + B_{\beta\beta}(\xi)C_2^2]$$

$$B_{\text{SV}}(\xi) = (Y^2\beta_0^2)^{-1}[B_{\rho\rho}(\xi)C_7^2 + 2B_{\beta\rho}(\xi)C_7C_8 + B_{\beta\beta}(\xi)C_8^2]$$

$$B_{\text{SH}}(\xi) = (Y^2\beta_0^2)^{-1}[B_{\rho\rho}(\xi)Y^4 + 2B_{\beta\rho}(\xi)C_{10}Y^2 + B_{\beta\beta}(\xi)C_{10}^2]$$

$$B_{\text{BB}}(\xi) = p^2\beta_0(4XY\alpha_0)^{-1}[B_{\rho\rho}(\xi)C_3^2 + 2B_{\beta\rho}(\xi)C_3C_4 + B_{\beta\beta}(\xi)C_4^2]$$

$$B_{\text{DD}}(\xi) = p^2\beta_0(4XY\alpha_0)^{-1}[B_{\rho\rho}(\xi)C_5^2 + 2B_{\beta\rho}(\xi)C_5C_6 + B_{\beta\beta}(\xi)C_6^2]$$

$$\hat{A}_{\text{P}} = (2X\alpha_0)^{-1}A_{\text{P}}, \; \hat{A}_{\text{SV}} = (2Y\beta_0)^{-1}A_{\text{SV}}, \; \hat{A}_{\text{SH}} = (2Y\beta_0)^{-1}A_{\text{SH}}$$

$$A_{\text{P}} = \sigma_{\rho\rho}^2(1 - 4p^2\beta_0^2 Z) + 2\sigma_{\alpha\rho}^2(1 - 4p^2\beta_0^2) + 8\sigma_{\beta\rho}^2 p^2\beta_0^2(1 - 2Z)$$

$$+ 3\sigma_{\alpha\alpha}^2 - 16p^2\beta_0^2\sigma_{\alpha\beta}^2 + 16p^2\beta_0^2\sigma_{\beta\beta}^2(1 - Z)$$

$$A_{\text{SV}} = \sigma_{\rho\rho}^2(1 - C_9) + \sigma_{\beta\rho}^2(2 - 4C_9) + \sigma_{\beta\beta}^2(3 - 4C_9)$$

$$A_{\text{SH}} = \sigma_{\rho\rho}^2 Y^2 + 2\sigma_{\beta\rho}^2(2Y^2 - 1) + \sigma_{\beta\beta}^2(4Y^2 - 1)$$

where

$$X = \sqrt{(1 - p^2\alpha_0^2)} \qquad Y = \sqrt{(1 - p^2\beta_0^2)} \qquad Z = p^2(\alpha_0^2 - \beta_0^2)$$

$$C_1 = X^2(1 - 4p^2\beta_0^2), \qquad C_2 = -8p^2\beta_0^2 X^2$$

$$C_3 = X(3 - 4\beta_0^2 p^2) - Y[(\alpha_0/\beta_0) + (2\beta_0/\alpha_0) - 4\alpha_0\beta_0 p^2]$$

$$C_4 = 4X(1 - 2\beta_0^2 p^2) - 4Y[(\beta_0/\alpha_0) - 2\alpha_0\beta_0 p^2]$$

$$C_5 = X(4\beta_0^2 p^2 - 3) - Y[(\alpha_0/\beta_0) + (2\beta_0/\alpha_0) - 4\alpha_0\beta_0 p^2]$$

$$C_6 = 4X(2\beta_0^2 p^2 - 1) - 4Y[(\beta_0/\alpha_0) - 2\alpha_0\beta_0 p^2]$$

$$C_7 = Y^2(1 - 4p^2\beta_0^2), \qquad C_8 = 1 - 8p^2\beta_0^2 Y^2$$

$$C_9 = 4Y^2 Z\beta_0^2/\alpha_0^2, \qquad C_{10} = 1 - 2p^2\beta_0^2$$

For multimode propagation (P–SV), neglecting higher-order terms restricts the range of applicable pathlengths L. This range is approximately given as

$$\max\{\lambda, a\} < L < Y^2 \max\{\lambda, a\}/\sigma^2$$

where λ is the wavelength, a is the correlation length of the medium, and σ^2 is the variance of the fluctuations. The equations are valid for the whole frequency range, and there is no restriction on the wavelength to correlation length ratio.

The angle- and frequency-dependent phase and group velocities are given by

$$c^{\text{phase}} = \frac{\omega}{\sqrt{k_x^2 + k_z^2}} = \frac{1}{\sqrt{p^2 + k_z^2/\omega^2}}$$

$$c^{\text{group}} = \sqrt{\left(\frac{\partial \omega}{\partial k_x}\right)^2 + \left(\frac{\partial \omega}{\partial k_z}\right)^2}$$

In the low- and high-frequency limits the phase and group velocities are the same. The low-frequency limit for pressure waves in acoustic media is

$$c_{\text{fluid}}^{\text{low freq}} \approx c_0 \left[1 - (\sigma_{\rho\rho}^2/2)\cos^2\theta - \sigma_{\rho\alpha}^2\right]$$

and for elastic waves

$$c_{\text{P}}^{\text{low freq}} = \alpha_0(1 - A_{\text{P}}/2)$$

$$c_{\text{SV}}^{\text{low freq}} = \beta_0(1 - A_{\text{SV}}/2)$$

$$c_{\text{SH}}^{\text{low freq}} = \beta_0(1 - A_{\text{SH}}/2)$$

These limits are the same as the result obtained from Backus averaging with higher-order terms in the medium fluctuations neglected. The high-frequency limit of phase and group velocities is

$$c_{\text{fluid}}^{\text{high freq}} = c_0\big[1 + (\sigma_{\alpha\alpha}{}^2/2\cos^2\theta)\big]$$

for fluids, and

$$c_{\text{P}}^{\text{high freq}} = \alpha_0\big[1 - \sigma_{\alpha\alpha}{}^2\big(1 - 3p^2\alpha_0{}^2/2\big)\big/\big(1 - p^2\alpha_0{}^2\big)\big]$$

$$c_{\text{SV,SH}}^{\text{high freq}} = \beta_0\big[1 - \sigma_{\beta\beta}{}^2\big(1 - 3p^2\beta_0{}^2/2\big)\big/\big(1 - p^2\beta_0{}^2\big)\big]$$

for elastic media, which are in agreement with ray theory predictions, again neglecting higher-order terms.

Shear-wave splitting or birefringence and its frequency dependence can be characterized by

$$S(\omega, p) = [c_{\text{SV}}(\omega, p) - c_{\text{SH}}(\omega, p)]/c_{\text{SV}}(\omega, p) \approx [c_{\text{SV}}(\omega, p) - c_{\text{SH}}(\omega, p)]/\beta_0$$

In the low-frequency limit $S^{\text{low freq}}(\omega, p) \approx (A_{\text{SH}} - A_{\text{SV}})/2$, which is the shear-wave splitting in the transversely isotropic medium obtained from a Backus average of the elastic moduli. In the high-frequency limit $S^{\text{high freq}}(\omega, p) = 0$.

For a medium with exponential correlation functions $\sigma^2 \exp(-\xi/a)$ (where a is the correlation length) for the velocity and density fluctuations with different variances σ^2 but the same correlation length a, the complete frequency dependence of the phase and group velocities is expressed as

$$c_{\text{fluid}}^{\text{phase}} = c_0\big[1 + \{2k_0^2 a^2 \sigma_{\alpha\alpha}{}^2 - \sigma_{\rho\alpha}{}^2 - (\cos^2\theta)\sigma_{\rho\rho}{}^2/2\}\big/\big(1 + 4k_0^2 a^2 \cos^2\theta\big)\big]$$

$$c_{\text{fluid}}^{\text{group}} = c_0\big[1 + N\big/\big(1 + 4k_0^2 a^2 \cos^2\theta\big)^2\big]$$

$$N = 2k_0^2 a^2 \sigma_{\alpha\alpha}{}^2\big(3 + 4k_0^2 a^2 \cos^2\theta\big)$$

$$+ \left(\frac{\sigma_{\rho\rho}{}^2 \cos^2\theta}{2} + \sigma_{\rho\alpha}{}^2\right)\big(4k_0^2 a^2 \cos^2\theta - 1\big)$$

for pressure waves in acoustic media. For elastic P, SV, and SH waves in a randomly layered medium with exponential spatial autocorrelation, the real part of the vertical wave number (obtained by Fourier sine transforms) is given as (Shapiro and Hubral, 1995)

$$k_z^{\text{P}} = \lambda_a + \omega A_{\text{P}} - \omega^2 a^2\big[B_{\text{P}}(0)\{2\lambda_a/(1 + 4a^2\lambda_a{}^2)\}$$

$$+ B_{\text{BB}}(0)\{\lambda_-/(1 + a^2\lambda_-{}^2)\} + B_{\text{DD}}(0)\{\lambda_+/(1 + a^2\lambda_+{}^2)\}\big]$$

$$k_z^{\text{SV}} = \lambda_b + \omega A_{\text{SV}} - \omega^2 a^2\big[B_{\text{SV}}(0)\{2\lambda_b/(1 + 4a^2\lambda_b{}^2)\}$$

$$- B_{\text{BB}}(0)\{\lambda_-/(1 + a^2\lambda_-{}^2)\} + B_{\text{DD}}(0)\{\lambda_+/(1 + a^2\lambda_+{}^2)\}\big]$$

$$k_z^{\text{SH}} = \lambda_b + \omega A_{\text{SH}} - \omega^2 a^2 B_{\text{SH}}(0)\{2\lambda_b/(1 + 4a^2\lambda_b{}^2)\}$$

and the shear-wave splitting for exponentially correlated randomly layered media is

$$S(\omega, p) \approx S^{\text{low freq}} + \omega a^2 \beta_0 Y \big[\{ 2\lambda_b / (1 + 4a^2 \lambda_b^2) \} \{ B_{SV}(0) - B_{SH}(0) \}$$
$$- \{ \lambda_+ / (1 + a^2 \lambda_+^2) \} B_{DD}(0) + \{ \lambda_- / (1 + a^2 \lambda_-^2) \} B_{BB}(0) \big]$$

These equations reveal the general feature that the *anisotropy* (change in velocity with angle) *depends on the frequency*, and the *dispersion* (change in velocity with frequency) *depends on the angle*.

USES

The equations described in this section can be used to estimate velocity dispersion and frequency-dependent anisotropy for plane-wave propagation at any angle in randomly layered one-dimensional media.

ASSUMPTIONS AND LIMITATIONS

The results described in this section are based on the following assumptions:

- Layers are isotropic, linear elastic with no lateral variation.
- The layered medium is statistically stationary with small fluctuations (<30%) in the velocity and density.
- The propagation path is very much longer than any characteristic correlation length of the medium.
- Incident plane-wave propagation is assumed.

3.11 SCALE-DEPENDENT SEISMIC VELOCITIES IN HETEROGENEOUS MEDIA

SYNOPSIS

Measurable travel times of seismic events propagating in heterogeneous media depend on the scale of the seismic wavelength relative to the scale of the geologic heterogeneities. In general, the velocity inferred from arrival times is slower when the wavelength, λ, is longer than the scale of the heterogeneity, a, and faster when the wavelength is shorter (Mukerji et al., 1995).

LAYERED (ONE-DIMENSIONAL) MEDIA

For normal-incidence propagation in stratified media, in the long-wavelength limit ($\lambda/a \gg 1$), where a is the scale of the layering, the stratified medium behaves as a homogeneous effective medium with a velocity given by effective medium theory as

$$V_{\text{EMT}} = \left(\frac{M_{\text{EMT}}}{\rho_{\text{ave}}} \right)^{1/2}$$

The effective modulus M_{EMT} is obtained from the Backus average. For normal-incidence plane-wave propagation, the effective modulus is given by the harmonic average

$$M_{\text{EMT}} = \left[\sum_k \frac{f_k}{M_k} \right]^{-1}$$

$$\frac{1}{\rho_{\text{ave}} V_{\text{EMT}}^2} = \sum_k \frac{f_k}{\rho_k V_k^2}$$

$$\rho_{\text{ave}} = \sum_k f_k \rho_k$$

where f_k, ρ_k, M_k, and V_k are the volume fractions, densities, moduli, and velocities of each constituent layer, respectively. The modulus M can be interpreted as C_{3333} or $K + 4\mu/3$ for P-waves and as C_{2323} or μ for S waves (where K and μ are the bulk and shear moduli, respectively).

In the short-wavelength limit ($\lambda/a \ll 1$), the travel time for plane waves traveling perpendicularly to the layers is given by ray theory as the sum of the travel times through each layer. The ray theory or short-wavelength velocity through the medium is, therefore,

$$\frac{1}{V_{\text{RT}}} = \sum_k \frac{f_k}{V_k}$$

The ray theory velocity involves averaging slownesses, whereas the effective medium velocity involves averaging compliances (slownesses squared). The result is that V_{RT} is always faster than V_{EMT}.

In **two- and three-dimensional heterogeneous media,** there is also the path effect as a result of Fermat's principle. Shorter wavelengths tend to find fast paths and diffract around slower inhomogeneities, thus biasing the travel times to lower values (Nolet, 1987; Müller, Roth, and Korn, 1992). This is sometimes referred to as the "Wielandt effect," "fast path effect," or "velocity shift." The velocity shift was quantified by Boyse (1986) and by Roth, Müller, and Sneider (1993) using an asymptotic ray-theoretical approach. The heterogeneous random medium is characterized by a spatially varying, statistically stationary slowness field

$$n(\mathbf{r}) = n_0 + \varepsilon n_1(\mathbf{r})$$

where r is the position vector, $n_1(r)$ is a zero mean small fluctuation superposed on the constant background slowness n_0, and $\varepsilon \ll 1$ is a small perturbation parameter. The spatial structure of the heterogeneities is described by the isotropic spatial autocorrelation function:

$$\langle \varepsilon n_1(r_1)\varepsilon n_1(r_2)\rangle = \varepsilon^2 \langle n_1^2\rangle N(|r_1 - r_2|)$$

and the coefficient of variation (normalized standard deviation) is given by

$$\sigma_n = \frac{\varepsilon\sqrt{\langle n_1^2\rangle}}{n_0}$$

where $\langle \cdot \rangle$ denotes the expectation operator. For short wavelength ($\lambda/a \ll 1$) initially plane waves traveling along the x-direction, the expected travel time (spatial average of the travel time over a plane normal to x) at distance X is given as (Boyse, 1986)

$$\langle T\rangle = n_0\left[X + \alpha\sigma_n^2\int_0^x (X - \xi)^2 \frac{N'(\xi)}{\xi}d\xi\right] + O(\varepsilon^3)$$

where $\alpha = 1, 1/2$, and 0 for three, two, and one dimension(s), respectively, and the prime denotes differentiation. When the wave has traveled a large distance compared with the correlation length a of the medium ($X \gg a$), and when the autocorrelation function is such that $N(\xi) \ll N(0)$ for $\xi > a$, then

$$\langle T\rangle \approx n_0\left[X - \alpha X^2\sigma_n^2 D\right]$$

$$D = -\int_0^\infty \frac{N'(\xi)}{\xi}d\xi > 0$$

The ray-theory slowness calculated from the average ray theory travel time is given by $n_{RT} = \langle T\rangle/X$. Scaling the distance by the correlation length, a, results in the following equation:

$$\frac{n_{RT}}{n_0} = 1 - \alpha\sigma_n^2\left(\frac{X}{a}\right)\hat{D}$$

where $\hat{D}$ is defined similarly to D but with the autocorrelation function of unit correlation length ($a = 1$). For a Gaussian autocorrelation function $\hat{D} = \sqrt{\pi}$. In one dimension $\alpha = 0$, and the ray-theory slowness is just the average slowness. In two and three dimensions the path effect is described by the term $\alpha\sigma_n^2(X/a)\hat{D}$. In this case, the wave arrivals are on average faster in the random medium than in a uniform medium having the same mean slowness. The expected travel time in three dimensions is less than that in two dimensions, which in turn is less than that in one dimension. This is because in higher dimensions more admissible paths are available to minimize the travel time.

Müller et al. (1992) have established ray-theoretical results relating $N(|r|)$ to the spatial autocorrelation function $\phi(|r|)$ of the travel time fluctuations around the mean travel time. Using first-order straight-ray theory, they show

$$\phi(\zeta) = 2X \int_\zeta^\infty N(\xi) \frac{\xi}{\sqrt{\xi^2 - \zeta^2}} d\xi$$

For a medium with a Gaussian slowness autocorrelation function, $N(r) = \exp(-r^2/a^2)$, the variance of travel time fluctuations $\phi(0)$ is related to the variance of the slowness fluctuations by

$$\phi(0) = \sqrt{\pi} X \, a \, n_0^2 \, \sigma_n^2$$

The expected travel time $\langle T \rangle$ of the wavefield $U(n)$ in the heterogeneous medium with random slowness n is distinct from the travel time T of the expected wavefield $\langle U(n) \rangle$ in the random medium. The expected wave, which is an ensemble average of the wavefield over all possible realizations of the random medium, travels slower than the wavefield in the average medium $\langle n \rangle$ (Keller, 1964). Thus,

$$\langle T(U(n)) \rangle \leq T(U(\langle n \rangle)) \leq T(\langle U(n) \rangle)$$

USES

The results described in this section can be used for the following purposes:

- To estimate velocity shift caused by fast path effects in heterogeneous media.
- To relate statistics of observed travel times to the statistics of the heterogeneities.

ASSUMPTIONS AND LIMITATIONS

The equations described in this section apply under the following conditions:

- Small fluctuations in the material properties of the heterogeneous medium.
- Isotropic spatial autocorrelation function of the slowness fluctuations.
- Ray theory results are valid only for wavelengths much smaller than the spatial correlation length of the media.

3.12 SCATTERING ATTENUATION

SYNOPSIS

The attenuation coefficient, $\gamma_s = \pi f/QV$, (where Q is the quality factor, V is the seismic velocity, and f is frequency) that results from elastic scattering depends on the ratio of seismic wavelength, λ, to the diameter, d_s, of the scattering heterogeneity. Roughly speaking there are three domains:

- Rayleigh scattering, where $\lambda > d_s$ and $\gamma_s \propto d_s^3 f^4$
- Stochastic/Mie scattering, where $\lambda \approx d_s$ and $\gamma_s \propto d_s f^2$
- Diffusion scattering, where $\lambda < d_s$ and $\gamma_s \propto 1/d_s$

When $\lambda \gg d_s$, the heterogeneous medium behaves like an effective homogeneous medium, and scattering effects may be negligible. At the other limit, when $\lambda \ll d_s$, the heterogeneous medium may be treated as a piecewise homogeneous medium.

Figure 3.12.1 shows schematically the general scale dependence (or, equivalently, frequency dependence) of wave velocity that is expected owing to scattering in heterogeneous media. At very long wavelengths ($\lambda \gg d_s$) the phase velocity is nondispersive and is close to the static effective medium result. As the wavelength decreases (frequency increases), scattering causes velocity dispersion. In the Rayleigh scattering domain ($\lambda/d_s \approx 2\pi$), the velocity shows a slight decrease with increasing frequency. This is usually followed by a rapid and much larger increase in phase velocity owing to resonant (or Mie) scattering ($\lambda \approx d_s$). When $\lambda \ll d_s$ (specular scattering or ray theory), the velocity is again nondispersive or weakly dispersive and is usually significantly higher than its long-wavelength limit.

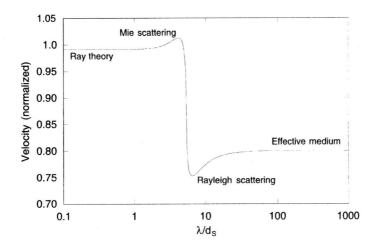

Figure 3.12.1

It is usually assumed that the long-wavelength Rayleigh limit is most appropriate for analyzing laboratory rock physics results because the seismic wavelength is often much larger than the grain size. However, Lucet and Zinszner (1992) and Blair (1990), among others, have shown that the scattering heterogeneities can be *clusters* of grains that are comparable to, or larger than, the wavelength. Certainly any of the domains are possible in situ.

Blair (1990) suggests a simple (ad hoc) expression that is consistent with both the Rayleigh and diffusion scattering limits:

$$\gamma_s(f) = \frac{C_s}{d_s} \frac{\left(\frac{f}{f_d}\right)^4}{\left(1 + \frac{f}{f_d}\right)^4}, \qquad f_d = \frac{k_s V}{d_s} = \frac{k_s f \lambda}{d_s}$$

where C_s and k_s are constants.

Many theoretical estimates of scattering effects on velocity and attenuation have appeared (see Mehta, 1983, and Berryman, 1992, for reviews). Most are in the long-wavelength limit, and most assume that the concentration of scatterers is small, and thus only single scattering is considered.

The attenuation of P-waves caused by a low concentration of small spherical inclusions is given by (Yamakawa, 1962; Kuster and Toksöz, 1974)

$$\gamma_{sph} = c \frac{3\omega}{4V_P} \left(\frac{\omega}{V_P}a\right)^3 \left[2B_0^2 + \frac{2}{3}(1 + 2\zeta^3)B_1^2 + \frac{(2 + 3\zeta^5)}{5}B_2^2\right]$$

where

$$B_0 = \frac{K - K'}{3K' + 4\mu}$$

$$B_1 = \frac{\rho - \rho'}{3\rho}$$

$$B_3 = \frac{20}{3} \frac{\mu(\mu' - \mu)}{6\mu'(K + 2\mu) + \mu(9K + 8\mu)}$$

$$\zeta = V_P/V_S$$

The terms V_P and V_S are the P and S velocities of the host medium, respectively, c is the volume concentration of the spheres, a is their radius, $2\pi f = \omega$ is the frequency, ρ is the density, K is the bulk modulus, and μ is the shear modulus. The unprimed moduli refer to the background host medium, and the primed moduli refer to the inclusions.

In the case of elastic spheres in a linear viscous fluid, with viscosity η, the attenuation is given by (Epstein, 1941; Epstein and Carhart, 1953; Kuster and Toksöz, 1974)

$$\gamma_{sph} = c \frac{\omega}{2V_P}(\rho - \rho') \, \text{Real}\left[\frac{i + b_0 - ib_0^2/3}{\rho(1 - ib_0) - (\rho + 2\rho')b_0^2/9}\right]$$

where

$$b_0 = (1 + i)a\sqrt{\frac{\pi f \rho}{\eta}}$$

Hudson (1981) gives the attenuation coefficient for elastic waves in cracked media (see Section 4.10 on Hudson). For aligned penny-shaped ellipsoidal cracks with normals along the 3-axis, the attenuation coefficients for P, SV, and SH waves are

$$\gamma_P = \frac{\omega}{V_S}\varepsilon\left(\frac{\omega a}{V_P}\right)^3 \frac{1}{30\pi}\left[AU_1^2 \sin^2 2\theta + BU_3^2\left(\frac{V_P^2}{V_S^2} - 2\sin^2\theta\right)^2\right]$$

$$\gamma_{SV} = \frac{\omega}{V_S}\varepsilon\left(\frac{\omega a}{V_S}\right)^3 \frac{1}{30\pi}\left[AU_1^2 \cos^2 2\theta + BU_3^2 \sin^2 2\theta\right]$$

$$\gamma_{SH} = \frac{\omega}{V_S}\varepsilon\left(\frac{\omega a}{V_S}\right)^3 \frac{1}{30\pi}\left[AU_1^2 \cos^2\theta\right]$$

$$A = \frac{3}{2} + \frac{V_S^5}{V_P^5}$$

$$B = 2 + \frac{15}{4}\frac{V_S}{V_P} - 10\frac{V_S^3}{V_P^3} + 8\frac{V_S^5}{V_P^5}$$

In these expressions, θ is the angle between the direction of propagation and the 3-axis (axis of symmetry), and ε is the crack density parameter:

$$\varepsilon = \frac{N}{V}a^3 = \frac{3\phi}{4\pi\alpha}$$

where N/V is the number of penny-shaped cracks of radius a per unit volume, ϕ is the crack porosity, and α is the crack aspect ratio.

U_1 and U_3 depend on the crack conditions. For dry cracks

$$U_1 = \frac{16(\lambda + 2\mu)}{3(3\lambda + 4\mu)}$$

$$U_3 = \frac{4(\lambda + 2\mu)}{3(\lambda + \mu)}$$

For "weak" inclusions (i.e., when $\mu\alpha/[K' + (4/3)\mu']$ is of the order 1 and is not small enough to be neglected)

$$U_1 = \frac{16(\lambda + 2\mu)}{3(3\lambda + 4\mu)}\frac{1}{(1 + M)}$$

$$U_3 = \frac{4(\lambda + 2\mu)}{3(\lambda + \mu)}\frac{1}{(1 + \kappa)}$$

where

$$M = \frac{4\mu'}{\pi\alpha\mu} \frac{(\lambda + 2\mu)}{(3\lambda + 4\mu)}$$

$$\kappa = \frac{[K' + (4/3)\mu'](\lambda + 2\mu)}{\pi\alpha\mu(\lambda + \mu)}$$

with K' and μ' equal to the bulk and shear moduli of the inclusion material. The criterion for an inclusion to be "weak" depends on its shape, or aspect ratio α, as well as on the relative moduli of the inclusion and matrix material. Dry cavities can be modeled by setting the inclusion moduli to zero. Fluid-saturated cavities are simulated by setting the inclusion shear modulus to zero. Remember that these give only the scattering losses and do not incorporate other viscous losses caused by the pore fluid.

Hudson also gives expressions for infinitely thin fluid-filled cracks:

$$U_1 = \frac{16(\lambda + 2\mu)}{3(3\lambda + 4\mu)}$$

$$U_3 = 0$$

These assume no discontinuity in the normal component of crack displacements and therefore predict no change in the compressional modulus with saturation. There is, however, a shear displacement discontinuity and a resulting effect on shear stiffness. This case should be used with care.

For randomly oriented cracks (isotropic distribution) the P and S attenuation coefficients are given as

$$\gamma_P = \frac{\omega}{V_S} \varepsilon \left(\frac{\omega a}{V_P}\right)^3 \frac{4}{(15)^2\pi} \left(AU_1^2 + \frac{1}{2}\frac{V_P^5}{V_S^5}B(B-2)U_3^2\right)$$

$$\gamma_S = \frac{\omega}{V_S} \varepsilon \left(\frac{\omega a}{V_S}\right)^3 \frac{1}{75\pi} \left(AU_1^2 + \frac{1}{3}BU_3^2\right)$$

The fourth-power dependence on ω is characteristic of Rayleigh scattering.

Random heterogeneous media with spatially varying velocity $c = c_0 + c'$ may be characterized by the autocorrelation function

$$N(r) = \frac{\langle \xi(r')\xi(r' + r)\rangle}{\langle \xi^2 \rangle}$$

where $\xi = -c'/c_0$ and c_0 denotes the mean background velocity. For small fluctuations, the fractional energy loss caused by scattering is given by (Aki and

Richards, 1980)

$$\frac{\Delta E}{E} = \frac{8\langle\xi^2\rangle k^4 a^3 L}{1 + 4k^2 a^2} \quad \text{for } N(r) = e^{-r/a}$$

$$\frac{\Delta E}{E} = \sqrt{\pi}\langle\xi^2\rangle k^2 a L \left(1 - e^{-k^2 a^2}\right) \quad \text{for } N(r) = e^{-r^2/a^2}$$

These expressions are valid for small $\Delta E/E$ values as they are derived under the Born approximation, which assumes that the primary incident waves are unchanged as they propagate through the heterogeneous medium.

Aki and Richards (1980) classify scattering phenomena in terms of two dimensionless numbers ka and kL, where $k = 2\pi/\lambda$ is the wavenumber, a is the characteristic scale of the heterogeneity, and L is the path length of the primary incident wave in the heterogeneous medium. Scattering effects are not very important for very small or very large ka, and they become increasingly important with increasing kL. Scattering problems may be classified on the basis of the fractional energy loss caused by scattering, $\Delta E/E$, and the wave parameter D defined by $D = 4L/ka^2$. The wave parameter is the ratio of the first Fresnel zone to the scale length of the heterogeneity. Ray theory is applicable when $D < 1$. In this case the inhomogeneities are smooth enough to be treated as piecewise homogeneous. Effective medium theories are appropriate when ka and $\Delta E/E$ are small. These domains are summarized in Figure 3.12.2.

Scattering becomes complex when heterogeneity scales are comparable with the wavelength and when the path lengths are long. Energy diffusion models are used for long path lengths and strong scattering.

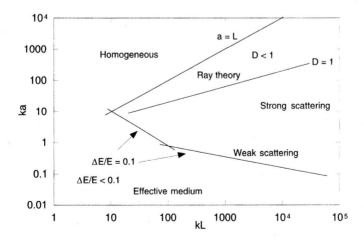

Figure 3.12.2

USES

The results described in this section can be used to estimate the seismic attenuation caused by scattering.

ASSUMPTIONS AND LIMITATIONS

The results described in this section have the following limitations:

- Formulas for spherical and ellipsoidal inclusions are limited to low pore concentrations and wavelengths much larger than the scatterer diameter.
- Formulas for fractional energy loss in random heterogeneous media are limited to weak scattering.

3.13 WAVES IN CYLINDRICAL RODS – THE RESONANT BAR

SYNOPSIS

Time-harmonic waves propagating in the axial direction along a circular cylindrical rod involve radial, circumferential, and axial components of displacement, u_r, u_θ, and u_z, respectively. Motions that depend on z but are independent of θ may be separated into **torsional** waves involving u_θ only and **longitudinal** waves involving u_r and u_z. **Flexural** waves consist of motions that depend on both z and θ.

TORSIONAL WAVES

Torsional waves involve purely circumferential displacements that are independent of θ. The dispersion relation (for free-surface boundary conditions) is of the form (Achenbach, 1984)

$$sa\, J_0(sa) - 2J_1(sa) = 0$$

$$s^2 = \frac{\omega^2}{V_s^2} - k^2$$

where

$$J_n(\ ) = \text{Bessel functions of the first kind of order } n$$
$$a = \text{radius of cylindrical rod}$$
$$V_S = \text{S-wave velocity}$$
$$k = \text{wavenumber for torsional waves}$$

For practical purposes, the lowest mode of each kind of motion is important. The lowest torsional mode consists of displacement proportional to the radius, and the motion is a rotation of each cross section of the cylinder about its center. The phase velocity of the lowest torsional mode is nondispersive and is given by

$$V_{\text{torsion}} = V_S = \sqrt{\frac{\mu}{\rho}}$$

where μ and ρ are the shear modulus and density of the rod, respectively.

LONGITUDINAL WAVES

Longitudinal waves are axially symmetric and have displacement components in the axial and radial directions. The dispersion relation (for free-surface boundary conditions), known as the **Pochhammer** equation, is (Achenbach, 1984)

$$2pa[(sa)^2 + (ka)^2]J_1(pa)J_1(sa) - [(sa)^2 - (ka)^2]^2 J_0(pa)J_1(sa)$$
$$- 4(ka)^2(pa)(sa)J_1(pa)J_0(sa) = 0$$

$$p^2 = \frac{\omega^2}{V_P^2} - k^2$$

where V_P is the P-wave velocity.

The phase velocity of the lowest longitudinal mode for small $ka(ka \ll 1)$ can be expressed as

$$V_{\text{long}} = \sqrt{\frac{E}{\rho}\left[1 - \frac{1}{4}v^2(ka)^2\right]} + O[(Ka)^4]$$

where

$$E = \text{Young's modulus of cylindrical rod}$$
$$v = \text{Poisson's ratio of cylindrical rod}$$

In the limit as $(ka) \to 0$, the phase velocity tends to the bar velocity or extensional velocity $V_E = \sqrt{E/\rho}$. For very large $ka(ka \gg 1)$, V_{long} approaches the Rayleigh wave velocity.

FLEXURAL WAVES

Flexural modes have all three displacement components – axial, radial, and circumferential and involve motion that depends on both z and θ. The phase velocity of the lowest flexural mode for small values of ka ($ka \ll 1$) may be written as

$$V_{\text{flex}} = \frac{1}{2}\sqrt{\frac{E}{\rho}}(ka) + O[(ka)^3]$$

The phase velocity of the lowest flexural mode goes to zero as $(ka) \to 0$ and approaches the Rayleigh wave velocity for large ka values.

BAR RESONANCE

Resonant modes (or standing waves) occur when the bar length is an integer number of half-wavelengths:

$$V = \lambda f = \frac{2Lf}{n}$$

where

$$V = \text{velocity}$$
$$\lambda = \text{wavelength}$$
$$f = \text{resonant frequency}$$
$$L = \text{bar length}$$
$$n = \text{positive integer}$$

In practice, the shear or extensional velocity is calculated from the observed resonant frequency, most often at the fundamental mode, where $n = 1$.

POROUS, FLUID-SATURATED RODS

Biot's theory has been used to extend Pochhammer's method of analysis for fluid-saturated porous rods (Gardner, 1962; Berryman, 1983). The dependence of the velocity and attenuation of longitudinal waves on the skeleton and fluid properties is rather complicated. The motions of the solid and the fluid are partly parallel to the axis of the cylinder and partly along the radius. The dispersion relations are obtained from plane-wave solutions of Biot's equations in cylindrical (r, θ, z) coordinates. For an open (unjacketed) surface boundary condition the ω–k_z dispersion relation is given by (in the notation of Berryman, 1983)

$$D_{\text{open}} = \begin{vmatrix} a_{11} & a_{12} & a_{13} \\ a_{21} & a_{22} & 0 \\ a_{31} & a_{32} & a_{33} \end{vmatrix} = 0$$

$$a_{11} = \frac{[(C\Gamma_- - H)k_+^2 + 2\mu_{\text{fr}}k_z^2]J_0(k_+a) + 2\mu_{\text{fr}}k_{r+}J_1(k_+a)/a}{(\Gamma_+ - \Gamma_-)}$$

$$a_{12} = \frac{\left[(H - C\Gamma_+)k_-^2 - 2\mu_{fr}k_z^2\right]J_0(k_-a) - 2\mu_{fr}k_{r-}J_1(k_-a)/a}{(\Gamma_+ - \Gamma_-)}$$

$$a_{13} = -2\mu_{fr}k_{sr}[k_{sr}J_0(k_{sr}a) - J_1(k_{sr}a)/a]$$

$$a_{21} = \frac{(M\Gamma_- - C)k_+^2 J_0(k_+a)}{(\Gamma_+ - \Gamma_-)}$$

$$a_{22} = \frac{(C - M\Gamma_+)k_-^2 J_0(k_-a)}{(\Gamma_+ - \Gamma_-)}$$

$$a_{23} = 0$$

$$a_{31} = \frac{-2i\mu_{fr}k_z k_{r+} J_1(k_+a)}{(\Gamma_+ - \Gamma_-)}$$

$$a_{32} = \frac{2i\mu_{fr}k_z k_{r-} J_1(k_-a)}{(\Gamma_+ - \Gamma_-)}$$

$$a_{33} = -\mu_{fr}\left(k_s^2 - 2k_z^2\right)k_{sr}J_1(k_{sr}a)/(ik_z)$$

$$k_{r\pm}^2 = k_\pm^2 - k_z^2, \quad k_{sr}^2 = k_s^2 - k_z^2$$

$$k_s^2 = \omega^2\left(\rho - \rho_{fl}^2/q\right)\mu_{fr}$$

$$k_+^2 = \frac{1}{2}[b + f - \sqrt{(b - f)^2 + 4cd}]$$

$$k_-^2 = \frac{1}{2}[b + f + \sqrt{(b - f)^2 + 4cd}]$$

$$b = \omega^2(\rho M - \rho_{fl}C)/\Delta$$

$$c = \omega^2(\rho_{fl}M - qC)/\Delta$$

$$d = \omega^2(\rho_{fl}H - \rho C)/\Delta$$

$$f = \omega^2(qH - \rho_{fl}C)/\Delta$$

$$\Delta = MH - C^2$$

$$\Gamma_\pm = d/\left(k_\pm^2 - b\right) = \left(k_\pm^2 - f\right)/c$$

$$H = K_{fr} + \frac{4}{3}\mu_{fr} + \frac{(K_0 - K_{fr})^2}{(D - K_{fr})}$$

$$C = \frac{(K_0 - K_{fr})K_0}{(D - K_{fr})}$$

$$M = \frac{K_0^2}{(D - K_{fr})}$$

$$D = K_0[1 + \phi(K_0/K_{fl} - 1)]$$

$$\rho = (1 - \phi)\rho_0 + \phi\rho_{fl}$$

$$q = \frac{\alpha\rho_{fl}}{\phi} - \frac{i\eta F(\zeta)}{\omega\kappa}$$

where

K_{fr}, μ_{fr} = effective bulk and shear moduli of rock frame: either the dry frame or the high-frequency unrelaxed "wet frame" moduli predicted by the Mavko–Jizba squirt theory

K_0 = bulk modulus of mineral material making up rock

K_{fl} = effective bulk modulus of pore fluid

ϕ = porosity

ρ_0 = mineral density

ρ_{fl} = fluid density

α = tortuosity parameter (always greater than 1)

η = viscosity of the pore fluid

κ = absolute permeability of the rock

ω = angular frequency of plane wave

The viscodynamic operator $F(\zeta)$ incorporates the frequency dependence of viscous drag and is defined by

$$F(\zeta) = \frac{1}{4}\frac{\zeta T(\zeta)}{1 + 2i T(\zeta)/\zeta}$$

$$T(\zeta) = \frac{\text{ber}'(\zeta) + i\,\text{bei}'(\zeta)}{\text{ber}(\zeta) + i\,\text{bei}(\zeta)} = \frac{e^{i3\pi/4}J_1(\zeta e^{-i\pi/4})}{J_0(\zeta e^{-i\pi/4})}$$

$$\zeta = (\omega/\omega_r)^{1/2} = \left(\frac{\omega h^2 \rho_{fl}}{\eta}\right)^{1/2}$$

where

ber( ), bei( ) = real and imaginary parts of the Kelvin function

$J_n($) = Bessel function of order n

h = pore-size parameter

The pore-size parameter h depends on both the dimensions and shape of the pore space. Stoll (1974) found that values between 1/6 and 1/7 of the mean grain diameter gave good agreement with experimental data from several investigators. For spherical grains, Hovem and Ingram (1979) obtained $h = \phi d/[3(1 - \phi)]$, where d is the grain diameter.

This dispersion relation gives the same results as Gardner (1962). When the surface pores are closed (jacketed) the resulting dispersion relation is

$$D_{\text{closed}} = \begin{vmatrix} a_{11} & a_{12} & a_{13} \\ a_{31} & a_{32} & a_{33} \\ a_{41} & a_{42} & a_{43} \end{vmatrix} = 0$$

$$a_{41} = \frac{k_{r+}\Gamma_- J_1(k_+a)}{(\Gamma_+ - \Gamma_-)}$$

$$a_{42} = \frac{-k_{r-}\Gamma_+ J_1(k_-a)}{(\Gamma_+ - \Gamma_-)}$$

$$a_{43} = k_{\text{sr}} J_1(k_{\text{sr}}a)\rho_{\text{fl}}/q$$

For open-pore surface conditions the vanishing of the fluid pressure at the surface of the cylinder causes strong radial motion of the fluid relative to the solid. This relative motion absorbs energy, causing greater attenuation than would be present in a plane longitudinal wave in an extended porous saturated medium (White, 1986). Narrow stop bands and sharp peaks in the attenuation can occur if the slow P-wave has wavelength $\lambda < 2.6a$. Such stop bands do not exist in the case of the jacketed, closed-pore surface. A slow extensional wave propagates under jacketed boundary conditions but not under the open-surface condition.

USES

The results described in this section can be used to model wave propagation and geometric dispersion in resonant bar experiments.

ASSUMPTIONS AND LIMITATIONS

The results described in this section assume the following:

- Isotropic, linear, homogeneous, and elastic–poroelastic rod of solid circular cross section.
- For elastic rods the cylindrical surface is taken to be free of tractions.
- For porous rods, unjacketed and jacketed surface boundary conditions are assumed.

PART 4

EFFECTIVE MEDIA

4.1 HASHIN–SHTRIKMAN BOUNDS

SYNOPSIS

If we wish to predict the effective elastic moduli of a mixture of grains and pores theoretically, we generally need to specify (1) the volume fractions of the various phases, (2) the elastic moduli of the various phases, and (3) the geometric details of how the phases are arranged relative to each other. If we specify only the volume fractions and the constituent moduli, the best we can do is to predict the upper and lower bounds (shown shematically in Figure 4.1.1).

At any given volume fraction of constituents the effective modulus will fall between the bounds (somewhere along the vertical dashed line in the plot of bulk modulus, Figure 4.1.1), but its precise value depends on the geometric details. We use, for example, terms like "stiff pore shapes" and "soft pore shapes." Stiffer shapes cause the value to be higher within the allowable range; softer shapes cause the value to be lower. The best bounds, defined as giving the narrowest possible range without specifying anything about the geometries of the constituents, are the Hashin–Shtrikman bounds (Hashin and Shtrikman, 1963), given by

$$K^{\text{HS}\pm} = K_1 + \frac{f_2}{(K_2 - K_1)^{-1} + f_1 \left(K_1 + \frac{4}{3}\mu_1\right)^{-1}}$$

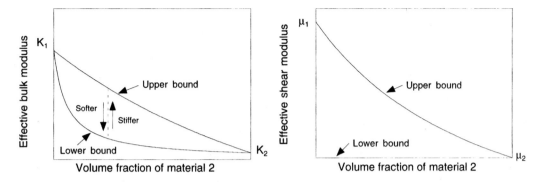

Figure 4.1.1. Schematic representation of the upper and lower bounds on the elastic bulk and shear moduli.

$$\mu^{HS\pm} = \mu_1 + \frac{f_2}{(\mu_2 - \mu_1)^{-1} + \frac{2f_1(K_1+2\mu_1)}{5\mu_1(K_1+\frac{4}{3}\mu_1)}}$$

where

$$K_1, K_2 = \text{bulk moduli of individual phases}$$
$$\mu_1, \mu_2 = \text{shear moduli of individual phases}$$
$$f_1, f_2 = \text{volume fractions of individual phases}$$

Upper and lower bounds are computed by interchanging which material is termed 1 and which is termed 2. Generally, the expressions give the upper bound when the stiffest material is termed 1 in the expressions above, and the lower bound when the softest material is termed 1.

The physical interpretation of the bounds for bulk modulus is shown schematically in Figure 4.1.2. The space is filled by an assembly of spheres of material 2, each surrounded by a shell of material 1. Each sphere and its shell has precisely the volume fractions f_1 and f_2. The upper bound is realized when the stiffer material forms the shell; the lower bound is realized when it is in the core.

A more general form of the bounds, which can be applied to more than two phases (Berryman, 1995), can be written as

$$K^{HS+} = \Lambda(\mu_{max}), \quad K^{HS-} = \Lambda(\mu_{min})$$
$$\mu^{HS+} = \Gamma(\zeta(K_{max}, \mu_{max})), \quad \mu^{HS-} = \Gamma(\zeta(K_{min}, \mu_{min}))$$

where

$$\Lambda(z) = \left\langle \frac{1}{K(r) + \frac{4}{3}z} \right\rangle^{-1} - \frac{4}{3}z$$

$$\Gamma(z) = \left\langle \frac{1}{\mu(r) + z} \right\rangle^{-1} - z$$

$$\zeta(K, \mu) = \frac{\mu}{6}\left(\frac{9K + 8\mu}{K + 2\mu}\right)$$

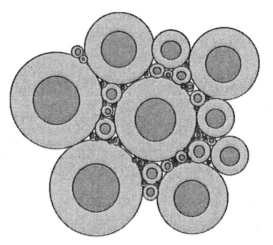

Figure 4.1.2. Physical interpretation of the Hashin–Shtrikman bounds for bulk modulus of a two-phase material.

The brackets $\langle \cdot \rangle$ indicate an average over the medium, which is the same as an average over the constituents weighted by their volume fractions.

EXAMPLE

Compute the Hashin–Shtrikman upper and lower bounds on the bulk and shear moduli for a mixture of quartz, calcite, and water. The porosity (water fraction) is 27 percent; quartz is 80 percent by volume of the solid fraction, and calcite is 20 percent by volume of the solid fraction. The moduli of the individual constituents are

$$K_{quartz} = 36 \, \text{GPa}, \quad K_{calcite} = 75 \, \text{GPa}, \quad K_{water} = 2.2 \, \text{GPa},$$

$$\mu_{quartz} = 45 \, \text{GPa}, \quad \mu_{calcite} = 31 \, \text{GPa}, \quad \text{and} \quad \mu_{water} = 0 \, \text{GPa}. \quad \text{Hence}$$

$$\mu_{min} = 0 \, \text{GPa}, \quad \mu_{max} = 45 \, \text{GPa}, \quad K_{min} = 2.2 \, \text{GPa}, \quad \text{and} \quad K_{max} = 75 \, \text{GPa}.$$

$$K^{HS-} = \Lambda(\mu_{min})$$

$$= \left[\frac{\phi}{2.2} + \frac{(1-\phi)(0.8)}{36.0} + \frac{(1-\phi)(0.2)}{75.0} \right]^{-1}$$

$$= 7.10 \, \text{GPa}$$

$$K^{HS+} = \Lambda(\mu_{max})$$

$$= \left[\frac{\phi}{2.2 + \left(\frac{4}{3}\right)45} + \frac{(1-\phi)(0.8)}{36.0 + \left(\frac{4}{3}\right)45} + \frac{(1-\phi)(0.2)}{75.0 + \left(\frac{4}{3}\right)45.0} \right]^{-1} - \left(\frac{4}{3}\right)45$$

$$= 26.9 \, \text{GPa}$$

$$\zeta(K_{max}, \mu_{max}) = \frac{45}{6}\left(\frac{9 \times 75 + 8 \times 45}{75 + 2 \times 45}\right) = 47.0 \text{ GPa}$$

$$\zeta(K_{min}, \mu_{min}) = 0 \text{ GPa}$$

$$\mu^{HS+} = \Gamma(\zeta(K_{max}, \mu_{max}))$$

$$= \left[\frac{\phi}{47.0} + \frac{(1-\phi)(0.8)}{45.0 + 47.0} + \frac{(1-\phi)(0.2)}{31.0 + 47.0}\right]^{-1} - 47.0$$

$$= 24.6 \text{ GPa}$$

$$\mu^{HS-} = \Gamma(\zeta(K_{min}, \mu_{min}))$$

$$= 0$$

The separation between the upper and lower bounds depends on how different the constituents are. As shown in Figure 4.1.3, the bounds are often quite similar when mixing solids, for the moduli of common minerals are usually within a factor of two of each other. Because many effective medium models (e.g., Biot, Gassmann, Kuster–Toksöz, etc.) assume a homogeneous mineral modulus, it is often useful (and adequate) to represent a mixed mineralogy with an "average mineral" equal to either one of the bounds or to their average $(M^{HS+} + M^{HS-})/2$. On the other hand, when the constituents are quite different (such as minerals and pore fluids), the bounds become quite separated, and we lose some of the predictive value.

Note that when $\mu_{min} = 0$, K^{HS-} is the same as the Reuss bound. In this case, the Reuss or Hashin–Shtrikman lower bound describes the moduli of a suspension of grains in a pore fluid exactly (see Section 4.2 on Voigt–Reuss bounds and also Section 4.3 on Wood's relation). These also describe the moduli of a mixture of fluids or gases, or both.

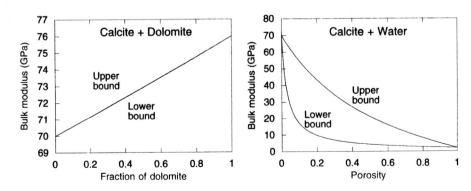

Figure 4.1.3

When all phases have the same shear modulus, $\mu = \mu_{\min} = \mu_{\max}$, the upper and lower bounds become identical and we get the expression by Hill (1963) for the effective bulk modulus of a composite with uniform shear modulus (see Section 4.5 on composites with uniform shear modulus).

USES

The bounds described in this section can be used for the following:

- To compute the estimated range of average mineral modulus for a mixture of mineral grains.
- To compute the upper and lower bounds for a mixture of mineral and pore fluid.

ASSUMPTIONS AND LIMITATIONS

The bounds described in this section apply under the following conditions:

- Each constituent is isotropic, linear, elastic.
- The rock is isotropic linear elastic.

4.2 VOIGT AND REUSS BOUNDS

SYNOPSIS

If we wish to predict the effective elastic moduli of a mixture of grains and pores theoretically, we generally need to specify (1) the volume fractions of the various phases, (2) the elastic moduli of the various phases, and (3) the geometric details of how the phases are arranged relative to each other. If we specify only the volume fractions and the constituent moduli, the best we can do is to predict the upper and lower bounds (shown schematically in Figure 4.2.1).

At any given volume fraction of constituents, the effective modulus will fall between the bounds (somewhere along the vertical dashed line in Figure 4.2.1), but its precise value depends on the geometric details. We use, for example, terms like "stiff pore shapes" and "soft pore shapes." Stiffer shapes cause the value to be higher within the allowable range; softer shapes cause the value to be lower. The simplest, but not necessarily the best, bounds are the Voigt and Reuss bounds. (See also Section 4.1 on Hashin–Shtrikman bounds, which are narrower.)

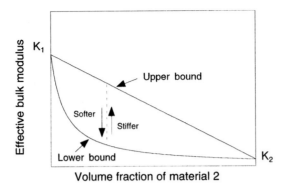

Figure 4.2.1

The **Voigt upper bound** of the effective elastic modulus, M_V, of N phases is

$$M_V = \sum_{i=1}^{N} f_i M_i$$

where

f_i = the volume fraction of the ith medium
M_i = the elastic modulus of the ith medium

The Voigt bound is sometimes called the **isostrain average** because it gives the ratio of average stress to average strain when all constituents are assumed to have the same strain.

The **Reuss lower bound** of the effective elastic modulus, M_R, is (Reuss, 1929)

$$\frac{1}{M_R} = \sum_{i=1}^{N} \frac{f_i}{M_i}$$

The Reuss bound is sometimes called the **isostress average** because it gives the ratio of average stress to average strain when all constituents are assumed to have the same stress.

When one of the constituents is a liquid or gas with zero shear modulus, the Reuss average bulk and shear moduli for the composite are exactly the same as given by the Hashin–Shtrikman lower bound.

The Reuss average exactly describes the effective moduli of a suspension of solid grains in a fluid. It also describes the moduli of "shattered" materials in which solid fragments are completely surrounded by the pore fluid.

When all constituents are gases or liquids, or both, with zero shear modulus, the Reuss average gives the effective moduli of the mixture exactly.

In contrast to the Reuss average, which describes a number of real physical systems, real isotropic mixtures can never be as stiff as the Voigt bound (except for the single phase end members).

Mathematically the M in the Reuss average formula can represent any modulus: K, μ, E, and so forth. However, it makes most sense to compute only the

Reuss averages of the shear modulus, $M = \mu$, and the bulk modulus, $M = K$, and then compute the other moduli from these.

USES

The methods described in this section can be used for the following purposes:

- To compute the estimated range of average mineral modulus for a mixture of mineral grains.
- To compute the upper and lower bounds for a mixture of mineral and pore fluid.

ASSUMPTIONS AND LIMITATIONS

The methods described in this section presuppose that each constituent is isotropic, linear, elastic.

4.3 WOOD'S FORMULA

SYNOPSIS

In a fluid suspension or fluid mixture, where the heterogeneities are small compared with a wavelength, the sound velocity is given exactly by Wood's (1955) relation

$$V = \sqrt{\frac{K_R}{\rho}}$$

where K_R is the Reuss (isostress) average of the composite

$$\frac{1}{K_R} = \sum_{i=1}^{N} \frac{f_i}{K_i}$$

and ρ is the average density defined by

$$\rho = \sum_{i=1}^{N} f_i \rho_i$$

The f_i, K_i, and ρ_i are the volume fractions, bulk moduli, and densities of the phases, respectively.

EXAMPLE

Use Wood's relation to estimate the speed of sound in a water-saturated suspension of quartz particles at atmospheric conditions. The quartz properties are $K_{quartz} = 36$ GPa and $\rho_{quartz} = 2.65$ g/cm^3. The water properties are $K_{water} = 2.2$ GPa and $\rho_{water} = 1.0$ g/cm^3. The porosity is $\phi = 0.40$.

The Reuss average bulk modulus of the suspension is given by

$$K_{Reuss} = \left(\frac{\phi}{K_{water}} + \frac{1-\phi}{K_{quartz}} \right)^{-1} = \left(\frac{0.4}{2.2} + \frac{0.6}{36} \right)^{-1} = 5.04 \text{ GPa}$$

The density of the suspension is

$$\rho = \phi \rho_{water} + (1-\phi)\rho_{quartz} = (0.4)(1.0) + (0.6)(2.65) = 1.99 \text{ g/cm}^3$$

This gives the sound speed of

$$V = \sqrt{K/\rho} = \sqrt{5.04/1.99} = 1.59 \text{ km/s}$$

EXAMPLE

Use Wood's relation to estimate the speed of sound in a suspension of quartz particles in water with 50 percent saturation of air at atmospheric conditions. The quartz properties are $K_{quartz} = 36$ GPa and $\rho_{quartz} = 2.65$ g/cm^3. The water properties are $K_{water} = 2.2$ GPa and $\rho_{water} = 1.0$ g/cm^3. The air properties are $K_{air} = 0.000131$ GPa and $\rho_{air} = 0.00119$ g/cm^3. The porosity is $\phi = 0.40$.

The Reuss average bulk modulus of the suspension is given by

$$K_{Reuss} = \left(\frac{0.5\phi}{K_{water}} + \frac{0.5\phi}{K_{air}} + \frac{1-\phi}{K_{quartz}} \right)^{-1}$$

$$= \left(\frac{(0.5)(0.4)}{2.2} + \frac{(0.5)(0.4)}{0.000131} + \frac{0.6}{36} \right)^{-1} = 0.00065 \text{ GPa}$$

The density of the suspension is

$$\rho = 0.5\phi\rho_{water} + 0.5\phi\rho_{air} + (1-\phi)\rho_{quartz}$$

$$= (0.5)(0.4)(1.0) + (0.5)(0.4)(0.00119) + (0.6)(2.65) = 1.79 \text{ g/cm}^3$$

This gives the sound speed of

$$V = \sqrt{K/\rho} = \sqrt{0.00065/1.79} = 0.019 \text{ km/s}$$

USES

Wood's formula may be used to estimate the velocity in suspensions.

ASSUMPTIONS AND LIMITATIONS

Wood's formula presupposes that composite rock and each of its components are isotropic, linear, and elastic.

4.4 HILL AVERAGE MODULI ESTIMATE

SYNOPSIS

The Voigt–Reuss–Hill average is simply the arithmetic average of the Voigt upper bound and the Reuss lower bound. (See the discussion of the Voigt–Reuss bounds in Section 4.2.) This average is expressed as

$$M_{VRH} = \frac{M_V + M_R}{2}$$

where

$$M_V = \sum_{i=1}^{N} f_i M_i$$

$$\frac{1}{M_R} = \sum_{i=1}^{N} \frac{f_i}{M_i}$$

The terms f_i and M_i are the volume fraction and modulus of the ith component, respectively. Although M can be any modulus, it makes most sense for it to be the shear modulus or the bulk modulus.

The Voigt–Reuss–Hill average is useful when an *estimate* of the moduli is needed, not just the allowable range of values. An obvious extension would be to average, instead, the Hashin–Shtrikman upper and lower bounds.

This resembles, but is not exactly the same as the average of the algebraic and harmonic means of velocity used by Greenberg and Castagna (1992) in their empirical V_P–V_S relation (see Section 7.8).

USES

The Voigt–Reuss–Hill average is used to estimate the effective elastic moduli of a rock in terms of its constituents and pore space.

ASSUMPTIONS AND LIMITATIONS

The following limitation and assumption apply to the Voigt–Reuss–Hill average:

- The result is strictly heuristic. Hill (1952) showed that the Voigt and Reuss averages are upper and low bounds, respectively. Several authors have shown that the average of these bounds can be a useful and sometimes accurate estimate of rock properties.
- The rock is isotropic.

4.5 COMPOSITE WITH UNIFORM SHEAR MODULUS

SYNOPSIS

Hill (1963) showed that when all of the phases or constituents in a composite have the same shear modulus, μ, the effective P-wave modulus, $M_{\text{eff}} = (K_{\text{eff}} + \frac{4}{3}\mu_{\text{eff}})$, is given exactly by

$$\frac{1}{\left(K_{\text{eff}} + \frac{4}{3}\mu_{\text{eff}}\right)} = \sum_{i=1}^{N} \frac{x_i}{\left(K_i + \frac{4}{3}\mu\right)} = \left\langle \frac{1}{K + \frac{4}{3}\mu} \right\rangle$$

where x_i is the volume fraction of the ith component, K_i is its bulk modulus, and $\langle \cdot \rangle$ refers to the volume average. Because $\mu_{\text{eff}} = \mu_i = \mu$, any of the effective moduli can then be easily obtained.

This is obviously the same as

$$\frac{1}{\left(\rho V_{\text{P}}^2\right)_{\text{eff}}} = \left\langle \frac{1}{\rho V_{\text{P}}^2} \right\rangle$$

This striking result states that the effective moduli of a composite with uniform shear modulus can be found *exactly* if one knows only the volume fractions of the constituents independent of the constituent geometries. There is no dependence, for example, on elllipsoids, spheres, or other idealized shapes.

Hill's equation follows simply from the expressions for Hashin–Shtrikman bounds (see Section 4.1 on Hashin–Shtrikman bounds) on the effective bulk modulus:

$$\frac{1}{K^{HS\pm} + \frac{4}{3}\mu_{\{^{max}_{min}\}}} = \left\langle \frac{1}{K + \frac{4}{3}\mu_{\{^{max}_{min}\}}} \right\rangle$$

where μ_{min} and μ_{max} are the minimum and maximum shear moduli of the various constituents, yielding, respectively, the lower and upper bounds on the bulk modulus, $K^{HS\pm}$. Any composite must have an effective bulk modulus that falls between the bounds. Because here $\mu = \mu_{min} = \mu_{max}$, the two bounds on the bulk modulus are equal and reduce to the Hill expression above.

In the case of a mixture of liquids or gases, or both, where $\mu = 0$ for all the constituents, the Hill's equation becomes the well-known isostress equation or Reuss average:

$$\frac{1}{K_{eff}} = \sum_{i=1}^{N} \frac{x_i}{K_i} = \left\langle \frac{1}{K} \right\rangle$$

A somewhat surprising result is that a finely layered medium, where each layer is isotropic and has the same shear modulus but a different bulk modulus, is *isotropic* with a bulk modulus given by Hill's equation. (See Section 4.12 on the Backus average.)

USES

Hill's equation can be used to calculate the effective low-frequency moduli for rocks with spatially nonuniform or *patchy* saturation. At low frequencies, Gassmann's relations predict no change in the shear modulus between dry and saturated patches, allowing this relation to be used to estimate K.

ASSUMPTIONS AND LIMITATIONS

Hill's equation applies when the composite rock and each of its components are isotropic and have the same shear modulus.

4.6 ROCK AND PORE COMPRESSIBILITIES AND SOME PITFALLS

SYNOPSIS

This section summarizes useful relations among the compressibilities of porous materials and addresses some commonly made mistakes.

A nonporous elastic solid has a single compressibility

$$\beta = \frac{1}{V}\frac{\partial V}{\partial \sigma}$$

where σ is the hydrostatic stress applied on the outer surface and V is the sample bulk volume. In contrast, compressibilities for porous media are more complicated. We have to account for at least two pressures (the external confining pressure and the internal pore pressure) and two volumes (bulk volume and pore volume). Therefore, we can define at least four compressibilities. Following Zimmerman's (1991) notation, in which the first subscript indicates the volume change (b for bulk, p for pore) and the second subscript denotes the pressure that is varied (c for confining, p for pore), these compressibilities are

$$\beta_{bc} = \frac{1}{V_b}\left(\frac{\partial V_b}{\partial \sigma_c}\right)_{\sigma_p}$$

$$\beta_{bp} = -\frac{1}{V_b}\left(\frac{\partial V_b}{\partial \sigma_p}\right)_{\sigma_c}$$

$$\beta_{pc} = \frac{1}{v_p}\left(\frac{\partial v_p}{\partial \sigma_c}\right)_{\sigma_p}$$

$$\beta_{pp} = -\frac{1}{v_p}\left(\frac{\partial v_p}{\partial \sigma_p}\right)_{\sigma_c}$$

Note that the signs are chosen to ensure that the compressibilities are positive when tensional stress is taken to be positive. Thus, for instance, β_{bp} is to be interpreted as the fractional change in the bulk volume with respect to change in the pore pressure while the confining pressure is held constant. These are the *dry* or *drained* bulk and pore compressibilities. The effective dry bulk modulus is $K_{dry} = 1/\beta_{bc}$, and the dry-pore-space stiffness is $K_\phi = 1/\beta_{pc}$. In addition, there is the *saturated* or *undrained* bulk compressibility when the mass of the pore fluid is kept constant as the confining pressure changes:

$$\beta_{\text{u}} = \frac{1}{K_{\text{sat low f}}} = \frac{1}{V_{\text{b}}} \left(\frac{\partial V_{\text{b}}}{\partial \sigma_{\text{c}}} \right)_{m_{\text{fluid}}}$$

This equation assumes that the pore pressure is equilibrated throughout the pore space, and the expression is therefore appropriate for very low frequencies. At high frequencies, with unequilibrated pore pressures, the appropriate bulk modulus is $K_{\text{sat hi f}}$ calculated from some high-frequency theory such as the squirt, Biot, or inclusion models, or some other viscoelastic model.

The moduli K_{dry}, $K_{\text{sat low f}}$, $K_{\text{sat hi f}}$, and K_{ϕ} are the ones most useful in wave propagation rock physics. The other compressibilities are used in calculations of subsidence caused by fluid withdrawal and reservoir compressibility analyses. Some of the compressibilities can be related to each other by linear superposition and reciprocity. The well-known Gassmann's equation relates K_{dry} to K_{sat} through the mineral and fluid bulk moduli K_0 and K_{fl}. A few other relations are (for simple derivations, see Zimmerman, 1991)

$$\beta_{\text{bp}} = \beta_{\text{bc}} - \frac{1}{K_0}$$

$$\beta_{\text{pc}} = \beta_{\text{bp}}/\phi$$

$$\beta_{\text{pp}} = \left[\beta_{\text{bc}} - (1 + \phi)\frac{1}{K_0} \right] \Big/ \phi$$

MORE ON DRY ROCK COMPRESSIBILITY

The effective dry rock compressibility of a homogeneous, linear, porous, elastic solid with any arbitrarily shaped pore space (sometimes called the "drained" or "frame" compressibility) can be written as

$$\frac{1}{K_{\text{dry}}} = \frac{1}{K_0} + \frac{1}{V_{\text{b}}} \frac{\partial v_{\text{p}}}{\partial \sigma_{\text{c}}} \Big|_{\sigma_{\text{p}}}$$

or

$$\frac{1}{K_{\text{dry}}} = \frac{1}{K_0} + \frac{\phi}{K_{\phi}} \tag{1}$$

where

$$\frac{1}{K_{\phi}} = \frac{1}{v_{\text{p}}} \frac{\partial v_{\text{p}}}{\partial \sigma_{\text{c}}} \Big|_{\sigma_{\text{p}}}$$

is defined as the dry pore space compressibility (K_ϕ is the dry pore space stiffness),

$$K_{\text{dry}} = 1/\beta_{bc} = \text{effective bulk modulus of dry porous solid}$$
$$K_0 = \text{bulk modulus of intrinsic mineral material}$$
$$V_b = \text{total bulk volume}$$
$$v_p = \text{pore volume}$$
$$\phi = v_p/V_b = \text{porosity}$$
$$\sigma_c, \sigma_p = \text{hydrostatic confining stress and pore stress}$$
$$\text{(pore pressure)}$$

We assume that no inelastic effects such as friction or viscosity are present. These equations are strictly true, regardless of pore geometry and pore concentration.

CAUTION: "Dry rock" is not the same as gas-saturated rock. The dry-frame modulus refers to the incremental bulk deformation resulting from an increment of applied confining pressure while pore pressure is held constant. This corresponds to a "drained" experiment in which pore fluids can flow freely in or out of the sample to ensure constant pore pressure. Alternatively, it can correspond to an undrained experiment in which the pore fluid has zero bulk modulus and thus the pore compressions do not induce changes in pore pressure, which is approximately the case for an air-filled sample at standard temperature and pressure. However, at reservoir conditions (high pore pressure), gas takes on a nonnegligible bulk modulus and should be treated as a saturating fluid.

CAUTION: The harmonic average of the mineral and dry pore moduli, which resembles equation (1) above, is incorrect:

$$\frac{1}{K_{\text{dry}}} \stackrel{?}{=} \frac{1-\phi}{K_0} + \frac{\phi}{K_\phi} \quad \text{(incorrect)}$$

This equation is sometimes "guessed" because it resembles the Reuss average, but it has no justification from elasticity analysis. It is also *incorrect* to write

$$\frac{1}{K_{\text{dry}}} \stackrel{?}{=} \frac{1}{K_0} + \frac{\partial \phi}{\partial \sigma_c} \quad \text{(incorrect)} \tag{2}$$

The correct expression is

$$\frac{1}{K_{\text{dry}}} = \frac{1}{(1-\phi)}\left(\frac{1}{K_0} + \frac{\partial \phi}{\partial \sigma_c}\right)$$

The incorrect equation (2) appears as an intermediate result in some of the classic literature of rock physics. The notable final results are still correct, for the actual derivations are done in terms of the pore volume change, $\partial v_p/\partial \sigma_c$, and not $\partial \phi/\partial \sigma_c$.

Not distinguishing between changes in differential (effective) pressure, $\sigma_d = \sigma_c - \sigma_p$, and confining pressure, σ_c, can lead to confusion. Changing σ_c while σ_p

is kept constant ($\delta\sigma_p = 0$) is not the same as changing σ_c with $\delta\sigma_p = \delta\sigma_c$ (i.e., the differential stress is kept constant). In the first situation the porous medium deforms with the effective dry modulus K_{dry}. The second situation is one of uniform hydrostatic pressure outside *and* inside the porous rock. For this stress state the rock deforms with the intrinsic mineral modulus K_0. Not understanding this can lead to the following erroneous results:

$$\frac{1}{K_0} \stackrel{?}{=} \frac{1}{K_{dry}} - \frac{1}{(1-\phi)}\frac{\partial\phi}{\partial\sigma_c} \quad \text{(incorrect)}$$

or

$$\frac{\partial\phi}{\partial\sigma_c} \stackrel{?}{=} (1-\phi)\left(\frac{1}{K_{dry}} - \frac{1}{K_0}\right) \quad \text{(incorrect)}$$

TABLE 4.6.1. Correct and incorrect versions of the fundamental equations.

Incorrect	Correct
$\dfrac{1}{K_{dry}} \stackrel{?}{=} \dfrac{1}{K_0} + \dfrac{\partial\phi}{\partial\sigma_c}$	$\dfrac{1}{K_{dry}} = \dfrac{1}{K_0} + \dfrac{1}{V_b}\dfrac{\partial v_p}{\partial\sigma_c}$
$\dfrac{1}{K_{dry}} \stackrel{?}{=} \dfrac{1-\phi}{K_0} + \dfrac{\phi}{K_\phi}$	$\dfrac{1}{K_{dry}} = \dfrac{1}{K_0} + \dfrac{\phi}{K_\phi}$
$\dfrac{1}{K_{dry}} \stackrel{?}{=} \dfrac{1}{K_0} + \dfrac{1}{(1-\phi)}\dfrac{\partial\phi}{\partial\sigma_c}$	$\dfrac{1}{K_{dry}} = \dfrac{1}{(1-\phi)}\left(\dfrac{1}{K_0} + \dfrac{\partial\phi}{\partial\sigma_c}\right)$
$\dfrac{\partial\phi}{\partial\sigma_c} \stackrel{?}{=} \left(\dfrac{1}{K_{dry}} - \dfrac{1}{K_0}\right)(1-\phi)$	$\dfrac{\partial\phi}{\partial\sigma_c} = \dfrac{1-\phi}{K_{dry}} - \dfrac{1}{K_0}$

ASSUMPTIONS AND LIMITATIONS

The following presuppositions apply to the equations presented in this section:

- They assume isotropic, linear, porous, elastic media.
- All derivations here are in the context of linear elasticity with infinitesimal, incremental strains and stresses. Hence Eulerian and Lagrangian formulations are equivalent.
- It is assumed that the temperature is always held constant as the pressure varies.
- Inelastic effects such as friction and viscosity are neglected.

4.7 KUSTER AND TOKSÖZ FORMULATION FOR EFFECTIVE MODULI

SYNOPSIS

Kuster and Toksöz (1974) derived expressions for P- and S- wave velocities by using a long-wavelength first-order scattering theory. A generalization of their expressions for the effective moduli K_{KT}^* and μ_{KT}^* for a variety of inclusion shapes can be written as (Kuster and Toksöz, 1974; Berryman, 1980b)

$$\left(K_{KT}^* - K_m\right)\frac{\left(K_m + \frac{4}{3}\mu_m\right)}{\left(K_{KT}^* + \frac{4}{3}\mu_m\right)} = \sum_{i=1}^{N} x_i(K_i - K_m)P^{mi}$$

$$\left(\mu_{KT}^* - \mu_m\right)\frac{\left(\mu_m + \zeta_m\right)}{\left(\mu_{KT}^* + \zeta_m\right)} = \sum_{i=1}^{N} x_i(\mu_i - \mu_m)Q^{mi}$$

where the summation is over the different inclusion types with volume concentration x_i, and

$$\zeta = \frac{\mu}{6}\frac{(9K + 8\mu)}{(K + 2\mu)}$$

The coefficients P^{mi} and Q^{mi} describe the effect of an inclusion of material i in a background medium m. For example, a two-phase material with a single type of inclusion embedded within a background medium has a single term on the right-hand side. Inclusions with different material properties or different shapes require separate terms in the summation. Each set of inclusions must be distributed randomly, and thus its effect is isotropic. These formulas are uncoupled and can be made explicit for easy evaluation. Table 4.7.1 gives expressions for P and Q for some simple inclusion shapes.

Dry cavities can be modeled by setting the inclusion moduli to zero. Fluid-saturated cavities are simulated by setting the inclusion shear modulus to zero.

CAUTION: Because the cavities are isolated with respect to flow, this approach simulates very high frequency saturated rock behavior appropriate to ultrasonic laboratory conditions. At low frequencies, when there is time for wave-induced pore pressure increments to flow and equilibrate, it is better to find the effective moduli for dry cavities and then saturate them with the Gassmann low-frequency relations (see Section 6.3). This should not be confused with the tendency to term this approach a low-frequency theory, for crack dimensions are assumed to be much smaller than a wavelength.

EXAMPLE

Calculate the effective bulk and shear moduli, K_{KT}^* and μ_{KT}^*, for a quartz matrix with spherical, water-filled inclusions of porosity 0.1.

$$K_m = 37\,\text{GPa}, \quad \mu_m = 44\,\text{GPa}, \quad K_i = 2.25\,\text{GPa}, \quad \mu_i = 0\,\text{GPa}.$$

Volume fraction of spherical inclusions $x_1 = 0.1$ and $N = 1$. The P and Q values for spheres are obtained from the table as follows:

$$P^{m1} = \frac{\left(37 + \frac{4}{3}44\right)}{\left(2.25 + \frac{4}{3}44\right)} = 1.57$$

$$\zeta_m = \frac{44}{6}\frac{(9 \times 37 + 8 \times 44)}{(37 + 2 \times 44)} = 40.2$$

$$Q^{m1} = \frac{(44 + 40.2)}{(0 + 40.2)} = 2.095$$

Substituting these in the Kuster–Toksöz equations gives:

$$K_{KT}^* = 31.84\,\text{GPa}, \quad \mu_{KT}^* = 35.7\,\text{GPa}$$

TABLE 4.7.1. Coefficients P and Q for some specific shapes. The subscripts m and i refer to the background and inclusion materials [from Berryman (1995)].

Inclusion Shape	P^{mi}	Q^{mi}
Spheres	$\dfrac{K_m + \frac{4}{3}\mu_m}{K_i + \frac{4}{3}\mu_m}$	$\dfrac{\mu_m + \zeta_m}{\mu_i + \zeta_m}$
Needles	$\dfrac{K_m + \mu_m + \frac{1}{3}\mu_i}{K_i + \mu_m + \frac{1}{3}\mu_i}$	$\dfrac{1}{5}\left(\dfrac{4\mu_m}{\mu_m + \mu_i} + 2\dfrac{\mu_m + \gamma_m}{\mu_i + \gamma_m} + \dfrac{K_i + \frac{4}{3}\mu_m}{K_i + \mu_m + \frac{1}{3}\mu_i}\right)$
Disks	$\dfrac{K_m + \frac{4}{3}\mu_i}{K_i + \frac{4}{3}\mu_i}$	$\dfrac{\mu_m + \zeta_i}{\mu_i + \zeta_i}$
Penny cracks	$\dfrac{K_m + \frac{4}{3}\mu_i}{K_i + \frac{4}{3}\mu_i + \pi\alpha\beta_m}$	$\dfrac{1}{5}\left(1 + \dfrac{8\mu_m}{4\mu_i + \pi\alpha(\mu_m + 2\beta_m)} + 2\dfrac{K_i + \frac{2}{3}(\mu_i + \mu_m)}{K_i + \frac{4}{3}\mu_i + \pi\alpha\beta_m}\right)$

$$\beta = \mu\frac{(3K + \mu)}{(3K + 4\mu)} \quad \gamma = \mu\frac{(3K + \mu)}{(3K + 7\mu)} \quad \zeta = \frac{\mu}{6}\frac{(9K + 8\mu)}{(K + 2\mu)}$$

The P and Q values for ellipsoidal inclusions with arbitrary aspect ratio are the same as given in Section 4.8 on self-consistent methods.

Note that for spherical inclusions, the Kuster–Toksöz expressions for bulk modulus are identical to the Hashin–Shtrikman upper bound even though the Kuster–Toksöz expressions are formally limited to low porosity.

ASSUMPTIONS AND LIMITATIONS

The following presuppositions and limitations apply to the Kuster–Toksöz formulations:

- They assume isotropic, linear, elastic media.
- They are limited to dilute concentrations of the inclusions.
- They assume idealized ellipsoidal inclusion shapes.

4.8 SELF-CONSISTENT APPROXIMATIONS OF EFFECTIVE MODULI

SYNOPSIS

Theoretical estimates of the effective moduli of composite or porous elastic materials generally depend on (1) the properties of the individual components of the composite, (2) the volume fractions of the components, and (3) the geometric details of the shapes and spatial distributions of the components. The bounding methods (see discussions of the Hashin–Shtrikman and Voigt–Reuss bounds, Sections 4.1 and 4.2) establish upper and lower bounds when only (1) and (2) are known with no geometric details. A second approach improves these estimates by adding statistical information about the phases (e.g., Beran and Molyneux, 1966; McCoy, 1970; Corson, 1974; Watt, Davies, and O'Connell, 1976). A third approach is to assume very specific inclusion shapes. Most methods use the solution for the elastic deformation of a single inclusion of one material in an infinite background medium of the second material and then use one scheme or another to estimate the effective moduli when there is a distribution of these inclusions. These estimates are generally limited to dilute distributions of inclusions owing to the difficulty of modeling or estimating the elastic interaction of inclusions in close proximity.

A relatively successful, and certainly popular, method to extend these specific geometry methods to slightly higher concentrations of inclusions is the

self-consistent approximation (Budiansky, 1965; Hill, 1965; Wu, 1966). In this approach one still uses the mathematical solution for the deformation of isolated inclusions, but the interaction of inclusions is approximated by replacing the background medium with the as-yet-unknown effective medium. These methods were made popular following a series of papers by O'Connell and Budiansky (see, for example, O'Connell and Budiansky, 1974). Their equations for effective bulk and shear moduli, K_{SC}^* and μ_{SC}^*, respectively, of a cracked medium with randomly oriented dry penny-shaped cracks (in the limiting case when the aspect ratio α goes to 0) are

$$\frac{K_{SC}^*}{K} = 1 - \frac{16}{9}\left(\frac{1 - v_{SC}^{*2}}{1 - 2v_{SC}^*}\right)\varepsilon$$

$$\frac{\mu_{SC}^*}{\mu} = 1 - \frac{32}{45}\frac{\left(1 - v_{SC}^*\right)\left(5 - v_{SC}^*\right)}{\left(2 - v_{SC}^*\right)}\varepsilon$$

where K and μ are the bulk and shear moduli, respectively, of the uncracked medium, and ε is the crack density parameter, which is defined as the number of cracks per unit volume times the crack radius cubed. The effective Poisson ratio v_{SC}^* is related to ε and the Poisson's ratio v of the uncracked solid by

$$\varepsilon = \frac{45}{16}\frac{\left(v - v_{SC}^*\right)\left(2 - v_{SC}^*\right)}{\left(1 - v_{SC}^{*2}\right)\left(10v - 3vv_{SC}^* - v_{SC}^*\right)}$$

This equation must first be solved for v_{SC}^* for a given ε, after which K_{SC}^* and μ_{SC}^* can be evaluated. The nearly linear dependence of v_{SC}^* on ε is well approximated by

$$v_{SC}^* = v\left(1 - \frac{16}{9}\varepsilon\right)$$

and this simplifies the calculation of the effective moduli. For fluid-saturated, infinitely thin penny-shaped cracks

$$\frac{K_{SC}^*}{K} = 1$$

$$\frac{\mu_{SC}^*}{\mu} = 1 - \frac{32}{15}\left(\frac{1 - v_{SC}^*}{2 - v_{SC}^*}\right)\varepsilon$$

$$\varepsilon = \frac{45}{32}\frac{\left(v_{SC}^* - v\right)\left(2 - v_{SC}^*\right)}{\left(1 - v_{SC}^{*2}\right)(1 - 2v)}$$

However, this result is inadequate for small aspect ratio cracks with soft-fluid saturation, such as when the parameter $\omega = K_{\text{fluid}}/(\alpha K)$ is of the order 1. Then the appropriate equations given by O'Connell and Budiansky are

$$\frac{K_{SC}^*}{K} = 1 - \frac{16}{9}\frac{\left(1 - v_{SC}^{*2}\right)}{\left(1 - 2v_{SC}^*\right)}D\varepsilon$$

$$\frac{\mu_{SC}^*}{\mu} = 1 - \frac{32}{45}\left(1 - v_{SC}^*\right)\left[D + \frac{3}{\left(2 - v_{SC}^*\right)}\right]\varepsilon$$

$$\varepsilon = \frac{45}{16}\frac{\left(v - v_{SC}^*\right)}{\left(1 - v_{SC}^{*\,2}\right)}\frac{\left(2 - v_{SC}^*\right)}{\left[D(1 + 3v)\left(2 - v_{SC}^*\right) - 2(1 - 2v)\right]}$$

$$D = \left[1 + \frac{4}{3\pi}\frac{\left(1 - v_{SC}^{*\,2}\right)}{\left(1 - 2v_{SC}^*\right)}\frac{K}{K_{SC}^*}\omega\right]^{-1}$$

Wu's self-consistent modulus estimates for two-phase composites may be expressed as (m = matrix, i = inclusion)

$$K_{SC}^* = K_m + x_i(K_i - K_m)P^{*i}$$

$$\mu_{SC}^* = \mu_m + x_i(\mu_i - \mu_m)Q^{*i}$$

Berryman (1980b, 1995) gives a more general form of the self-consistent approximations for N-phase composites:

$$\sum_{i=1}^{N} x_i\left(K_i - K_{SC}^*\right)P^{*i} = 0$$

$$\sum_{i=1}^{N} x_i\left(\mu_i - \mu_{SC}^*\right)Q^{*i} = 0$$

where i refers to the ith material, x_i is its volume fraction, P and Q are geometric factors given in Table 4.8.1, and the superscript $*i$ on P and Q indicates that the factors are for an inclusion of material i in a background medium with self-consistent effective moduli K_{SC}^* and μ_{SC}^*. These equations are coupled and must be solved by simultaneous iteration. Although Berryman's self-consistent method does not converge for fluid disks ($\mu_2 = 0$), the formulas for penny-shaped fluid-filled cracks are generally not singular and converge rapidly. However, his estimates for needles, disks, and penny cracks should be used cautiously for fluid-saturated composite materials.

Dry cavities can be modeled by setting the inclusion moduli to zero. Fluid-saturated cavities are simulated by setting the inclusion shear modulus to zero.

CAUTION: Because the cavities are isolated with respect to flow, this approach simulates very-high-frequency saturated rock behavior appropriate to ultrasonic laboratory conditions. At low frequencies, when there is time for wave-induced pore pressure increments to flow and equilibrate, it is better to find the effective moduli for dry cavities and then saturate them with the Gassmann low-frequency relations. This should not be confused with the tendency to term this approach a low-frequency theory, for crack dimensions are assumed to be much smaller than a wavelength.

TABLE 4.8.1. Coefficients P and Q for some specific shapes. The subscripts m and i refer to the background and inclusion materials [from Berryman (1995)].

Inclusion Shape	P^{mi}	Q^{mi}
Spheres	$\dfrac{K_m + \frac{4}{3}\mu_m}{K_i + \frac{4}{3}\mu_m}$	$\dfrac{\mu_m + \zeta_m}{\mu_i + \zeta_m}$
Needles	$\dfrac{K_m + \mu_m + \frac{1}{3}\mu_i}{K_i + \mu_m + \frac{1}{3}\mu_i}$	$\dfrac{1}{5}\left(\dfrac{4\mu_m}{\mu_m + \mu_i} + 2\dfrac{\mu_m + \gamma_m}{\mu_i + \gamma_m} + \dfrac{K_i + \frac{4}{3}\mu_m}{K_i + \mu_m + \frac{1}{3}\mu_i}\right)$
Disks	$\dfrac{K_m + \frac{4}{3}\mu_i}{K_i + \frac{4}{3}\mu_i}$	$\dfrac{\mu_m + \zeta_i}{\mu_i + \zeta_i}$
Penny cracks	$\dfrac{K_m + \frac{4}{3}\mu_i}{K_i + \frac{4}{3}\mu_i + \pi\alpha\beta_m}$	$\dfrac{1}{5}\left[1 + \dfrac{8\mu_m}{4\mu_i + \pi\alpha(\mu_m + 2\beta_m)} + 2\dfrac{K_i + \frac{2}{3}(\mu_i + \mu_m)}{K_i + \frac{4}{3}\mu_i + \pi\alpha\beta_m}\right]$

$$\beta = \mu\frac{(3K + \mu)}{(3K + 4\mu)} \quad \gamma = \mu\frac{(3K + \mu)}{(3K + 7\mu)} \quad \zeta = \frac{\mu}{6}\frac{(9K + 8\mu)}{(K + 2\mu)}$$

EXAMPLE

Calculate the self-consistent effective bulk and shear moduli, K^*_{SC}, and μ^*_{SC}, for a water-saturated rock consisting of spherical quartz grains (aspect ratio $\alpha = 1$) and total porosity 0.3. The pore space consists of spherical pores ($\alpha = 1$) and thin penny-shaped cracks ($\alpha = 10^{-2}$). The thin cracks have a porosity of 0.01, whereas the remaining porosity (0.29) is made up of the spherical pores.

The total number of phases, N, is 3.

$$K_1(\text{quartz}) = 37 \text{ GPa}, \; \mu_1(\text{quartz}) = 44 \text{ GPa},$$
$$\alpha_1 = 1, x_1 \text{ (volume fraction)} = 0.7$$
$$K_2(\text{water, spherical pores}) = 2.25 \text{ GPa},$$
$$\mu_2(\text{water, spherical pores}) = 0 \text{ GPa},$$
$$\alpha_2(\text{spherical pores}) = 1, x_2 \text{ (volume fraction)} = 0.29$$
$$K_3(\text{water, thin cracks}) = 2.25 \text{ GPa},$$
$$\mu_3(\text{water, thin cracks}) = 0 \text{ GPa},$$
$$\alpha_3(\text{thin cracks}) = 10^{-2}, x_3(\text{volume fraction}) = 0.01$$

The coupled equations for K^*_{SC} and μ^*_{SC} are

$$x_1(K_1 - K^*_{SC})P^{*1} + x_2(K_2 - K^*_{SC})P^{*2} + x_3(K_3 - K^*_{SC})P^{*3} = 0$$

$$x_1(\mu_1 - \mu^*_{SC})Q^{*1} + x_2(\mu_2 - \mu^*_{SC})Q^{*2} + x_3(\mu_3 - \mu^*_{SC})Q^{*3} = 0$$

The P's and Q's are obtained from Table 4.8.1 or from the more general equation for ellipsoids of arbitrary aspect ratio. In the equations for P's and Q's, K_m and μ_m are replaced everywhere by K_{SC}^* and μ_{SC}^*, respectively. The coupled equations are solved iteratively, starting from some initial guess for K_{SC}^* and μ_{SC}^*. The Voigt average may be taken as the starting point. The converged solutions (known as the fixed points of the coupled equations) are $K_{SC}^* = 16.8$ GPa and $\mu_{SC}^* = 11.6$ GPa.

The coefficients P and Q for ellipsoidal inclusions of arbitrary aspect ratio are given by

$$P = \frac{1}{3} T_{iijj}$$

$$Q = \frac{1}{5}\left(T_{ijij} - \frac{1}{3}T_{iijj}\right)$$

where the tensor T_{ijkl} relates the uniform far-field strain field to the strain within the ellipsoidal inclusion (Wu, 1966). Berryman (1980b) gives the pertinent scalars required for computing P and Q as

$$T_{iijj} = 3F_1/F_2$$

$$T_{ijij} - \frac{1}{3}T_{iijj} = \frac{2}{F_3} + \frac{1}{F_4} + \frac{F_4 F_5 + F_6 F_7 - F_8 F_9}{F_2 F_4}$$

where

$$F_1 = 1 + A\left[\frac{3}{2}(f+\theta) - R\left(\frac{3}{2}f + \frac{5}{2}\theta - \frac{4}{3}\right)\right]$$

$$F_2 = 1 + A\left[1 + \frac{3}{2}(f+\theta) - (R/2)(3f+5\theta)\right] + B(3-4R)$$

$$+ (A/2)(A+3B)(3-4R)[f + \theta - R(f - \theta + 2\theta^2)]$$

$$F_3 = 1 + A\left[1 - \left(f + \frac{3}{2}\theta\right) + R(f+\theta)\right]$$

$$F_4 = 1 + (A/4)[f + 3\theta - R(f - \theta)]$$

$$F_5 = A\left[-f + R\left(f + \theta - \frac{4}{3}\right)\right] + B\theta(3-4R)$$

$$F_6 = 1 + A[1 + f - R(f+\theta)] + B(1-\theta)(3-4R)$$

$$F_7 = 2 + (A/4)[3f + 9\theta - R(3f+5\theta)] + B\theta(3-4R)$$

$$F_8 = A[1 - 2R + (f/2)(R - 1) + (\theta/2)(5R - 3)] + B(1 - \theta)(3 - 4R)$$

$$F_9 = A[(R - 1)f - R\theta] + B\theta(3 - 4R)$$

with A, B, and R given by

$$A = \mu_i/\mu_m - 1$$

$$B = \frac{1}{3}(K_i/K_m - \mu_i/\mu_m)$$

and

$$R = [(1 - 2\nu_m)/2(1 - \nu_m)]$$

The functions θ and f are given by

$$\theta = \begin{cases} \frac{\alpha}{(\alpha^2 - 1)^{3/2}}\left[\alpha(\alpha^2 - 1)^{1/2} - \cosh^{-1}\alpha\right] \\ \frac{\alpha}{(1 - \alpha^2)^{3/2}}\left[\cos^{-1}\alpha - \alpha(1 - \alpha^2)^{1/2}\right] \end{cases}$$

for prolate and oblate spheroids, respectively, and

$$f = \frac{\alpha^2}{1 - \alpha^2}(3\theta - 2)$$

Note that $\alpha < 1$ for oblate spheroids, and $\alpha > 1$ for prolate spheroids.

ASSUMPTIONS AND LIMITATIONS

The approach described in this section has the following presuppositions:

- Idealized ellipsoidal inclusion shapes.
- Isotropic, linear, elastic media.
- Cracks are isolated with respect to fluid flow. Pore pressures are unequilibrated and adiabatic. Appropriate for high-frequency laboratory conditions. For low-frequency field situations use dry inclusions and then saturate by using Gassmann relations. This should not be confused with the tendency to term this approach a low-frequency theory, for crack dimensions are assumed to be much smaller than a wavelength.

4.9 DIFFERENTIAL EFFECTIVE MEDIUM MODEL

SYNOPSIS

The differential effective medium (DEM) theory models two-phase composites by incrementally adding inclusions of one phase (phase 2) to the matrix phase (Cleary et al., 1980; Norris, 1985; Zimmerman, 1991). The matrix begins as phase 1 (when concentration of phase 2 is zero) and is changed at each step as a new increment of phase 2 material is added. The process is continued until the desired proportion of the constituents is reached. The DEM formulation does not treat each constituent symmetrically. There is a preferred matrix or host material, and the effective moduli depend on the construction path taken to reach the final composite. Starting with material 1 as the host and incrementally adding inclusions of material 2 will not, in general, lead to the same effective properties as starting with phase 2 as the host. For multiple inclusion shapes or multiple constituents, the effective moduli depend not only on the final volume fractions of the constituents but also on the order in which the incremental additions are done. The process of incrementally adding inclusions to the matrix is really a thought experiment and should not be taken to provide an accurate description of the true evolution of rock porosity in nature.

The coupled system of ordinary differential equations for the effective bulk and shear moduli, K^* and μ^*, respectively, are (Berryman, 1992)

$$(1 - y)\frac{d}{dy}[K^*(y)] = (K_2 - K^*)P^{(*2)}(y)$$

$$(1 - y)\frac{d}{dy}[\mu^*(y)] = (\mu_2 - \mu^*)Q^{(*2)}(y)$$

with initial conditions $K^*(0) = K_1$ and $\mu^*(0) = \mu_1$, where

$K_1, \mu_1 =$ bulk and shear moduli of the initial host material
(phase 1)
$K_2, \mu_2 =$ bulk and shear moduli of the incrementally added
inclusions (phase 2)
$y =$ concentration of phase 2

For fluid inclusions and voids, y equals the porosity, ϕ. The terms P and Q are geometric factors given in Table 4.9.1, and the superscript ^{*2} on P and

TABLE 4.9.1. Coefficients P and Q for some specific shapes. The subscripts m and i refer to the background and inclusion materials [from Berryman (1995)].

Inclusion Shape	P^{mi}	Q^{mi}
Spheres	$\dfrac{K_m + \frac{4}{3}\mu_m}{K_i + \frac{4}{3}\mu_m}$	$\dfrac{\mu_m + \zeta_m}{\mu_i + \zeta_m}$
Needles	$\dfrac{K_m + \mu_m + \frac{1}{3}\mu_i}{K_i + \mu_m + \frac{1}{3}\mu_i}$	$\dfrac{1}{5}\left(\dfrac{4\mu_m}{\mu_m + \mu_i} + 2\dfrac{\mu_m + \gamma_m}{\mu_i + \gamma_m} + \dfrac{K_i + \frac{4}{3}\mu_m}{K_i + \mu_m + \frac{1}{3}\mu_i} \right)$
Disks	$\dfrac{K_m + \frac{4}{3}\mu_i}{K_i + \frac{4}{3}\mu_i}$	$\dfrac{\mu_m + \zeta_i}{\mu_i + \zeta_i}$
Penny cracks	$\dfrac{K_m + \frac{4}{3}\mu_i}{K_i + \frac{4}{3}\mu_i + \pi\alpha\beta_m}$	$\dfrac{1}{5}\left[1 + \dfrac{8\mu_m}{4\mu_i + \pi\alpha(\mu_m + 2\beta_m)} + 2\dfrac{K_i + \frac{2}{3}(\mu_i + \mu_m)}{K_i + \frac{4}{3}\mu_i + \pi\alpha\beta_m} \right]$

$$\beta = \mu\frac{(3K + \mu)}{(3K + 4\mu)} \qquad \gamma = \mu\frac{(3K + \mu)}{(3K + 7\mu)} \qquad \zeta = \frac{\mu}{6}\frac{(9K + 8\mu)}{(K + 2\mu)}$$

Q indicates that the factors are for an inclusion of material 2 in a background medium with effective moduli K^* and μ^*. Dry cavities can be modeled by setting the inclusion moduli to zero. Fluid-saturated cavities are simulated by setting the inclusion shear modulus to zero.

CAUTION: Because the cavities are isolated with respect to flow, this approach simulates very-high-frequency saturated rock behavior appropriate to ultrasonic laboratory conditions. At low frequencies, when there is time for wave-induced pore pressure increments to flow and equilibrate, it is better to find the effective moduli for dry cavities and then saturate them with the Gassmann low-frequency relations. This should not be confused with the tendency to term this approach a low-frequency theory, for inclusion dimensions are assumed to be much smaller than a wavelength.

The P and Q for ellipsoidal inclusions with arbitrary aspect ratio are the same as given in Section 4.8 for the self-consistent methods.

Norris et al. (1985) have shown that the DEM is realizable and therefore is always consistent with the Hashin–Shtrikman upper and lower bounds.

The derivation of the DEM equations as given above (Norris, 1985; Berryman, 1992) assumes that, as each new inclusion (or pore) is introduced, it displaces on average either the host matrix material or the inclusion material with probabilities $(1 - y)$ and y, respectively. A slightly different derivation by Zimmerman (1984) assumed that when a new inclusion is introduced, it always displaces the host material alone. This leads to similar differential equations with $dy/(1 - y)$ replaced by dy. The effective moduli predicted by the Zimmerman version of DEM are always slightly stiffer (for the same inclusion geometry and concentration) than the DEM equations given above. They both predict the same first-order terms in y but begin to diverge at concentrations above 10 percent. The dependence of effective moduli on concentration goes as $e^{-2y} = (1 - 2y + 2y^2 - ...)$ for Zimmerman's equations, whereas it behaves as $(1 - y)^2 = (1 - 2y + 2y^2 - ...)$ for the Norris version. In general, for a fixed inclusion geometry and porosity, the Zimmerman DEM effective moduli are close to the Kuster–Toksöz effective moduli and are stiffer than the Norris–Berryman DEM predictions, which in turn are stiffer than the Berryman self-consistent effective moduli. For spherical inclusions, the Zimmerman estimates fall above the Hashin–Shtrikman upper bound for high concentrations.

An important conceptual difference between the DEM and self-consistent schemes for calculating effective moduli of composites is that the DEM scheme identifies one of the constituents as a host or matrix material in which inclusions of the other constituent(s) are embedded, whereas the self-consistent scheme does not identify any specific host material but treats the composite as an aggregate of all the constituents.

MODIFIED DEM WITH CRITICAL POROSITY CONSTRAINTS

In the usual DEM model, starting from a solid initial host, a porous material stays intact at all porosities and falls apart only at the very end when $y = 1$ (100-percent porosity). This is because the solid host remains connected and therefore load bearing.

Although DEM is a good model for materials such as glass foam (Berge, Berryman, and Bonner, 1993) and oceanic basalts (Berge, Fryer, and Wilkens, 1992), most reservoir rocks fall apart at a critical porosity, ϕ_c, significantly less than 1.0 and are not represented very well by the conventional DEM theory. The modified DEM model (Mukerji et al., 1995) incorporates percolation behavior at any desired ϕ_c by redefining the phase 2 end member. The inclusions are now no longer made up of pure fluid (the original phase 2 material) but are composite inclusions of the critical phase at ϕ_c with elastic moduli (K_c, μ_c). With this definition, y denotes the concentration of the critical phase in the matrix. The total porosity is given by $\phi = y\phi_c$.

The computations are implemented by replacing (K_2, μ_2) with (K_c, μ_c) everywhere in the equations. Integrating along the reverse path, from $\phi = \phi_c$ to $\phi = 0$ gives lower moduli, for now the softer critical phase is the matrix. The moduli of the critical phase may be taken as the Reuss average value at ϕ_c of the pure end member moduli. Because the critical phase consists of grains just barely touching each other, better estimates of K_c and μ_c may be obtained from measurements

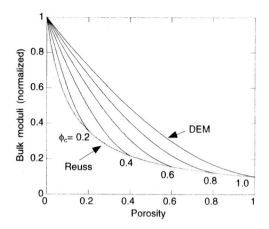

Figure 4.9.1

on loose sands or from models of granular material. For porosities greater than ϕ_c, the material is a suspension and is best characterized by the Reuss average (or Wood's equation).

Figure 4.9.1 shows normalized bulk moduli curves for the conventional DEM theory (percolation at $\phi = 1$) and for the modified DEM (percolation at $\phi_c < 1$) for a range of ϕ_c values. When $\phi_c = 1$, the modified DEM coincides with the conventional DEM curve. The shapes of the inclusions were taken to be spheres. The path was from 0 to ϕ_c and (K_c, μ_c) were taken as the Reuss average values at ϕ_c. For this choice of (K_c, μ_c), estimates along the reversed path coincide with the Reuss curve.

USES

The purpose of the differential effective medium model is to estimate the effective elastic moduli of a rock in terms of its constituents and pore space.

ASSUMPTIONS AND LIMITATIONS

The following assumptions and limitations apply to the differential effective medium model:

- The rock is isotropic, linear, elastic.
- The process of incrementally adding inclusions to the matrix is a thought experiment and should not be taken to provide an accurate description of the true evolution of rock porosity in nature.

- Idealized ellipsoidal inclusion shapes are assumed.
- Cracks are isolated with respect to fluid flow. Pore pressures are unequilibrated and adiabatic. The model is appropriate for high-frequency laboratory conditions. For low-frequency field situations, use dry inclusions and then saturate by using Gassmann relations. This should not be confused with the tendency to term this approach a low-frequency theory, for crack dimensions are assumed to be much smaller than a wavelength.

4.10 HUDSON'S MODEL FOR CRACKED MEDIA

SYNOPSIS

Hudson's model is based on a scattering theory analysis of the mean wave field in an elastic solid with thin, penny-shaped ellipsoidal cracks or inclusions (Hudson, 1980, 1981). The effective moduli c_{ij}^{eff} are given as

$$c_{ij}^{\text{eff}} = c_{ij}^0 + c_{ij}^1 + c_{ij}^2$$

where c_{ij}^0 are the isotropic background moduli, and c_{ij}^1, c_{ij}^2 are the first- and second-order corrections, respectively. (See Section 2.2 on anisotropy for the two-index notation of elastic moduli. Note also that Hudson uses a slightly different definition and that there is an extra factor of 2 in his c_{44}, c_{55}, and c_{66}. This makes the equations given in his paper for c_{44}^1 and c_{44}^2 slightly different from those given below, which are consistent with the more standard notation described in Section 2.2 on anisotropy.)

For a single crack set with crack normals aligned along the 3-axis, the cracked media show transverse isotropic symmetry, and the corrections are

$$c_{11}^1 = -\frac{\lambda^2}{\mu}\varepsilon U_3$$

$$c_{13}^1 = -\frac{\lambda(\lambda + 2\mu)}{\mu}\varepsilon U_3$$

$$c_{33}^1 = -\frac{(\lambda + 2\mu)^2}{\mu}\varepsilon U_3$$

$$c_{44}^1 = -\mu\varepsilon U_1$$

$$c_{66}^1 = 0$$

and (superscripts on the c_{ij} denote second order, not quantities squared)

$$c_{11}^2 = \frac{q}{15} \frac{\lambda^2}{(\lambda + 2\mu)} (\varepsilon U_3)^2$$

$$c_{13}^2 = \frac{q}{15} \lambda (\varepsilon U_3)^2$$

$$c_{33}^2 = \frac{q}{15} (\lambda + 2\mu)(\varepsilon U_3)^2$$

$$c_{44}^2 = \frac{2}{15} \frac{\mu(3\lambda + 8\mu)}{\lambda + 2\mu} (\varepsilon U_1)^2$$

$$c_{66}^2 = 0$$

where

$$q = 15\frac{\lambda^2}{\mu^2} + 28\frac{\lambda}{\mu} + 28$$

$$\varepsilon = \frac{N}{V} a^3 = \frac{3\phi}{4\pi\alpha} = \text{crack density}$$

The isotropic background elastic moduli are λ and μ, and a and α are the crack radius and aspect ratio, respectively. The corrections c_{ij}^1 and c_{ij}^2 obey the usual symmetry properties for transverse isotropy or hexagonal symmetry (see Section 2.2 on anisotropy). The terms U_1 and U_3 depend on the crack conditions. For dry cracks

$$U_1 = \frac{16(\lambda + 2\mu)}{3(3\lambda + 4\mu)}$$

$$U_3 = \frac{4(\lambda + 2\mu)}{3(\lambda + \mu)}$$

For "weak" inclusions (i.e., when $\mu\alpha/[K' + (4/3)\mu']$ is of the order 1 and is not small enough to be neglected)

$$U_1 = \frac{16(\lambda + 2\mu)}{3(3\lambda + 4\mu)} \frac{1}{(1 + M)}$$

$$U_3 = \frac{4(\lambda + 2\mu)}{3(\lambda + \mu)} \frac{1}{(1 + \kappa)}$$

where

$$M = \frac{4\mu'}{\pi\alpha\mu} \frac{(\lambda + 2\mu)}{(3\lambda + 4\mu)}$$

$$\kappa = \frac{[K' + (4/3)\mu'](\lambda + 2\mu)}{\pi\alpha\mu(\lambda + \mu)}$$

with K' and μ' the bulk and shear modulus of the inclusion material. The criteria for an inclusion to be "weak" depend on its shape or aspect ratio α as well as on the relative moduli of the inclusion and matrix material. Dry cavities can be modeled by setting the inclusion moduli to zero. Fluid-saturated cavities are simulated by setting the inclusion shear modulus to zero.

CAUTION: Because the cavities are isolated with respect to flow, this approach simulates very-high-frequency behavior appropriate to ultrasonic laboratory conditions. At low frequencies, when there is time for wave-induced pore pressure increments to flow and equilibrate, it is better to find the effective moduli for dry cavities and then saturate them with the Brown and Korringa low-frequency relations. This should not be confused with the tendency to term this approach a low-frequency theory, for crack dimensions are assumed to be much smaller than a wavelength.

Hudson also gives expressions for infinitely thin fluid-filled cracks:

$$U_1 = \frac{16(\lambda + 2\mu)}{3(3\lambda + 4\mu)}$$
$$U_3 = 0$$

These assume no discontinuity in the normal component of crack displacements and therefore predict no change in the compressional modulus with saturation. There is, however, a shear displacement discontinuity and a resulting effect on shear stiffness. This case should be used with care.

The first-order changes λ_1 and μ_1 in the isotropic elastic moduli λ and μ of a material containing randomly oriented inclusions are given by

$$\mu_1 = -\frac{2\mu}{15}\varepsilon(3U_1 + 2U_3)$$
$$3\lambda_1 + 2\mu_1 = -\frac{(3\lambda + 2\mu)^2}{3\mu}\varepsilon U_3$$

These results agree with the self-consistent results of Budiansky and O'Connell (1976).

For two or more crack sets aligned in different directions, corrections for each crack set are calculated separately in a crack-local coordinate system with the 3-axis normal to the crack plane and then rotated or transformed back (see Section 1.4 on coordinate transformations) into the coordinates of c_{ij}^{eff}; finally the results are added to get the overall correction. Thus, for three crack sets with crack densities ε_1, ε_2, and ε_3 with crack normals aligned along the 1-, 2- and 3-axis, respectively, the overall first-order corrections to c_{ij}^0, $c_{ij}^{1(3\text{sets})}$ may be given in terms of linear combinations of the corrections for a single set with the appropriate crack densities as follows (where we have taken into account the symmetry properties of c_{ij}^1):

$$c_{11}^{1(3\text{sets})} = c_{33}^1(\varepsilon_1) + c_{11}^1(\varepsilon_2) + c_{11}^1(\varepsilon_3)$$
$$c_{12}^{1(3\text{sets})} = c_{13}^1(\varepsilon_1) + c_{13}^1(\varepsilon_2) + c_{12}^1(\varepsilon_3)$$

$$c_{13}^{1(3\text{sets})} = c_{13}^{1}(\varepsilon_1) + c_{12}^{1}(\varepsilon_2) + c_{13}^{1}(\varepsilon_3)$$

$$c_{22}^{1(3\text{sets})} = c_{11}^{1}(\varepsilon_1) + c_{33}^{1}(\varepsilon_2) + c_{11}^{1}(\varepsilon_3)$$

$$c_{23}^{1(3\text{sets})} = c_{12}^{1}(\varepsilon_1) + c_{13}^{1}(\varepsilon_2) + c_{13}^{1}(\varepsilon_3)$$

$$c_{33}^{1(3\text{sets})} = c_{11}^{1}(\varepsilon_1) + c_{11}^{1}(\varepsilon_2) + c_{33}^{1}(\varepsilon_3)$$

$$c_{44}^{1(3\text{sets})} = c_{44}^{1}(\varepsilon_2) + c_{44}^{1}(\varepsilon_3)$$

$$c_{55}^{1(3\text{sets})} = c_{44}^{1}(\varepsilon_1) + c_{44}^{1}(\varepsilon_3)$$

$$c_{66}^{1(3\text{sets})} = c_{44}^{1}(\varepsilon_1) + c_{44}^{1}(\varepsilon_2)$$

Note that $c_{66}^{1} = 0$ and $c_{12}^{1} = c_{11}^{1} - 2c_{66}^{1} = c_{11}^{1}$.

Hudson (1981) also gives the attenuation coefficient ($\gamma = \omega Q^{-1}/2V$) for elastic waves in cracked media. For aligned cracks with normals along the 3-axis, the attenuation coefficients for P-, SV-, and SH-waves are

$$\gamma_P = \frac{\omega}{V_S}\varepsilon\left(\frac{\omega a}{V_P}\right)^3 \frac{1}{30\pi}\left[AU_1^2 \sin^2 2\theta + BU_3^2\left(\frac{V_P^2}{V_S^2} - 2\sin^2\theta\right)^2\right]$$

$$\gamma_{SV} = \frac{\omega}{V_S}\varepsilon\left(\frac{\omega a}{V_S}\right)^3 \frac{1}{30\pi}\left(AU_1^2 \cos^2 2\theta + BU_3^2 \sin^2 2\theta\right)$$

$$\gamma_{SH} = \frac{\omega}{V_S}\varepsilon\left(\frac{\omega a}{V_S}\right)^3 \frac{1}{30\pi}\left(AU_1^2 \cos^2\theta\right)$$

$$A = \frac{3}{2} + \frac{V_S^5}{V_P^5}$$

$$B = 2 + \frac{15}{4}\frac{V_S}{V_P} - 10\frac{V_S^3}{V_P^3} + 8\frac{V_S^5}{V_P^5}$$

In the preceding expressions V_P and V_S are the P and S velocities in the uncracked isotropic background matrix, ω is the angular frequency, and θ is the angle between the direction of propagation and the 3-axis (axis of symmetry).

For randomly oriented cracks (isotropic distribution), the P and S attenuation coefficients are given as

$$\gamma_P = \frac{\omega}{V_S}\varepsilon\left(\frac{\omega a}{V_P}\right)^3 \frac{4}{(15)^2\pi}\left[AU_1^2 + \frac{1}{2}\frac{V_P^5}{V_S^5}B(B-2)U_3^2\right]$$

$$\gamma_S = \frac{\omega}{V_S}\varepsilon\left(\frac{\omega a}{V_S}\right)^3 \frac{1}{75\pi}\left(AU_1^2 + \frac{1}{3}BU_3^2\right)$$

The fourth-power dependence on ω is characteristic of Rayleigh scattering.

Hudson (1990) gives results for overall elastic moduli of material with various distributions of penny-shaped cracks. If conditions at the cracks are taken to be uniform so that U_1 and U_3 do not depend on the polar and azimuthal angles θ and ϕ, the first-order correction is given as

$$c^1_{ijpq} = -\frac{A}{\mu}U_3[\lambda^2\delta_{ij}\delta_{pq} + 2\lambda\mu(\delta_{ij}\tilde{\varepsilon}_{pq} + \delta_{pq}\tilde{\varepsilon}_{ij}) + 4\mu^2\tilde{\varepsilon}_{ijpq}]$$
$$- A\mu U_1(\delta_{jq}\tilde{\varepsilon}_{ip} + \delta_{jp}\tilde{\varepsilon}_{iq} + \delta_{iq}\tilde{\varepsilon}_{jp} + \delta_{ip}\tilde{\varepsilon}_{jq} - 4\tilde{\varepsilon}_{ijpq})$$

where

$$A = \int_0^{2\pi}\int_0^{\pi/2} \varepsilon(\theta,\phi)\sin\theta\,d\theta\,d\phi$$

$$\tilde{\varepsilon}_{ij} = \frac{1}{A}\int_0^{2\pi}\int_0^{\pi/2} \varepsilon(\theta,\phi)n_i n_j \sin\theta\,d\theta\,d\phi$$

$$\tilde{\varepsilon}_{ijpq} = \frac{1}{A}\int_0^{2\pi}\int_0^{\pi/2} \varepsilon(\theta,\phi)n_i n_j n_p n_q \sin\theta\,d\theta\,d\phi$$

and n_i are the components of the unit vector along the crack normal, $n = (\sin\theta\cos\phi, \sin\theta\sin\phi, \cos\theta)$, whereas $\varepsilon(\theta,\phi)$ is the crack density distribution function so that $\varepsilon(\theta,\phi)\sin\theta\,d\theta\,d\phi$ is the density of cracks with normals lying in the solid angle between $(\theta, \theta + d\theta)$ and $(\phi, \phi + d\phi)$.

SPECIAL CASES OF CRACK DISTRIBUTIONS

a) When cracks with total crack density ε_t have all their normals aligned along $\theta = \theta_0, \phi = \phi_0$:

$$\varepsilon(\theta,\phi) = \varepsilon_t\frac{\delta(\theta - \theta_0)}{\sin\theta}\delta(\phi - \phi_0)$$

and then

$$A = \varepsilon_t$$
$$\tilde{\varepsilon}_{ij} = n^0_i n^0_j$$
$$\tilde{\varepsilon}_{ijpq} = n^0_i n^0_j n^0_p n^0_q$$

where $n^0_1 = \sin\theta_0\cos\phi_0$, $n^0_2 = \sin\theta_0\sin\phi_0$, $n^0_3 = \cos\theta_0$.

b) Rotationally symmetric crack distributions with normals symmetrically distributed about $\theta = 0$, that is, ε is a function of θ only:

$$A = 2\pi\int_0^{\pi/2} \varepsilon(\theta)\sin\theta\,d\theta$$

$$\tilde{\varepsilon}_{12} = \tilde{\varepsilon}_{23} = \tilde{\varepsilon}_{31} = 0$$

$$\tilde{\varepsilon}_{11} = \tilde{\varepsilon}_{22} = \frac{\pi}{A}\int_0^{\pi/2} \varepsilon(\theta)\sin^3\theta\,d\theta = \frac{1}{2}(1 - \tilde{\varepsilon}_{33})$$

$$\tilde{\varepsilon}_{1111} = \tilde{\varepsilon}_{2222} = \frac{3\pi}{4A} \int_0^{\pi/2} \varepsilon(\theta) \sin^5 \theta \, d\theta = 3\tilde{\varepsilon}_{1122} = 3\tilde{\varepsilon}_{1212}, \text{ etc.}$$

$$\tilde{\varepsilon}_{3333} = \frac{8}{3}\tilde{\varepsilon}_{1111} - 4\tilde{\varepsilon}_{11} + 1$$

$$\tilde{\varepsilon}_{1133} = \tilde{\varepsilon}_{11} - \frac{4}{3}\tilde{\varepsilon}_{1111} = \tilde{\varepsilon}_{2233} = \tilde{\varepsilon}_{1313} = \tilde{\varepsilon}_{2323}, \text{ etc.}$$

Elements other than those related to the preceding elements by symmetry are zero. A particular rotationally symmetric distribution is the Fisher distribution, for which $\varepsilon(\theta)$ is

$$\varepsilon(\theta) = \frac{\varepsilon_t}{2\pi} \frac{e^{(\cos\theta)/\sigma^2}}{\sigma^2(e^{1/\sigma^2} - 1)}$$

For small σ^2, this is approximately a model for a Gaussian distribution on the sphere

$$\varepsilon(\theta) \approx \varepsilon_t \frac{e^{-\theta^2/2\sigma^2}}{2\pi\sigma^2}$$

The proportion of crack normals outside the range $0 \le \theta \le 2\sigma$ is approximately $1/e^2$. For this distribution,

$$A = \varepsilon_t$$

$$\tilde{\varepsilon}_{11} = \frac{-1 + 2\sigma^2 e^{1/\sigma^2} - 2\sigma^4(e^{1/\sigma^2} - 1)}{2(e^{1/\sigma^2} - 1)} \approx \sigma^2$$

$$\tilde{\varepsilon}_{1111} = \frac{3}{8}\left[\frac{-1 + 4\sigma^4(2e^{1/\sigma^2} + 1) - 24\sigma^6 e^{1/\sigma^2} + 24\sigma^8(e^{1/\sigma^2} - 1)}{(e^{1/\sigma^2} - 1)}\right] \approx 3\sigma^4$$

This distribution is suitable when crack normals are oriented randomly with a small variance about a mean direction along the 3-axis.

c) Cracks with normals randomly distributed at a fixed angle from the 3-axis forming a cone. In this case ε is independent of ϕ and is zero unless $\theta = \theta_0, 0 \le \phi \le 2\pi$.

$$\varepsilon(\theta) = \varepsilon_t \frac{\delta(\theta - \theta_0)}{2\pi \sin\theta}$$

which gives

$$A = \varepsilon_t$$

$$\tilde{\varepsilon}_{11} = \frac{1}{2} \sin^2 \theta_0$$

$$\tilde{\varepsilon}_{1111} = \frac{3}{8} \sin^4 \theta_0$$

and the first-order corrections are

$$c^1_{1111} = \frac{-\varepsilon_t}{2\mu}[U_3(2\lambda^2 + 4\lambda\mu\sin^2\theta_0 + 3\mu^2\sin^4\theta_0) + U_1\mu^2\sin^2\theta_0(4 - 3\sin^2\theta_0)]$$

$$= c^1_{2222}$$

$$c^1_{3333} = \frac{-\varepsilon_t}{\mu}[U_3(\lambda + 2\mu\cos^2\theta_0)^2 + U_1\mu^2 4\cos^2\theta_0\sin^2\theta_0]$$

$$c^1_{1122} = \frac{-\varepsilon_t}{2\mu}[U_3(2\lambda^2 + 4\lambda\mu\sin^2\theta_0 + \mu^2\sin^4\theta_0) - U_1\mu^2\sin^4\theta_0]$$

$$c^1_{1133} = \frac{-\varepsilon_t}{\mu}[U_3(\lambda + \mu\sin^2\theta_0)(\lambda + 2\mu\cos^2\theta_0) - U_1\mu^2 2\sin^2\theta_0\cos^2\theta_0]$$

$$= c^1_{2233}$$

$$c^1_{2323} = \frac{-\varepsilon_t}{2}\mu[U_3 4\sin^2\theta_0\cos^2\theta_0 + U_1(\sin^2\theta_0 + 2\cos^2\theta_0 - 4\sin^2\theta_0\cos^2\theta_0)]$$

$$= c^1_{1313}$$

$$c^1_{1212} = \frac{-\varepsilon_t}{2}\mu[U_3\sin^4\theta_0 + U_1\sin^2\theta_0(2 - \sin^2\theta_0)]$$

USES

Hudson's model is used to estimate the effective elastic moduli and attenuation of a rock in terms of its constituents and pore space.

ASSUMPTIONS AND LIMITATIONS

The use of Hudson's model requires the following considerations:

- Idealized crack shape (penny-shaped) with small aspect ratios and crack density are assumed. Crack radius and distance between cracks are much smaller than a wavelength. The formal limit quoted by Hudson for both first- and second-order terms is ε less than 0.1.
- The second-order expansion is not a uniformly converging series and predicts increasing moduli with crack density beyond the formal limit (Cheng, 1993). Better results will be obtained by using just the first-order correction rather than inappropriately using the second-order correction. Cheng gives a new expansion based on Padé approximation, which avoids this problem.
- Cracks are isolated with respect to fluid flow. Pore pressures are unequilibrated and adiabatic. The model is appropriate for high-frequency laboratory conditions. For low-frequency field situations use Hudson's dry equations and

then saturate by using Brown and Korringa relations. This should not be confused with the tendency to think of this approach as a low-frequency theory, because crack dimensions are assumed to be much smaller than a wavelength.

- Sometimes a single crack set may not be an adequate representation of crack-induced anisotropy. In this case we need to superpose several crack sets with angular distributions.

4.11 ESHELBY–CHENG MODEL FOR CRACKED ANISOTROPIC MEDIA

SYNOPSIS

Cheng (1978, 1993) has given a model for the effective moduli of cracked transversely isotropic rocks based on Eshelby's (1957) static solution for the strain inside an ellipsoidal inclusion in an isotropic matrix. The effective moduli c_{ij}^{eff} for a rock containing fluid-filled ellipsoidal cracks with their normals aligned along the 3-axis are given as

$$c_{ij}^{\text{eff}} = c_{ij}^0 - \phi c_{ij}^1$$

where ϕ is the porosity and c_{ij}^0 are the moduli of the uncracked isotropic rock. The corrections c_{ij}^1 are

$$c_{11}^1 = \lambda(S_{31} - S_{33} + 1) + \frac{2\mu E}{D(S_{12} - S_{11} + 1)}$$

$$c_{33}^1 = \frac{(\lambda + 2\mu)(-S_{12} - S_{11} + 1) + 2\lambda S_{13} + 4\mu C}{D}$$

$$c_{13}^1 = \frac{(\lambda + 2\mu)(S_{13} + S_{31}) - 4\mu C + \lambda(S_{13} - S_{12} - S_{11} - S_{33} + 2)}{2D}$$

$$c_{44}^1 = \frac{\mu}{1 - 2S_{1313}}$$

$$c_{66}^1 = \frac{\mu}{1 - 2S_{1212}}$$

with

$$C = \frac{K_{\text{fl}}}{3(K - K_{\text{fl}})}$$

$$D = S_{33}S_{11} + S_{33}S_{12} - 2S_{31}S_{13} - (S_{11} + S_{12} + S_{33} - 1 - 3C)$$
$$\quad - C[S_{11} + S_{12} + 2(S_{33} - S_{13} - S_{31})]$$

$$E = S_{33}S_{11} - S_{31}S_{13} - (S_{33} + S_{11} - 2C - 1)$$
$$\quad + C(S_{31} + S_{13} - S_{11} - S_{33})$$

$$S_{11} = QI_{aa} + RI_a$$

$$S_{33} = Q\left(\frac{4\pi}{3} - 2I_{ac}\alpha^2\right) + I_c R$$

$$S_{12} = QI_{ab} - RI_a$$

$$S_{13} = QI_{ac}\alpha^2 - RI_a$$

$$S_{31} = QI_{ac} - RI_c$$

$$S_{1212} = QI_{ab} + RI_a$$

$$S_{1313} = \frac{Q(1 + \alpha^2)I_{ac}}{2} + \frac{R(I_a + I_c)}{2}$$

$$I_a = \frac{2\pi\alpha(\cos^{-1}\alpha - \alpha S_a)}{S_a^3}$$

$$I_c = 4\pi - 2I_a$$

$$I_{ac} = \frac{I_c - I_a}{3S_a^2}$$

$$I_{aa} = \pi - \frac{3I_{ac}}{4}$$

$$I_{ab} = \frac{I_{aa}}{3}$$

$$\sigma = \frac{3K - 2\mu}{6K + 2\mu}$$

$$S_a = \sqrt{1 - \alpha^2}$$

$$R = \frac{1 - 2\sigma}{8\pi(1 - \sigma)}$$

$$Q = \frac{3R}{1 - 2\sigma}$$

In the preceding equations K and μ are the bulk and shear modulus of the isotropic matrix, respectively; K_{fl} is the bulk modulus of the fluid; and α is the crack aspect ratio. Dry cavities can be modeled by setting the inclusion moduli to zero. Do not confuse the S's with the anisotropic compliance tensor. This model is valid for arbitrary aspect ratios, unlike the Hudson model (see Section 4.10 on Hudson's model), which assumes very small aspect ratio cracks. The results of the two models are essentially the same for small aspect ratios and low crack densities (<0.1) as long as the "weak inclusion" form of Hudson's theory is used (Cheng, 1993).

USES

The Eshelby–Cheng model is used to obtain effective anisotropic stiffness tensor for transversely isotropic cracked rocks.

ASSUMPTIONS AND LIMITATIONS

The following presuppositions and limitations apply to the Eshelby–Cheng model:

- The model assumes an isotropic, homogeneous, elastic background matrix and an idealized ellipsoidal crack shape.
- The model assumes low crack concentrations but can handle all aspect ratios.
- Because the cavities are isolated with respect to flow, this approach simulates very-high-frequency behavior appropriate to ultrasonic laboratory conditions. At low frequencies, when there is time for wave-induced pore pressure increments to flow and equilibrate, it is better to find the effective moduli for dry cavities and then saturate them with the Brown and Korringa low-frequency relations. This should not be confused with the tendency to term this approach a low-frequency theory, for crack dimensions are assumed to be much smaller than a wavelength.

EXTENSIONS

The model has been extended to a transversely isotropic background by Nishizawa (1982).

4.12 ELASTIC CONSTANTS IN FINELY LAYERED MEDIA – BACKUS AVERAGE

SYNOPSIS

A transversely isotropic medium with the symmetry axis in the x_3 direction has an elastic stiffness tensor that can be written in the condensed matrix form (see

Section 2.2 on anisotropy):

$$\begin{bmatrix} a & b & f & 0 & 0 & 0 \\ b & a & f & 0 & 0 & 0 \\ f & f & c & 0 & 0 & 0 \\ 0 & 0 & 0 & d & 0 & 0 \\ 0 & 0 & 0 & 0 & d & 0 \\ 0 & 0 & 0 & 0 & 0 & m \end{bmatrix}, \quad m = \frac{1}{2}(a - b)$$

where a, b, c, d, and f are five independent elastic constants. Backus (1962) showed that in the long-wavelength limit a stratified medium composed of **layers of transversely isotropic materials** (each with its symmetry axis normal to the strata) is also effectively anisotropic with effective stiffness as follows:

$$\begin{bmatrix} A & B & F & 0 & 0 & 0 \\ B & A & F & 0 & 0 & 0 \\ F & F & C & 0 & 0 & 0 \\ 0 & 0 & 0 & D & 0 & 0 \\ 0 & 0 & 0 & 0 & D & 0 \\ 0 & 0 & 0 & 0 & 0 & M \end{bmatrix}, \quad M = \frac{1}{2}(A - B)$$

where

$$A = \langle a - f^2 c^{-1} \rangle + \langle c^{-1} \rangle^{-1} \langle f c^{-1} \rangle^2$$

$$B = \langle b - f^2 c^{-1} \rangle + \langle c^{-1} \rangle^{-1} \langle f c^{-1} \rangle^2$$

$$C = \langle c^{-1} \rangle^{-1}$$

$$F = \langle c^{-1} \rangle^{-1} \langle f c^{-1} \rangle$$

$$D = \langle d^{-1} \rangle^{-1}$$

$$M = \langle m \rangle$$

The brackets $\langle \cdot \rangle$ indicate averages of the enclosed properties weighted by their volumetric proportions. This is often called the *Backus average*.

If the **individual layers are isotropic**, the effective medium is still transversely isotropic, but the number of independent constants needed to describe each individual layer is reduced to 2:

$$a = c = \lambda + 2\mu, \quad b = f = \lambda, \quad d = m = \mu$$

giving for the effective medium

$$A = \left\langle \frac{4\mu(\lambda + \mu)}{\lambda + 2\mu} \right\rangle + \left\langle \frac{1}{\lambda + 2\mu} \right\rangle^{-1} \left\langle \frac{\lambda}{\lambda + 2\mu} \right\rangle^2$$

$$B = \left\langle \frac{2\mu\lambda}{\lambda + 2\mu} \right\rangle + \left\langle \frac{1}{\lambda + 2\mu} \right\rangle^{-1} \left\langle \frac{\lambda}{\lambda + 2\mu} \right\rangle^2$$

$$C = \left\langle \frac{1}{\lambda + 2\mu} \right\rangle^{-1}$$

$$F = \left\langle \frac{1}{\lambda + 2\mu} \right\rangle^{-1} \left\langle \frac{\lambda}{\lambda + 2\mu} \right\rangle$$

$$D = \left\langle \frac{1}{\mu} \right\rangle^{-1}$$

$$M = \langle \mu \rangle$$

In terms of the P- and S-wave velocities and densities in the isotropic layers (Levin, 1979),

$$a = \rho V_P^2$$

$$d = \rho V_S^2$$

$$f = \rho \left(V_P^2 - V_S^2 \right)$$

the effective parameters can be rewritten as

$$A = \left\langle 4\rho V_S^2 \left[1 - \frac{V_S^2}{V_P^2} \right] \right\rangle + \left\langle 1 - 2\frac{V_S^2}{V_P^2} \right\rangle^2 \langle (\rho V_P^2)^{-1} \rangle^{-1}$$

$$B = \left\langle 2\rho V_S^2 \left[1 - \frac{2V_S^2}{V_P^2} \right] \right\rangle + \left\langle 1 - 2\frac{V_S^2}{V_P^2} \right\rangle^2 \langle (\rho V_P^2)^{-1} \rangle^{-1}$$

$$C = \langle (\rho V_P^2)^{-1} \rangle^{-1}$$

$$F = \left\langle 1 - 2\frac{V_S^2}{V_P^2} \right\rangle \langle (\rho V_P^2)^{-1} \rangle^{-1}$$

$$D = \langle (\rho V_S^2)^{-1} \rangle^{-1}$$

$$M = \langle \rho V_S^2 \rangle$$

The P- and S-wave velocities in the effective anisotropic medium can be written as

$$V_{SH,h} = \sqrt{M/\rho}$$

$$V_{SH,v} = V_{SV,h} = V_{SV,v} = \sqrt{D/\rho}$$

$$V_{P,h} = \sqrt{A/\rho}$$

$$V_{P,v} = \sqrt{C/\rho}$$

where ρ is the average density; $V_{P,v}$ is the vertically propagating P wave; $V_{P,h}$ is the horizontally propagating P wave; $V_{SH,h}$ is the horizontally propagating, horizontally polarized S wave; $V_{SV,h}$ is the horizontally propagating, vertically polarized S wave; and $V_{SV,v}$ and $V_{SH,v}$ are the vertically propagating S waves of any polarization (vertical is defined as normal to the layering).

EXAMPLE

Calculate the effective anisotropic elastic constants and the velocity anisotropy for a thinly layered sequence of dolomite and shale with the following layer properties:

$$V_{P(1)} = 5{,}200 \text{ m/s}, \quad V_{S(1)} = 2{,}700 \text{ m/s}, \quad \rho_1 = 2{,}450 \text{ kg/m}^3, \quad d_1 = 0.75 \text{ m}$$

$$V_{P(2)} = 2{,}900 \text{ m/s}, \quad V_{S(2)} = 1{,}400 \text{ m/s}, \quad \rho_2 = 2{,}340 \text{ kg/m}^3, \quad d_2 = 0.5 \text{ m}$$

The volumetric fractions are

$$f_1 = d_1/(d_1 + d_2) = 0.6, \quad f_2 = d_2/(d_1 + d_2) = 0.4$$

If one takes the volumetric weighted averages of the appropriate properties,

$$A = f_1 4\rho_1 V_{S(1)}{}^2 \left(1 - \frac{V_{S(1)}{}^2}{V_{P(1)}{}^2}\right) + f_2 4\rho_2 V_{S(2)}{}^2 \left(1 - \frac{V_{S(2)}{}^2}{V_{P(2)}{}^2}\right)$$

$$+ \left[f_1 \left(1 - 2\frac{V_{S(1)}{}^2}{V_{P(1)}{}^2}\right) + f_2 \left(1 - 2\frac{V_{S(2)}{}^2}{V_{P(2)}{}^2}\right)\right]^2 \frac{1}{\frac{f_1}{\rho_1 V_{P(1)}{}^2} + \frac{f_2}{\rho_2 V_{P(2)}{}^2}}$$

$$A = 45.1 \text{ GPa}$$

Similarly, computing the other averages we get

$$C = 34.03 \text{ GPa}, \quad D = 8.28 \text{ GPa}, \quad M = 12.55 \text{ GPa}, \quad F = 16.7 \text{ GPa}$$

$$B = A - 2M = 20 \text{ GPa}$$

The average density $\rho = f_1 \rho_1 + f_2 \rho_2 = 2{,}406 \text{ kg/m}^3$. The anisotropic velocities are

$$V_{SH,h} = \sqrt{M/\rho} = 2{,}284.0 \text{ m/s}$$

$$V_{SH,v} = V_{SV,h} = V_{SV,v} = \sqrt{D/\rho} = 1{,}854.8$$

$$V_{P,h} = \sqrt{A/\rho} = 4{,}329.5 \text{ m/s}$$

$$V_{P,v} = \sqrt{C/\rho} = 3{,}761.0 \text{ m/s}$$

$$\text{P-wave anisotropy} = (4{,}329.5 - 3{,}761.0)/3{,}761.0 \approx 15\%$$

$$\text{S-wave anisotropy} = (2{,}284.0 - 1{,}854.8)/1{,}854.8 \approx 23\%$$

Finally, consider the case in which **each layer is isotropic with the same shear modulus** but with a different bulk modulus. This might be the situation, for example, for a massive, homogeneous rock with fine layers of different fluids or saturations. Then, the elastic constants of the medium become

$$A = C = \left\langle \frac{1}{\rho V_P^2} \right\rangle^{-1} = \left\langle \frac{1}{K + \frac{4}{3}\mu} \right\rangle^{-1}$$

$$B = F = \left\langle \frac{1}{\rho V_P^2} \right\rangle^{-1} - 2\mu = \left\langle \frac{1}{K + \frac{4}{3}\mu} \right\rangle^{-1} - 2\mu = A - 2\mu$$

$$D = M = \mu$$

A finely layered medium of isotropic layers, all having the same shear modulus, is isotropic.

USES

The Backus average is used to model a finely stratified medium as a single homogeneous medium.

ASSUMPTIONS AND LIMITATIONS

The following presuppositions and conditions apply to the Backus average:

- All materials are linear elastic.
- There are no sources of intrinsic energy dissipation such as friction or viscosity.
- Layer thickness must be much smaller than the seismic wavelength. How small is still a question of disagreement and research, but a rule of thumb is that the wavelength must be at least ten times the layer thickness.

PART 5

PART 5

GRANULAR MEDIA

5.1 PACKING OF SPHERES – GEOMETRIC RELATIONS

SYNOPSIS

Spheres are often used as idealized models for pores or grains. Table 5.1.1 gives some geometric properties of various packings of identical spheres (summarized in part from Bourbié et al., 1987). Note that complementary interpretations are possible when the grains are considered spheres versus when the pores are considered spheres.

USES

These results can be used to estimate the geometric relations of the packing of granular materials.

ASSUMPTIONS AND LIMITATIONS

The preceding results assume idealized, identical spheres.

TABLE 5.1.1. Geometric properties of sphere packs.

Packing type	Porosity (nonspheres)	Solid fraction (spheres)	Specific surface area[a]	Number of contacts per sphere	Radius of maximum inscribable sphere[b,c]	Radius of maximum sphere fitting in narrowest channels[b,c]
Simple cubic	$1 - \pi/6 = 0.476$	$\pi/6 = 0.524$	$\pi/2R$	6	0.732	0.414
Simple hexagonal	$1 - 4\pi \cos(\pi/6)/18 = 0.395$	$4\pi \cos(\pi/6)/18 = 0.605$	$2\pi \cos(\pi/6)/3R$	8	0.528	0.414 and 0.155
Hexagonal close pack	0.259	0.741	$2.22/R$	12	0.225 and 0.414	0.155
Dense random pack	~0.36	~0.64	~$1.92/R$	~9		

[a] Specific surface area, S, is defined as the pore surface area in a sample divided by the total volume of the sample. If the grains are spherical, $S = 3(1 - \phi)/R$.
[b] Expressed in units of the radius of the packed spheres.
[c] Note that if the pore space is modeled as a packing of spherical pores, the inscribable spheres always have radius equal to 1.

5.2 RANDOM SPHERICAL GRAIN PACKINGS – CONTACT MODELS AND EFFECTIVE MODULI

SYNOPSIS

CONTACT STIFFNESSES AND EFFECTIVE MODULI

The effective elastic properties of packings of spherical particles depend on *normal and tangential contact stiffnesses of a two-particle combination*. The normal stiffness of two identical spheres is defined as the ratio of a confining force increment to the shortening of a sphere radius. The tangential stiffness of two identical spheres is the ratio of a tangential force increment to the increment of the tangential displacement of the center relative to the contact region:

$$S_n = \partial F / \partial \delta, \quad S_\tau = \partial T / \partial \tau$$

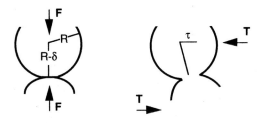

Figure 5.2.1

For a random sphere packing, effective bulk and shear moduli can be expressed through porosity, ϕ; coordination number, C (the average number of contacts per sphere); sphere radius, R; and normal and tangential stiffnesses, S_n and S_τ, respectively, of a two-sphere combination by

$$K_{eff} = \frac{C(1 - \phi)}{12 \pi R} S_n$$

$$G_{eff} = \frac{C(1 - \phi)}{20 \pi R} (S_n + 1.5 S_\tau)$$

The effective P- and S-wave velocities are (Winkler, 1983)

$$V_P^2 = \frac{3C}{20\pi R\rho}\left(S_n + \frac{2}{3}S_\tau\right)$$

$$V_S^2 = \frac{C}{20\pi R\rho}\left(S_n + \frac{3}{2}S_\tau\right)$$

$$\frac{V_P^2}{V_S^2} = 2\frac{3S_n/S_\tau + 2}{2S_n/S_\tau + 3}$$

where ρ is the grain-material density. Wang and Nur (1992) summarize some of the existing granular media models.

COORDINATION NUMBER

For a random dense pack of identical spheres the average number of contacts per grain C is about 9. An approximate C versus porosity curve, shown in Figure 5.2.2, is based on the summary in Murphy (1982).

THE HERTZ–MINDLIN MODEL

In the **Hertz model** of normal compression of two identical spheres, the radius of the contact area, a, and the normal displacement, δ, are

$$a = \sqrt[3]{\frac{3FR}{8G}(1 - v)}, \quad \delta = \frac{a^2}{R}$$

where G and v are the shear modulus and Poisson's ratio of the grain material, respectively.

If a hydrostatic confining pressure P is applied to a random identical sphere packing, a confining force acting between two particles is

$$F = \frac{4\pi R^2 P}{C(1 - \phi)}$$

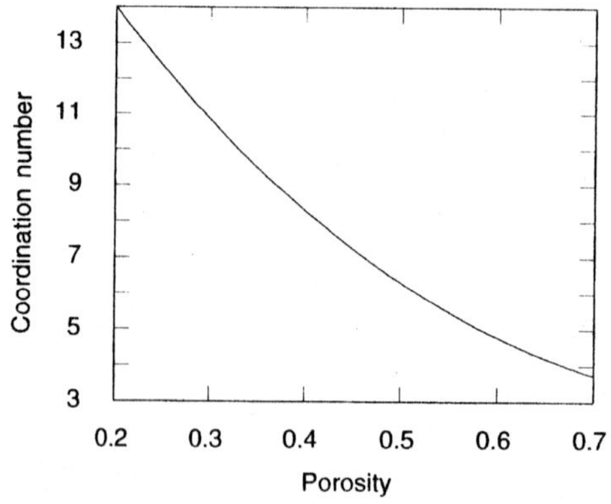

Porosity	C
0.20	14.007
0.25	12.336
0.30	10.843
0.35	9.5078
0.40	8.3147
0.45	7.2517
0.50	6.3108
0.55	5.4878
0.60	4.7826
0.65	4.1988
0.7	3.7440

Figure 5.2.2

Then

$$a = R \sqrt[3]{\frac{3\pi(1 - v)}{2C(1 - \phi)G} P}$$

The normal stiffness is

$$S_n = \frac{4Ga}{1 - v}$$

The effective bulk modulus of a dry random identical sphere packing is, then,

$$K_{eff} = \sqrt[3]{\frac{C^2(1 - \phi)^2 G^2}{18\pi^2(1 - v)^2} P}$$

Mindlin (1949) shows that if the spheres are first pressed together and a tangential force is applied *afterward*, the shear and normal stiffnesses are (the latter as in the Hertz solution)

$$S_\tau = \frac{8aG}{2 - v}, \quad S_n = \frac{4aG}{1 - v}$$

where v and G are the Poisson's ratio and shear modulus of the solid grains, respectively. The effective shear modulus of a dry random identical sphere packing is, then,

$$G_{eff} = \frac{5 - 4v}{5(2 - v)} \sqrt[3]{\frac{3C^2(1 - \phi)^2 G^2}{2\pi^2(1 - v)^2} P}$$

In the Mindlin formulas given above, it is assumed that there is no slip at the contact surface between two particles. In fact such slip will occur at the edges of the contact region. The no-slip assumption results in a small error if only acoustic wave propagation is concerned and can be safely used in estimating the effective moduli of granular materials. *The Hertz–Mindlin model can be used to describe the properties of precompacted granular rocks.*

THE WALTON MODEL

It is assumed in the Walton model (Walton, 1987) that *normal and shear deformation of a two-grain combination occur simultaneously*. This assumption leads to results somewhat different from those given by the Hertz–Mindlin model. Specifically, there is no partial slip in the contact area. The slip occurs across the whole area once applied tractions exceed friction resistance. The results discussed in the following paragraphs are given for two special cases: infinitely rough spheres (friction coefficient is very large) and ideally smooth spheres (friction coefficient is zero).

Under *hydrostatic pressure P*, an identical sphere packing is isotropic. Its effective bulk and shear moduli for the rough spheres case (dry pack) are described by

$$K_{\text{eff}} = \frac{1}{6} \sqrt[3]{\frac{3(1-\phi)^2 C^2 P}{\pi^4 B^2}}, \quad G_{\text{eff}} = \frac{3}{5} K_{\text{eff}} \frac{5B+A}{2B+A}$$

$$A = \frac{1}{4\pi} \left(\frac{1}{G} - \frac{1}{G+\lambda} \right), \quad B = \frac{1}{4\pi} \left(\frac{1}{G} + \frac{1}{G+\lambda} \right)$$

where λ is Lamé's coefficient of the grain material. For the smooth spheres case (dry pack)

$$G_{\text{eff}} = \frac{1}{10} \sqrt[3]{\frac{3(1-\phi)^2 C^2 P}{\pi^4 B^2}}, \quad K_{\text{eff}} = \frac{5}{3} G_{\text{eff}}$$

It is clear that the effective density of the aggregate is

$$\rho_{\text{eff}} = (1-\phi)\rho$$

Under *uniaxial pressure* σ_1 a dry identical sphere packing is transversely isotropic, and if the spheres are infinitely rough, it can be described by the following five constants:

$$C_{11} = 3(\alpha + 2\beta), \quad C_{12} = \alpha - 2\beta, \quad C_{13} = 2C_{12}$$
$$C_{33} = 8(\alpha + \beta), \quad C_{44} = \alpha + 7\beta$$

where

$$\alpha = \frac{(1-\phi)Ce}{32\pi^2 B}, \quad \beta = \frac{(1-\phi)Ce}{32\pi^2(2B+A)}$$

$$e = \sqrt[3]{\frac{24\pi^2 B(2B+A)\sigma_1}{(1-\phi)AC}}$$

THE DIGBY MODEL

The Digby model gives effective moduli for a dry random packing of identical elastic spherical particles. Neighboring particles are initially firmly bonded across small, flat, circular regions of the radius a. Outside these adhesion surfaces, the shape of each particle is assumed to be ideally smooth (with a continuous first derivative). Notice that this condition differs from that of Hertz, where the shape of a particle is not smooth at the intersection of the spherical surface and the plane of contact. Digby's normal and shear stiffnesses under hydrostatic pressure P are (Digby, 1981)

$$S_{\text{n}} = \frac{4Gb}{1-v}, \quad S_{\tau} = \frac{8Ga}{2-v}$$

where v and G are the Poisson's ratio and shear modulus of the grain material respectively. Parameter b can be found from the relation

$$\frac{b}{R} = \left[(d)^2 + \left(\frac{a}{R} \right)^2 \right]^{1/2}$$

where d satisfies the cubic equation

$$d^3 + \frac{3}{2} \left(\frac{a}{R} \right)^2 d - \frac{3\pi(1-v)}{2C(1-\phi)} \frac{P}{G} = 0$$

EXAMPLE

Use the Digby model to estimate the effective bulk and shear moduli for a dry random pack of spherical grains under a confining pressure of 10 MPa. The ratio of the radius of the initially bonded area to the grain radius a/R is 0.01. The bulk and shear moduli of the grain material are $K = 37$ GPa and $G = 44$ GPa, respectively. The porosity of the grain pack is 0.36.

The Poisson's ratio v for the grain material is calculated from K and G:

$$v = \frac{3K - 2G}{2(3K + G)} = 0.07$$

The coordination number $C = 9$. Solving the cubic equation for d

$$d^3 + \frac{3}{2} \left(\frac{a}{R} \right)^2 d - \frac{3\pi(1-v)}{2C(1-\phi)} \frac{P}{G} = 0$$

and taking the real root, neglecting the pair of complex conjugate roots, we get $d = 0.0547$. Next we calculate b/R as

$$b/R = \sqrt{d^2 + \left(\frac{a}{R} \right)^2} = 0.0556$$

The values of a/R and b/R are used to compute S_n/R and S_τ/R:

$$S_n/R = \frac{4G(b/R)}{1 - v} = 10.5, \quad S_\tau/R = \frac{8G(a/R)}{2 - v} = 1.8$$

which then finally give us

$$K_{\text{eff}} = \frac{C(1 - \phi)}{12\pi}(S_n/R) = 1.6 \, \text{GPa}$$

and

$$G_{\text{eff}} = \frac{C(1 - \phi)}{20\pi}[(S_n/R) + 1.5(S_\tau/R)] = 1.2 \, \text{GPa}$$

THE BRANDT MODEL

The Brandt model allows one to calculate the bulk modulus of randomly packed elastic spheres of identical mechanical properties but of different sizes. This packing is subject to external hydrostatic and internal hydrostatic pressures. The effective pressure P is the difference between these two pressures. The effective bulk modulus is (Brandt, 1955)

$$K_{\text{eff}} = \frac{2P^{1/3}}{9\phi}\left[\frac{E}{1.75(1-\nu^2)}\right]^{2/3} Z - 1.5PZ$$

$$Z = \frac{(1+30.75z)^{5/3}}{1+46.13z}, \quad z = \frac{K^{3/2}(1-\nu^2)}{E\sqrt{P}}$$

In this case E is the mineral Young's modulus, and K is the fluid bulk modulus.

THE CEMENTED SAND MODEL

The cemented sand model allows one to calculate the bulk and shear moduli of dry sand in which cement is deposited *at grain contacts*. The cement is elastic and its properties may differ from those of the spheres.

It is assumed that the starting framework of cemented sand is a dense, random pack of identical spherical grains with porosity $\phi_0 \approx 0.36$ and the average number of contacts per grain $C = 9$. Adding cement to the grains acts to reduce porosity and to increase the effective elastic moduli of the aggregate. Then, these effective dry-rock bulk and shear moduli are (Dvorkin and Nur, 1996)

$$K_{\text{eff}} = \frac{1}{6}C(1-\phi_0)M_c\hat{S}_n$$

$$G_{\text{eff}} = \frac{3}{5}K_{\text{eff}} + \frac{3}{20}C(1-\phi_0)G_c\hat{S}_\tau$$

$$M_c = \rho_c V_{Pc}^2$$

$$G_c = \rho_c V_{Sc}^2$$

where ρ_c is the cement's density and V_{Pc} and V_{Sc} are its P- and S-wave velocities. Parameters $\hat{S}_n$ and $\hat{S}_\tau$ are proportional to the normal and shear stiffness, respectively, of a cemented two-grain combination. They depend on the amount of the contact cement and on the properties of the cement and the grains as defined in the following relations:

$$\hat{S}_n = A_n\alpha^2 + B_n\alpha + C_n$$

$$A_n = -0.024153\Lambda_n^{-1.3646}$$

$$B_n = 0.20405\Lambda_n^{-0.89008}$$

$$C_n = 0.00024649\Lambda_n^{-1.9864}$$

$$\hat{S}_\tau = A_\tau \alpha^2 + B_\tau \alpha + C_\tau$$

$$A_\tau = -10^{-2}(2.26\nu^2 + 2.07\nu + 2.3)\Lambda_\tau^{0.079\nu^2 + 0.1754\nu - 1.342}$$

$$B_\tau = (0.0573\nu^2 + 0.0937\nu + 0.202)\Lambda_\tau^{0.0274\nu^2 + 0.0529\nu - 0.8765}$$

$$C_\tau = -10^{-4}(9.654\nu^2 + 4.945\nu + 3.1)\Lambda_\tau^{0.01867\nu^2 + 0.4011\nu - 1.8186}$$

$$\Lambda_n = \frac{2G_c}{\pi G}\frac{(1-\nu)(1-\nu_c)}{(1-2\nu_c)}, \quad \Lambda_\tau = \frac{G_c}{\pi G}, \quad \alpha = \frac{a}{R}$$

where G and ν are the shear modulus and the Poisson's ratio of the grains, respectively; G_c and ν_c are the shear modulus and the Poisson's ratio of the cement, respectively; a is the radius of the contact cement layer; and R is the grain radius.

The amount of the contact cement can be expressed through the ratio α of the radius of the cement layer a to the grain radius R:

$$\alpha = a/R$$

The radius of the contact cement layer a is not necessarily directly related to the total amount of cement; part of the cement may be deposited away from the intergranular contacts. However, by assuming that porosity reduction in sands is due to cementation only and by adopting certain schemes of cement deposition, we can relate parameter α to the current porosity of cemented sand ϕ. For example, we can use Scheme 1 (see Figure 5.2.3 below) in which all cement is deposited at grain contacts to get the formula

$$\alpha = 2\left[\frac{\phi_0 - \phi}{3C(1-\phi_0)}\right]^{0.25} = 2\left[\frac{S\phi_0}{3C(1-\phi_0)}\right]^{0.25}$$

or we can use Scheme 2 in which cement is evenly deposited on the grain surface:

$$\alpha = \left[\frac{2(\phi_0 - \phi)}{3(1-\phi_0)}\right]^{0.5} = \left[\frac{2S\phi_0}{3(1-\phi_0)}\right]^{0.5}$$

In these formulas S is the cement saturation of the pore space. It is the fraction of the pore space (of the uncemented sand) occupied by cement (in the cemented sand).

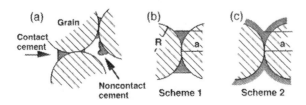

Figure 5.2.3

If the cement's properties are identical to those of the grains, the cementation theory gives results that are very close to those of the Digby model. The cementation theory allows one to **diagnose** a rock by determining what type of cement prevails. For example, the theory helps to distinguish between quartz and clay cement. Generally, V_P predictions are much better than V_S predictions.

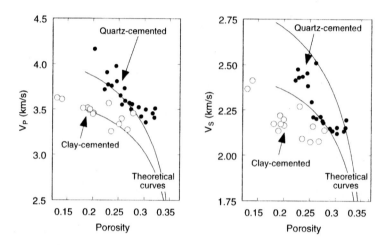

Figure 5.2.4. Predictions of V_P and V_S using the Scheme 2 model for quartz and clay cement compared with data from quartz and clay cemented rocks from the North Sea.

THE UNCEMENTED SAND MODEL

The uncemented sand model allows one to calculate the bulk and shear moduli of dry sand in which cement is deposited *away from grain contacts*. It is assumed that the starting framework of uncemented sand is a dense random pack of identical spherical grains with porosity $\phi_0 = 0.36$ and the average number of contacts per grain $C = 9$. At this porosity, the contact Hertz–Mindlin theory gives the following expressions for the effective bulk (K_{HM}) and shear (G_{HM}) moduli of a dry dense random pack of identical spherical grains subject to a hydrostatic pressure P:

$$K_{HM} = \left[\frac{C^2(1 - \phi_0)^2 G^2}{18\pi^2(1 - v)^2} P \right]^{1/3}$$

$$G_{HM} = \frac{5 - 4v}{5(2 - v)} \left[\frac{3C^2(1 - \phi_0)^2 G^2}{2\pi^2(1 - v)^2} P \right]^{1/3}$$

where v is the grain Poisson's ratio and G is the grain shear modulus.

To find the effective moduli (K_{eff} and G_{eff}) at a different porosity ϕ, a heuristic modified Hashin–Strikman lower bound is used:

$$K_{\text{eff}} = \left[\frac{\phi/\phi_0}{K_{\text{HM}} + \frac{4}{3}G_{\text{HM}}} + \frac{1 - \phi/\phi_0}{K + \frac{4}{3}G_{\text{HM}}} \right]^{-1} - \frac{4}{3}G_{\text{HM}}$$

$$G_{\text{eff}} = \left[\frac{\phi/\phi_0}{G_{\text{HM}} + \frac{G_{\text{HM}}}{6}\left(\frac{9K_{\text{HM}} + 8G_{\text{HM}}}{K_{\text{HM}} + 2G_{\text{HM}}}\right)} + \frac{1 - \phi/\phi_0}{G + \frac{G_{\text{HM}}}{6}\left(\frac{9K_{\text{HM}} + 8G_{\text{HM}}}{K_{\text{HM}} + 2G_{\text{HM}}}\right)} \right]^{-1}$$
$$- \frac{G_{\text{HM}}}{6}\left(\frac{9K_{\text{HM}} + 8G_{\text{HM}}}{K_{\text{HM}} + 2G_{\text{HM}}}\right)$$

where K is the grain bulk modulus.

EXAMPLE

Calculate V_P and V_S in uncemented dry quartz sand of porosity 0.3 at 40 MPa overburden and 20 MPa pore pressure. Use the uncemented sand model. For pure quartz, $G = 45$ GPa, $K = 36.6$ GPa, and $\nu = 0.06$. Then, for effective pressure 20 MPa $= 0.02$ GPa,

$$K_{\text{HM}} = \left[\frac{9^2(1 - 0.36)^2 45^2}{18 \cdot 3.14^2(1 - 0.06)^2} 0.02 \right]^{1/3} = 2 \text{ GPa}$$

$$G_{\text{HM}} = \frac{5 - 4 \cdot 0.06}{5(2 - 0.06)} \left[\frac{3 \cdot 9^2(1 - 0.36)^2 45^2}{2 \cdot 3.14^2(1 - 0.06)^2} 0.02 \right]^{1/3} = 3 \text{ GPa}$$

Next

$$K_{\text{eff}} = \left(\frac{0.3/0.36}{2 + \frac{4}{3}3} + \frac{1 - 0.3/0.36}{36.6 + \frac{4}{3}3} \right)^{-1} - \frac{4}{3}3 = 3 \text{ GPa}$$

$$G_{\text{eff}} = \left(\frac{0.3/0.36}{3 + 0.29} + \frac{1 - 0.3/0.36}{45 + 0.29} \right)^{-1} - 0.29 = 3.6 \text{ GPa}$$

Pure quartz density is 2.65 g/cm^3; then, the sandstone's density is

$$2.65 \cdot (1 - 0.3) = 1.855 \text{ g/cm}^3$$

The P-wave velocity is

$$V_P = \sqrt{\frac{3 + \frac{4}{3}3.6}{1.855}} = 2.05 \text{ km/s}$$

and the S-wave velocity is

$$V_S = \sqrt{\frac{3.6}{1.855}} = 1.39 \text{ km/s}$$

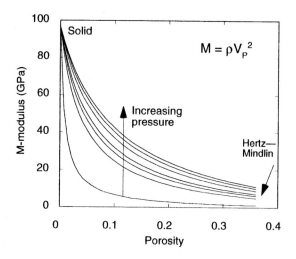

Figure 5.2.5. Illustration of the modified lower Hashin–Shtrikman bound for various effective pressures. The pressure dependence follows from the Hertz–Mindlin theory incorporated into the right end member.

This model connects two end members; one has zero porosity and the modulus of the solid phase and the other has high porosity and a pressure-dependent modulus as given by the Hertz–Mindlin theory. This contact theory allows one to describe the noticeable pressure dependence normally observed in sands.

The high-porosity end member does not necessarily have to be calculated from the Hertz–Mindlin theory. The end member can be measured experimentally on high-porosity sands from a given reservoir. Then, to estimate the moduli of sands

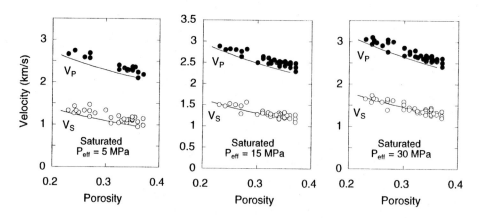

Figure 5.2.6. Prediction of V_P and V_S using the lower Hashin–Shtrikman bound compared with measured velocities from unconsolidated North Sea samples.

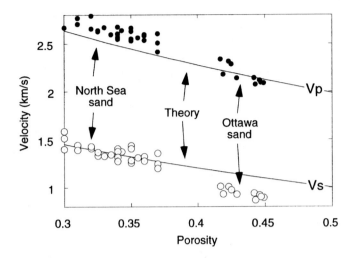

Figure 5.2.7

of different porosities, the modified Hashin–Strikman lower bound formulas can be used, where K_{HM} and G_{HM} are set at the measured values.

This method provides accurate estimates for velocities in uncemented sands. In Figures 5.2.6 and 5.2.7 the curves are from the theory.

This method can also be used for estimating velocities in sands of porosities exceeding 0.36.

USES

The methods can be used to model granular high-porosity rocks.

ASSUMPTIONS AND LIMITATIONS

The grain contact models presuppose the following:

- The strains are small.
- Identical homogeneous, isotropic, elastic spherical grains are assumed.

EXTENSIONS

To calculate the effective elastic moduli of saturated rocks (and low-frequency acoustic velocities), Gassmann's formula should be applied.

5.3 ORDERED SPHERICAL GRAIN PACKINGS – EFFECTIVE MODULI

SYNOPSIS

Ordered packings of identical spherical particles are generally anisotropic. Their effective elastic properties can be thus described through stiffness matrices.

SIMPLE CUBIC PACKING

The coordination number is 6, and the porosity is 47 percent. The stiffness matrix is

$$
\begin{pmatrix}
C_{11} & C_{12} & C_{12} & 0 & 0 & 0 \\
C_{12} & C_{11} & C_{12} & 0 & 0 & 0 \\
C_{12} & C_{12} & C_{11} & 0 & 0 & 0 \\
0 & 0 & 0 & \frac{1}{2}(C_{11} - C_{12}) & 0 & 0 \\
0 & 0 & 0 & 0 & \frac{1}{2}(C_{11} - C_{12}) & 0 \\
0 & 0 & 0 & 0 & 0 & \frac{1}{2}(C_{11} - C_{12})
\end{pmatrix}
$$

where

$$
c_{11} = c_0, \quad c_{12} = \frac{v}{2(2 - v)}c_0, \quad c_0 = \left[\frac{3G^2 P}{2(1 - v)^2}\right]^{1/3}
$$

P is the hydrostatic pressure, and G and v are the grain material shear modulus and Poisson's ratio, respectively.

HEXAGONAL CLOSE PACKING

The coordination number is 12, and the porosity is about 26 percent. The stiffness matrix is

$$
\begin{pmatrix}
C_{11} & C_{12} & C_{13} & 0 & 0 & 0 \\
C_{12} & C_{11} & C_{13} & 0 & 0 & 0 \\
C_{13} & C_{13} & C_{33} & 0 & 0 & 0 \\
0 & 0 & 0 & C_{44} & 0 & 0 \\
0 & 0 & 0 & 0 & C_{44} & 0 \\
0 & 0 & 0 & 0 & 0 & C_{66}
\end{pmatrix}
$$

where

$$c_{11} = \frac{1152 - 1848v + 725v^2}{24(2 - v)(12 - 11v)}c_0$$

$$c_{12} = \frac{v(120 - 109v)}{24(2 - v)(12 - 11v)}c_0$$

$$c_{13} = \frac{v}{3(2 - v)}c_0$$

$$c_{33} = \frac{4(3 - 2v)}{3(2 - v)}c_0$$

$$c_{44} = c_{55} = \frac{6 - 5v}{4(2 - v)}c_0$$

$$c_{66} = \frac{576 - 948v + 417v^2}{24(2 - v)(12 - 11v)}c_0$$

FACE-CENTERED CUBIC PACKING

The coordination number is 6, and the porosity is about 47 percent. The stiffness matrix is

$$\begin{pmatrix} C_{11} & C_{12} & C_{12} & 0 & 0 & 0 \\ C_{12} & C_{11} & C_{12} & 0 & 0 & 0 \\ C_{12} & C_{12} & C_{11} & 0 & 0 & 0 \\ 0 & 0 & 0 & C_{44} & 0 & 0 \\ 0 & 0 & 0 & 0 & C_{44} & 0 \\ 0 & 0 & 0 & 0 & 0 & C_{44} \end{pmatrix}$$

where

$$c_{11} = 2c_{44} = \frac{4 - 3v}{2 - v}c_0$$

$$c_{12} = \frac{v}{2(2 - v)}c_0$$

USES

The results of this section are sometimes used to estimate the elastic properties of granular materials.

ASSUMPTIONS AND LIMITATIONS

These models assume identical elastic spherical particles under small strain conditions.

PART 6

FLUID EFFECTS ON WAVE PROPAGATION

6.1 BIOT'S VELOCITY RELATIONS

SYNOPSIS

Biot (1956) derived theoretical formulas for predicting the frequency-dependent velocities of saturated rocks in terms of the dry rock properties. His formulation incorporates some, but not all, of the mechanisms of viscous and inertial interaction between the pore fluid and the mineral matrix of the rock. The low-frequency limiting velocities, V_{P0} and V_{S0}, are the same as predicted by Gassmann's relations (see the discussion of Gassmann in Section 6.3). The high-frequency limiting velocities, $V_{P\infty}$ and $V_{S\infty}$, are given by (cast in the notation of Johnson and Plona, 1982)

$$V_{P\infty}(\text{fast, slow}) = \left\{ \frac{\Delta \pm \left[\Delta^2 - 4\left(\rho_{11}\rho_{22} - \rho_{12}{}^2\right)\left(PR - Q^2\right)\right]^{1/2}}{2\left(\rho_{11}\rho_{22} - \rho_{12}{}^2\right)} \right\}^{1/2}$$

$$V_{S\infty} = \left(\frac{\mu_{\text{fr}}}{\rho - \phi\rho_{\text{fl}}\alpha^{-1}} \right)^{1/2}$$

$$\Delta = P\rho_{22} + R\rho_{11} - 2Q\rho_{12}$$

$$P = \frac{(1-\phi)(1-\phi-K_{\text{fr}}/K_0)K_0 + \phi K_0 K_{\text{fr}}/K_{\text{fl}}}{1-\phi-K_{\text{fr}}/K_0 + \phi K_0/K_{\text{fl}}} + \frac{4}{3}\mu_{\text{fr}}$$

$$Q = \frac{(1 - \phi - K_{fr}/K_0)\phi K_0}{1 - \phi - K_{fr}/K_0 + \phi K_0/K_{fl}}$$

$$R = \frac{\phi^2 K_0}{1 - \phi - K_{fr}/K_0 + \phi K_0/K_{fl}}$$

$$\rho_{11} = (1 - \phi)\rho_0 - (1 - \alpha)\phi\rho_{fl}$$

$$\rho_{22} = \alpha\phi\rho_{fl}$$

$$\rho_{12} = (1 - \alpha)\phi\rho_{fl}$$

$$\rho = \rho_0(1 - \phi) + \rho_{fl}\phi$$

where

K_{fr}, μ_{fr} = effective bulk and shear moduli of rock frame: either the
dry frame or the high-frequency unrelaxed "wet frame"
moduli predicted by the Mavko–Jizba squirt theory (Section 6.7)

K_0 = bulk modulus of mineral material making up the rock

K_{fl} = effective bulk modulus of pore fluid

ϕ = porosity

ρ_0 = mineral density

ρ_{fl} = fluid density

α = tortuosity parameter, always greater than 1

The term ρ_{12} describes the induced mass resulting from inertial drag caused by the relative acceleration of the solid frame and the pore fluid. The tortuosity, α, (sometimes called the structure factor) is a purely geometrical factor independent of the solid or fluid densities. Berryman (1981) obtained the relation

$$\alpha = 1 - r(1 - 1/\phi)$$

where $r = 1/2$ for spheres and lies between 0 and 1 for other ellipsoids. For uniform cylindrical pores with axes parallel to the pore pressure gradient, α equals 1 (the minimum possible value), whereas for a random system of pores with all possible orientations, $\alpha = 3$ (Stoll, 1977). The high-frequency limiting velocities depend quite strongly on α, with higher fast P-wave velocities for lower α values.

The two solutions given above for the high-frequency limiting P-wave velocity, designated by $\pm$, correspond to the "fast" and "slow" waves. The fast wave is the compressional body-wave most easily observed in the laboratory and the field, and it corresponds to overall fluid and solid motions that are in phase. The slow wave is a highly dissipative wave in which the overall solid and fluid motions are out of phase.

Another approximate expression for the high-frequency limit of the fast P-wave velocity is (Geertsma and Smit, 1961)

$$
V_{P\infty} = \left\{ \frac{1}{\rho_0(1-\phi) + \phi\rho_{fl}(1-\alpha^{-1})} \left[\left(K_{fr} + \frac{4}{3}\mu_{fr}\right) \right. \right.
$$
$$
\left. \left. + \frac{\phi\frac{\rho}{\rho_{fl}}\alpha^{-1} + \left(1-\frac{K_{fr}}{K_0}\right)\left(1-\frac{K_{fr}}{K_0} - 2\phi\alpha^{-1}\right)}{\left(1-\frac{K_{fr}}{K_0}-\phi\right)\frac{1}{K_0} + \frac{\phi}{K_{fl}}} \right] \right\}^{1/2}
$$

CAUTION: This form predicts velocities that are too high ($\approx$3–6%) compared with the actual high-frequency limit.

The complete frequency dependence can be obtained from the roots of the dispersion relations (Biot, 1956; Stoll, 1977; Berryman, 1980a):

$$
\begin{vmatrix} H/V_P^2 - \rho & \rho_{fl} - C/V_P^2 \\ C/V_P^2 - \rho_{fl} & q - M/V_P^2 \end{vmatrix} = 0
$$

$$
\begin{vmatrix} \rho - \mu_{fr}/V_S^2 & \rho_{fl} \\ \rho_{fl} & q \end{vmatrix} = 0
$$

The complex roots are

$$
\frac{1}{V_P^2}
$$
$$
= \frac{-(Hq + M\rho - 2C\rho_{fl}) \pm \sqrt{(Hq + M\rho - 2C\rho_{fl})^2 - 4(C^2 - MH)(\rho_{fl}^2 - \rho q)}}{2(C^2 - MH)}
$$

$$
\frac{1}{V_S^2} = \frac{q\rho - \rho_{fl}^2}{q\mu_{fr}}
$$

The real and imaginary parts of the roots give the velocity and attenuation, respectively. Again, the two solutions correspond to the fast and slow P-waves. The various terms are

$$
H = K_{fr} + \frac{4}{3}\mu_{fr} + \frac{(K_0 - K_{fr})^2}{(D - K_{fr})}
$$

$$
C = \frac{(K_0 - K_{fr})K_0}{(D - K_{fr})}
$$

$$
M = \frac{K_0^2}{(D - K_{fr})}
$$

$$
D = K_0[1 + \phi(K_0/K_{fl} - 1)]
$$

$$
\rho = (1 - \phi)\rho_0 + \phi\rho_{fl}
$$

$$
q = \frac{\alpha\rho_{fl}}{\phi} - \frac{i\eta F(\zeta)}{\omega\kappa}
$$

where

η = viscosity of the pore fluid
κ = absolute permeability of the rock
ω = angular frequency of the plane wave

The viscodynamic operator $F(\zeta)$ incorporates the frequency dependence of viscous drag and is defined by

$$F(\zeta) = \frac{1}{4} \frac{\zeta T(\zeta)}{1 + 2iT(\zeta)/\zeta}$$

$$T(\zeta) = \frac{\mathrm{ber}'(\zeta) + i\,\mathrm{bei}'(\zeta)}{\mathrm{ber}(\zeta) + i\,\mathrm{bei}(\zeta)} = \frac{e^{i3\pi/4} J_1\left(\zeta e^{-i\pi/4}\right)}{J_0\left(\zeta e^{-i\pi/4}\right)}$$

$$\zeta = (\omega/\omega_r)^{1/2} = \left(\frac{\omega a^2 \rho_{\mathrm{fl}}}{\eta}\right)^{1/2}$$

where

$\mathrm{ber}(\ )$, $\mathrm{bei}(\ )$ = real and imaginary parts of the Kelvin function
$J_n(\ )$ = Bessel function of order n
a = pore-size parameter

The pore-size parameter a depends on both the dimensions and the shape of the pore space. Stoll (1974) found that values between 1/6 and 1/7 of the mean grain diameter gave good agreement with experimental data from several investigators. For spherical grains, Hovem and Ingram (1979) obtained $a = \phi d/[3(1 - \phi)]$, where d is the grain diameter. The velocity dispersion curve for fast P-waves can be closely approximated by a standard linear solid viscoelastic model when $\kappa/a^2 \geq 1$. (See Sections 3.5 and 6.11.) However, for most consolidated crustal rocks, κ/a^2 is usually less than 1.

At very low frequencies $F(\zeta) \to 1$ and at very high frequencies (large ζ) the asymptotic values are $T(\zeta) \to (1 + i)/\sqrt{2}$ and $F(\zeta) \to (\kappa/4)(1 + i)/\sqrt{2}$.

The reference frequency, f_c, which determines the low-frequency range, $f \ll f_c$, and the high-frequency range, $f \gg f_c$, is given by

$$f_c = \frac{\phi\eta}{2\pi\rho_{\mathrm{fl}}\kappa}$$

One interpretation of this relation is that it is the frequency where viscous forces acting on the pore fluid approximately equal the inertial forces acting on it. In the high-frequency limit, the fluid motion is dominated by inertial effects; in the low-frequency limit, the fluid motion is dominated by viscous effects.

As mentioned above, Biot's theory predicts the existence of a slow, highly attenuated P-wave in addition to the usual fast P- and S-waves. The slow P-wave has been observed in the laboratory, and it is sometimes invoked to explain diffusional loss mechanisms.

USES

Biot's theory can be used for the following purposes:

- Estimating saturated rock velocities from dry rock velocities.
- Estimating frequency dependence of velocities.
- Estimating reservoir compaction caused by pumping using quasi-static limit of Biot's poroelasticity theory.

ASSUMPTIONS AND LIMITATIONS

The use of Biot's equations presented in this section requires the following considerations:

- The rock is isotropic.
- All minerals making up the rock have the same bulk and shear moduli.
- Fluid-bearing rock is completely saturated.
- **CAUTION: For most crustal rocks the amount of squirt dispersion (which is not included in Biot's formulation) is comparable with or greater than Biot's dispersion, and thus using Biot's theory alone will lead to poor predictions of high-frequency saturated velocities.** Exceptions: very high permeability materials such as ocean sediments and glass beads at very high effective pressure or near open boundaries such as at a borehole or at the surfaces of a laboratory sample. The recommended procedure is to use the Mavko–Jizba squirt theory (Section 6.7) first to estimate the high-frequency wet frame moduli and then substitute them into Biot's equations.
- Wavelength, even in the high-frequency limit, is much larger than the grain or pore scale.

EXTENSIONS

Biot's theory has been extended to anisotropic media (Biot, 1962).

6.2 GEERTSMA–SMIT APPROXIMATIONS OF BIOT'S RELATIONS

SYNOPSIS

Low- and middle-frequency approximations (Geertsma and Smit, 1961) of Biot's theoretical formulas for predicting the frequency-dependent velocities of saturated rocks in terms of the dry rock properties (see also Biot's relations, Section 6.1) may be expressed as

$$V_P^2 = \frac{V_{P\infty}^4 + V_{P0}^4\left(\frac{f_c}{f}\right)^2}{V_{P\infty}^2 + V_{P0}^2\left(\frac{f_c}{f}\right)^2}$$

where

V_P = frequency-dependent P-wave velocity of saturated rock
V_{P0} = Biot–Gassmann low-frequency limiting P-wave velocity
$V_{P\infty}$ = Biot high-frequency limiting P-wave velocity
f = frequency
f_c = Biot's reference frequency, f_c, which determines the low-frequency range, $f \ll f_c$, and the high-frequency range, $f \gg f_c$, given by

$$f_c = \frac{\phi\eta}{2\pi\rho_{fl}\kappa}$$

where

ϕ = porosity
ρ_{fl} = fluid density
η = viscosity of the pore fluid
κ = absolute permeability of the rock

USES

The Geertsma–Smit approximations can be used for the following:

- Estimating saturated rock velocities from dry rock velocities.
- Estimating the frequency dependence of velocities.

ASSUMPTIONS AND LIMITATIONS

The use of the Geertsma–Smit approximations presented in this section require the following considerations:

- Mathematical approximations are valid at moderate-to-low seismic frequencies, so that $f < f_c$. This generally means moderate-to-low permeabilities, but it is in this range of permeabilities that squirt dispersion may dominate the Biot effect.
- The rock is isotropic.
- All minerals making up the rock have the same bulk and shear moduli.
- Fluid-bearing rock is completely saturated.
- **CAUTION:** For most crustal rocks the amount of squirt dispersion (not included in Biot's theory) is comparable with or greater than Biot's

dispersion, and thus using Biot's theory alone will lead to poor predictions of high-frequency saturated velocities. Exceptions: very high permeability materials such as ocean sediments and glass beads, or at very high effective pressure. The recommended procedure is to use the Mavko–Jizba squirt theory (Section 6.7) first to estimate the high-frequency wet frame moduli and then substitute them into the Biot or Geertsma–Smit equations.

6.3 GASSMANN'S RELATIONS

SYNOPSIS

One of the most important problems in the rock physics analysis of logs, cores, and seismic data is the prediction of seismic velocities in rocks saturated with one fluid from rocks saturated with a second fluid – or, equivalently, saturated rock velocities from dry rock velocities, and vice versa. This is the *fluid substitution problem.*

Generally, when a rock is loaded under an increment of compression, such as from a passing seismic wave, an increment of pore pressure change is induced, which resists the compression and therefore stiffens the rock. The low-frequency Gassmann (1951)–Biot (1956) theory predicts the resulting increase in effective bulk modulus, K_{sat}, of the saturated rock through the following equation:

$$\frac{K_{sat}}{K_0 - K_{sat}} = \frac{K_{dry}}{K_0 - K_{dry}} + \frac{K_{fl}}{\phi(K_0 - K_{fl})}, \qquad \mu_{sat} = \mu_{dry}$$

where

$$K_{dry} = \text{effective bulk modulus of dry rock}$$
$$K_{sat} = \text{effective bulk modulus of the rock with pore fluid}$$
$$K_0 = \text{bulk modulus of mineral material making up rock}$$
$$K_{fl} = \text{effective bulk modulus of pore fluid}$$
$$\phi = \text{porosity}$$
$$\mu_{dry} = \text{effective shear modulus of dry rock}$$
$$\mu_{sat} = \text{effective shear modulus of rock with pore fluid}$$

Gassmann's equation assumes a homogeneous mineral modulus and statistical isotropy of the pore space but is free of assumptions about the pore geometry. Most importantly, it is valid only at sufficiently **low frequencies** such that the induced pore pressures are equilibrated throughout the pore space (i.e., there is sufficient time for the pore fluid to flow and eliminate wave-induced pore pressure gradients). This limitation to low frequencies explains why Gassmann's relation works best for very low frequency in situ seismic data (< 100 Hz) and may perform less well as frequencies increase toward sonic logging ($\approx 10^4$ Hz) and laboratory ultrasonic measurements ($\approx 10^6$ Hz).

CAUTION: "Dry rock" is not the same as gas-saturated rock. The "dry rock" or "dry frame" modulus refers to the incremental bulk deformation resulting from an increment of applied confining pressure with pore pressure held constant. This corresponds to a "drained" experiment in which pore fluids can flow freely in or out of the sample to ensure constant pore pressure. Alternatively, the "dry frame" modulus can correspond to an undrained experiment in which the pore fluid has zero bulk modulus and in which pore compressions therefore do not induce changes in pore pressure. This is approximately the case for an air-filled sample at standard temperature and pressure. However, at reservoir conditions (high pore pressure), gas takes on a nonnegligible bulk modulus and should be treated as a saturating fluid.

CAUTION: Laboratory measurements on very dry rocks, such as those prepared in a vacuum oven, are sometimes *too dry*. Several investigators have found that the first few percent of fluid saturation added to an extremely dry rock will lower the frame moduli probably as a result of disrupting surface forces acting on the pore surfaces. It is this slightly wet or moist rock modulus that should be used as the "dry rock" modulus in Gassmann's relations. A more thorough discussion is given in Section 6.12.

Although we often describe Gassmann's relations as allowing us to predict saturated rock moduli from dry rock moduli, and vice versa, the most common in situ problem is to predict the changes from one fluid to another. One procedure is simply to apply the equation twice: transform the moduli from the initial fluid saturation to the dry state and then immediately transform from the dry moduli to the new fluid-saturated state. Equivalently, we can algebraically eliminate the dry rock moduli from the equation and relate the saturated rock moduli $K_{sat\,1}$ and $K_{sat\,2}$ in terms of the two fluid bulk moduli K_{fl1} and K_{fl2} as follows:

$$\frac{K_{sat1}}{K_0 - K_{sat1}} - \frac{K_{fl1}}{\phi(K_0 - K_{fl1})} = \frac{K_{sat\,2}}{K_0 - K_{sat\,2}} - \frac{K_{fl2}}{\phi(K_0 - K_{fl2})}$$

CAUTION: It is NOT CORRECT simply to replace K_{dry} in Gassmann's equation by $K_{sat\,2}$.

A few more explicit, but entirely equivalent, forms of the equation are

$$K_{sat} = K_{dry} + \frac{\left(1 - \frac{K_{dry}}{K_0}\right)^2}{\frac{\phi}{K_{fl}} + \frac{1-\phi}{K_0} - \frac{K_{dry}}{K_0^2}}$$

$$K_{sat} = \frac{\phi\left(\frac{1}{K_0} - \frac{1}{K_{fl}}\right) + \frac{1}{K_0} - \frac{1}{K_{dry}}}{\frac{\phi}{K_{dry}}\left(\frac{1}{K_0} - \frac{1}{K_{fl}}\right) + \frac{1}{K_0}\left(\frac{1}{K_0} - \frac{1}{K_{dry}}\right)}$$

$$\frac{1}{K_{\text{sat}}} = \frac{1}{K_0} + \frac{\phi}{K_\phi + \frac{K_0 K_\text{fl}}{K_0 - K_\text{fl}}}$$

$$K_{\text{dry}} = \frac{K_{\text{sat}}\left(\frac{\phi K_0}{K_\text{fl}} + 1 - \phi\right) - K_0}{\frac{\phi K_0}{K_\text{fl}} + \frac{K_{\text{sat}}}{K_0} - 1 - \phi}$$

EXAMPLE

Use Gassmann's relation to compare the bulk modulus of a dry quartz sandstone having the properties $\phi = 0.20$, $K_{\text{dry}} = 12$ GPa, and $K_0 = 36$ GPa with the bulk moduli when the rock is saturated with gas and with water at $T = 80°C$ and $P = 300$ bar.

Calculate the bulk moduli and density for gas and for water by using the Batzle–Wang formulas discussed in Section 6.15. A gas with gravity 1 will have the properties $K_{\text{gas}} = 0.133$ GPa and $\rho_{\text{gas}} = 0.336$ g/cm^3, and water with salinity 50,000 ppm will have properties $K_{\text{water}} = 3.013$ GPa and $\rho_{\text{water}} = 1.055$ g/cm^3.

Next, substitute these into Gassmann's relations

$$\frac{K_{\text{sat-gas}}}{36 - K_{\text{sat-gas}}} = \frac{12}{36 - 12} + \frac{0.133}{0.20(36 - 0.133)}$$

to yield a value of $K_{\text{sat-gas}} = 12.29$ GPa for the gas-saturated rock. Similarly,

$$\frac{K_{\text{sat-water}}}{36 - K_{\text{sat-water}}} = \frac{12}{36 - 12} + \frac{3.013}{0.20(36 - 3.013)}$$

which yields a value of $K_{\text{sat-water}} = 17.6$ GPa for the water-saturated rock.

REUSS AVERAGE FORM

An equivalent form can be written as

$$\frac{K_{\text{sat}}}{K_0 - K_{\text{sat}}} = \frac{K_{\text{dry}}}{K_0 - K_{\text{dry}}} + \frac{K_R}{K_0 - K_R}$$

where

$$K_R = \left(\frac{\phi}{K_\text{fl}} + \frac{1 - \phi}{K_0}\right)^{-1}$$

is the Reuss average modulus of the fluid and mineral at porosity ϕ. This is consistent with the obvious result that when the dry frame modulus of the rock goes to zero, the fluid-saturated sample will behave as a suspension and lie on the Reuss bound.

LINEAR FORM

A particularly useful exact *linear* form of Gassmann's relation follows from the simple graphical construction shown in Figure 6.3.1 (Mavko and Mukerji, 1995). Draw a straight line from the mineral modulus on the left axis (at $\phi = 0$) through the data point (A) corresponding to the rock modulus with the initial pore fluid (in this case, air). The line intersects the Reuss average for that pore fluid at some porosity, ϕ_R, which is a measure of the pore space stiffness. Then the rock modulus (point A') for a new pore fluid (in this case, water) falls along a second straight line from the mineral modulus, intersecting the Reuss average for the new pore fluid at ϕ_R. Then we can write, *exactly*:

$$\Delta K_{Gass}(\phi) = \frac{\phi}{\phi_R} \Delta K_R(\phi_R)$$

where $\Delta K_{Gass} = K_{sat\,2} - K_{sat\,1}$ is the Gassmann-predicted change of saturated rock bulk modulus between any two pore fluids (including gas), and $\Delta K_R(\phi_R)$ is the difference in the Reuss average for the two fluids *evaluated at the intercept porosity* ϕ_R.

Because pore fluid moduli are usually much less than mineral moduli, we can approximate the Reuss average as

$$K_R(\phi_R) = \frac{K_{fl} K_0}{\phi_R K_0 + (1 - \phi_R) K_{fl}} \approx \frac{K_{fl}}{\phi_R}$$

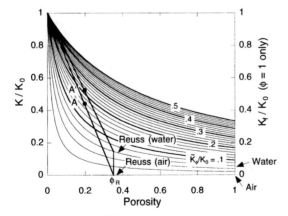

Figure 6.3.1. ϕ_R **is the porosity where a straight line drawn from the mineral modulus on the left axis, through the data point (A or A'), intersects the Reuss average for that corresponding pore fluid. The change in rock bulk modulus between any two pore fluids (A–A') is proportional to the change in Reuss average for the same two fluids evaluated at the intercept porosity ϕ_R.**

Then, the linear form of Gassmann's relations can be approximated as

$$\Delta K_{\text{Gass}}(\phi) \approx \frac{\phi}{\phi_{\text{R}}^2} \Delta K_{\text{fl}}$$

P-WAVE MODULUS FORM

Because Gassmann's relations predict no change in the shear modulus, we can also write the linear form of Gassmann's relation as

$$\Delta M_{\text{Gass}}(\phi) = \Delta K_{\text{Gass}}(\phi) \approx \frac{\phi}{\phi_{\text{R}}^2} \Delta K_{\text{fl}}$$

where $\Delta M_{\text{Gass}}(\phi)$ is the predicted change in the P-wave modulus $M = K + (4/3)\mu$.

VELOCITY FORM

Murphy, Schwartz, and Hornby (1991) suggested a velocity form of Gassmann's relation

$$\rho_{\text{sat}} V_{\text{Psat}}^2 = K_{\text{p}} + K_{\text{dry}} + \frac{4}{3}\mu$$

$$\rho_{\text{sat}} V_{\text{Ssat}}^2 = \mu$$

which is easily written as

$$\frac{V_{\text{Psat}}^2}{V_{\text{Ssat}}^2} = \frac{K_{\text{p}}}{\mu} + \frac{K_{\text{dry}}}{\mu} + \frac{4}{3}$$

where

$$\begin{aligned}
\rho_{\text{sat}} &= \text{density of the saturated rock} \\
V_{\text{Psat}} &= \text{P-wave saturated rock velocity} \\
V_{\text{Ssat}} &= \text{S-wave saturated rock velocity} \\
K_{\text{dry}} &= \text{dry rock bulk modulus} \\
\mu &= \mu_{\text{dry}} = \mu_{\text{sat}} = \text{rock shear modulus}
\end{aligned}$$

and

$$K_{\text{p}} = \frac{\left(1 - \frac{K_{\text{dry}}}{K_0}\right)^2}{\frac{\phi}{K_{\text{fl}}} + \frac{1-\phi}{K_0} - \frac{K_{\text{dry}}}{K_0^2}}$$

PORE STIFFNESS INTERPRETATION

One can write the dry rock compressibility at constant pore pressure, $K_{dry}{}^{-1}$, as (Walsh, 1965; Zimmerman, 1991)

$$\frac{1}{K_{dry}} = \frac{1}{K_0} + \frac{\phi}{K_\phi}$$

where

$$\frac{1}{K_\phi} = \frac{1}{v_p}\frac{\partial v_p}{\partial \sigma}\bigg|_P$$

is the effective dry rock pore space compressibility, which is defined as the ratio of the fractional change in pore volume, v_p, to an increment of applied external hydrostatic stress, σ, at *constant pore pressure*. (See Section 2.6.) This is related to another pore compressibility, $K_{\phi P}{}^{-1}$, that is more familiar to reservoir engineers and hydrogeologists:

$$\frac{1}{K_{\phi P}} = -\frac{1}{v_p}\frac{\partial v_p}{\partial P}\bigg|_\sigma$$

which is the ratio of the fractional change in pore volume to an increment of applied pore pressure, at *constant confining pressure*, by (Zimmerman, 1991)

$$\frac{1}{K_{\phi p}} = \frac{1}{K_\phi} - \frac{1}{K_0}$$

Figure 6.3.2 shows a plot of normalized dry bulk modulus, K_{dry}/K_0, versus porosity computed for various values of normalized pore space stiffness, K_ϕ/K_0.

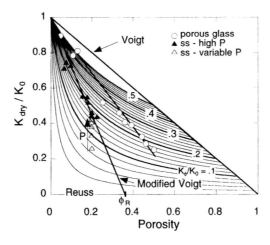

Figure 6.3.2

The data points plotted in Figure 6.3.2 are dynamic bulk moduli (calculated from ultrasonic velocities) for (1) ten clean sandstones, all at a pressure of 40 MPa (Han, 1986); (2) a single clean sandstone at pressures ranging from 5 to 40 MPa (Han, 1986); and (3) porous glass (Walsh, Brace, and England, 1965). Effective pore space compressibilities for each point can be read directly from the contours.

Similarly, the *saturated* rock compressibility, K_{sat}^{-1}, can be written as

$$\frac{1}{K_{\text{sat}}} = \frac{1}{K_0} + \frac{\phi}{\tilde{K}_\phi}$$

where

$$\tilde{K}_\phi = K_\phi + \frac{K_0 K_{\text{fl}}}{K_0 - K_{\text{fl}}} \approx K_\phi + K_{\text{fl}}$$

Here K_{fl} is the pore fluid bulk modulus, and K_ϕ is the same *dry* pore space stiffness defined above. The functional form for K_{sat} is exactly the same as for K_{dry}. The difference is only in the term $\tilde{K}_\phi$, which is equal to the dry pore stiffness, K_ϕ, incremented by a fluid term $F = (K_0 K_{\text{fl}})/(K_0 - K_{\text{fl}}) \approx K_{\text{fl}}$. Hence, the only effect of a change in fluid is a change in the modified pore stiffness $\tilde{K}_\phi$.

Therefore, fluid substitution, as predicted by Gassmann's equation, can be applied by computing this change, $\Delta \tilde{K}_\phi$, and then jumping the appropriate number of contours, as illustrated in Figure 6.3.3.

For the example shown, the starting point A was one of Han's (1986) data points for an effective dry rock bulk modulus, $K_{\text{dry}}/K_0 = 0.44$, and porosity, $\phi = 0.20$. Because the rock is dry, $F^{(1)}/K_0 \approx 0$. To saturate with water ($K_{\text{f}}/K_0 \approx 0.056$), we move up the amount $(F^{(2)} - F^{(1)})/K_0 = 0.06$, or 3 contours. The water-saturated modulus can be read off directly as $K_{\text{sat}}/K_0 = 0.52$, point A'. Obviously, the technique works equally well for going from dry to saturated, saturated to dry, or from one fluid to another. All we do is compute $\Delta F/K_0$ and count the contours.

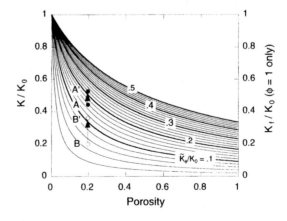

Figure 6.3.3

The second example shown (points B–B') is for the same two pore fluids and the same porosity. However, the change in rock stiffness during fluid substitution is much larger.

This illustrates the important point that a softer rock will have a larger sensitivity to fluid substitution than a stiffer rock at the same porosity. In fact, anything that causes the velocity input to Gassmann's equation to be higher (such as a stiffer rock, a measurement error, or velocities contaminated by velocity dispersion) will result in a smaller predicted sensitivity to changes in pore fluids. Similarly, lower input velocities will lead to a larger predicted sensitivity.

BIOT COEFFICIENT

The dry rock modulus can be written as

$$K_{\text{dry}} = K_0(1 - \beta)$$

where K_{dry} and K_0 are the bulk moduli of the dry rock and the mineral, and β is sometimes called the **Biot coefficient**, which is defined as the ratio of pore volume change Δv_p to bulk volume change ΔV at constant pore pressure:

$$\beta = \frac{\Delta v_p}{\Delta V}\bigg|_{\text{dry}} = \frac{\phi K_{\text{dry}}}{K_\phi} = 1 - \frac{K_{\text{dry}}}{K_0}$$

Then, Gassmann's equation can be expressed as

$$K_{\text{sat}} = K_{\text{dry}} + \beta^2 M$$

where

$$\frac{1}{M} = \frac{(\beta - \phi)}{K_0} + \frac{\phi}{K_{\text{fl}}}$$

V_P BUT NO V_S

In practice, fluid substitution is performed by starting with compressional and shear wave velocities measured on rocks saturated with the *initial* pore fluid (or gas) and then extracting the bulk and shear moduli. Then the bulk modulus of the rock saturated with the *new* pore fluid is calculated by using Gassmann's relation and the velocities are reconstructed.

A practical problem arises when we wish to estimate the change of V_P during fluid substitution but the shear velocity is unknown, which is almost always the case in situ. Then, strictly speaking, K cannot be extracted from V_P, and Gassmann's relations cannot be applied. To get around this problem, a common approach is to estimate V_S from an empirical V_S–V_P relation or to assume a dry rock Poisson's ratio (Castagna, Batzle, and Eastwood, 1985; Greenberg and Castagna, 1992). (See Section 7.8.)

Mavko, Chan and Mukerji (1995) have suggested an approximate method that operates directly on the P-wave modulus, $M = \rho V_P^2$. The method is equivalent to replacing the bulk moduli of the rock and mineral in Gassmann's relation with the corresponding P-wave moduli. For example,

$$\frac{M_{sat}}{M_0 - M_{sat}} \approx \frac{M_{dry}}{M_0 - M_{dry}} + \frac{M_{fl}}{\phi(M_0 - M_{fl})}$$

where M_{sat}, M_{dry}, M_0, and M_{fl} are the P-wave moduli of the saturated rock, the dry rock, the mineral, and the pore fluid, respectively. The approximate method performs the same operation with the P-wave modulus, M, as is done with the bulk modulus, K, in any of the various exact forms of Gassmann's relations listed above.

USES

Gassmann's relations are used to estimate the change of low-frequency elastic moduli of porous media caused by a change of pore fluids.

ASSUMPTIONS AND LIMITATIONS

The following considerations apply to the use of Gassmann's relations.

- **Low seismic frequencies** are assumed so that pore pressures are equilibrated throughout the pore space. In situ seismic conditions should generally be acceptable. Ultrasonic laboratory conditions will generally **not** be described well with Gassmann's equation. Sonic logging frequencies may or may not be within the range of validity, depending on the rock type and fluid viscosity.
- The rock is isotropic.
- All minerals making up the rock have the same bulk and shear moduli.
- Fluid-bearing rock is completely saturated.

EXTENSIONS

Gassmann's relations can be extended in the following ways:

- For mixed mineralogy, one can usually use an effective average modulus for K_0.
- For clay-filled rocks, it sometimes works best to consider the "soft" clay to be part of the pore-filling phase rather than part of the mineral matrix. Then the pore fluid is "mud," and its modulus can be estimated with an isostress calculation, as in the next item.

- For partially saturated rocks at sufficiently low frequencies, one can usually use an effective modulus for the pore fluid that is an isostress average of the moduli of the liquid and gaseous phases (see Section 6.12):

$$\frac{1}{K_{fl}} = \frac{S}{K_L} + \frac{1-S}{K_G}$$

where

K_L = the bulk modulus of the liquid phase
K_G = the bulk modulus of the gas phase
S = the saturation

- For anisotropic rocks use Brown and Korringa's extension of Gassmann's relations (see Section 6.5).
- Berryman and Milton have extended Gassmann's relation to include multiple porous constituents (see Section 6.6).

6.4 BAM – MARION'S BOUNDING AVERAGE METHOD

SYNOPSIS

Marion (1990) developed a heuristic method based on theoretical bounds for estimating the change in elastic moduli and velocities that result from substituting one pore-filling phase for another. The Hashin–Shtrikman (1963) bounds define the range of elastic moduli (velocities) possible for a given volume mix of two phases, either liquid or solid (see Figure 6.4.1). At any given volume fraction of constituents, the effective modulus will fall between the bounds (somewhere along the vertical dashed line in the top figure), but its precise value depends on the geometric details of the grains and pores. We use, for example, terms like "stiff pore shapes" and "soft pore shapes." Stiffer shapes cause the value to be higher within the allowable range; softer shapes cause the value to be lower.

Marion reasoned that the fractional vertical position within the bounds, $w = d/D$, where $0 \le w \le 1$, is therefore a measure of the pore geometry and is independent of the pore-filling properties – a reasonable assumption but one not proved. Because changing the pore-filling material does not change the geometry, w should remain constant, $w = d/D = d'/D'$, with any change in pore

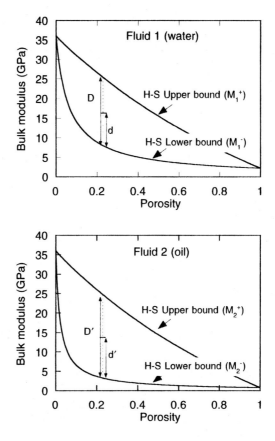

Figure 6.4.1

fluids. His method is as follows:

1) Begin with measurements of the modulus with the first pore-filling material (liquid, gas, or solid). Calculate the theoretical upper bound, M_1^+, and lower bound, M_1^-, corresponding to this state. (The subscript 1 refers to this first state with the first pore-filling material.) Plot the measured data value, M_1, and measure w relative to these bounds.

$$w = \frac{M_1 - M_1^-}{M_1^+ - M_1^-}$$

2) Recalculate the theoretical upper and lower bounds, M_2^+ and M_2^-, corresponding to the second pore-filling material of interest. Plot a point at the same position w relative to the new bounds. This is the new estimated modulus, M_2.

$$M_2 = M_2^- + w\left(M_2^+ - M_2^-\right)$$

Marion and others (Marion and Nur, 1991; Marion et al., 1992) showed that this method works quite well for several examples: predicting water-saturated rock velocities from dry rock velocities and predicting frozen rock (ice-filled) velocities from water-saturated velocities.

USES

Marion's bounding average method is applicable to the fluid substitution problem.

ASSUMPTIONS AND LIMITATIONS

Marion's bounding average method is primarily heuristic. Therefore, it needs to be tested empirically. A stronger theoretical basis would also be desirable. Nevertheless, it looks quite promising and extremely flexible.

EXTENSIONS

The simplicity of the method suggests that it be tried for comparing the effects of water-filled pores with clay-filled pores, altered clay grains versus the original crystalline grain, and so forth.

6.5 FLUID SUBSTITUTION IN ANISOTROPIC ROCKS: BROWN AND KORRINGA'S RELATIONS

SYNOPSIS

Brown and Korringa (1975) derived theoretical formulas relating the effective elastic moduli of an *anisotropic* dry rock to the effective moduli of the same rock containing fluid. This is simply the anisotropic version of Gassmann's relations. (Brown and Korringa also made a generalization to allow for a different pore compressibility relative to pore pressure rather than confining pressure. That generalization is not shown here.)

$$S_{ijkl}^{(\text{dry})} - S_{ijkl}^{(\text{sat})} = \frac{\left(S_{ij\alpha\alpha}^{(\text{dry})} - S_{ij\alpha\alpha}^{0}\right)\left(S_{kl\alpha\alpha}^{(\text{dry})} - S_{kl\alpha\alpha}^{0}\right)}{\left(S_{\alpha\alpha\beta\beta}^{(\text{dry})} - S_{\alpha\alpha\beta\beta}^{0}\right) + (\beta_{\text{fl}} - \beta_{0})\phi}$$

where

$S_{ijkl}^{(\text{dry})}$ = effective elastic compliance tensor of dry rock

$S_{ijkl}^{(\text{sat})}$ = effective elastic compliance tensor of rock saturated with pore fluid

S_{ijkl}^{0} = effective elastic compliance tensor of mineral material making up rock

β_{fl} = compressibility of pore fluid

β_{0} = compressibility of mineral material = $S_{\alpha\alpha\beta\beta}^{0}$

ϕ = porosity

USES

Brown and Korringa's relations are applicable to the fluid substitution problem in anisotropic rocks.

ASSUMPTIONS AND LIMITATIONS

The following considerations apply to the use of Brown and Korringa's relations:

- **Low seismic frequencies** (equilibrated pore pressures). In situ seismic frequencies generally should be acceptable. Ultrasonic laboratory conditions will generally **not** be described well with Brown and Korringa's equations. Sonic logging frequencies may or may not be within the range of validity, depending on the rock type and fluid viscosity.
- All minerals making up rock have the same moduli.
- Fluid-bearing rock is completely saturated.

EXTENSIONS

The following extensions of Brown and Korringa's relations can be made:

- For mixed mineralogy, one can usually use an effective average set of mineral compliances for S_{ijkl}^{0}.
- For clay-filled rocks, it often works best to consider the "soft" clay to be part of the pore-filling phase rather than part of the mineral matrix. Then, the pore fluid is "mud," and its modulus can be estimated with an isostress calculation, as in the next item.
- For partially saturated rocks at sufficiently low frequencies, one can usually use an effective modulus for the pore fluid that is an isostress average of the moduli of the liquid and gaseous phases:

$$\beta_{\text{fl}} = S\beta_{\text{L}} + (1 - S)\beta_{\text{G}}$$

where

β_L = the compressibility of the liquid phase
β_G = the compressibility of the gas phase
S = the saturation

6.6 GENERALIZED GASSMANN'S EQUATIONS FOR COMPOSITE POROUS MEDIA

SYNOPSIS

The generalized Gassmann's equation (Berryman and Milton, 1991) describes the static or low-frequency effective bulk modulus of a fluid-filled porous medium when the porous medium is a composite of two porous phases, each of which could be separately described by the more conventional Gassmann's relations. This is an improvement over the usual Gassmann's equation, which assumes that the porous medium is composed of a single, statistically homogeneous porous constituent with a single pore space stiffness and a single solid mineral. Like Gassmann's equation, the generalized Gassmann's formulation is completely independent of the pore geometry. The generalized formulation assumes that the two porous constituents are bonded at points of contact and fill all the space of the composite porous medium. Furthermore, like the Gassmann's formulation, it is assumed that the frequency is low enough that viscous and inertial effects are negligible and that any stress-induced increments of pore pressure are uniform within each constituent, although they could be different from one constituent to another (Berryman and Milton extend this to include dynamic poroelasticity, which is not presented here).

The pore microstructure within each phase is statistically homogeneous and much smaller than the size of the inclusions of each phase, which in turn are smaller than the size of the macroscopic sample. The inclusions of each porous phase are large enough to have effective dry frame bulk moduli, $K_{dry}^{(1)}$ and $K_{dry}^{(2)}$; porosities, $\phi^{(1)}$ and $\phi^{(2)}$; and solid mineral moduli, $K_0^{(1)}$ and $K_0^{(2)}$, respectively. The volume fractions of the two porous phases are $f^{(1)}$ and $f^{(2)}$, where $f^{(1)} + f^{(2)} = 1$. The generalized Gassmann's equation relates the effective saturated bulk modulus of the macroscopic sample, K_{sat}^*, to its dry frame bulk modulus, K_{dry}^*, through two

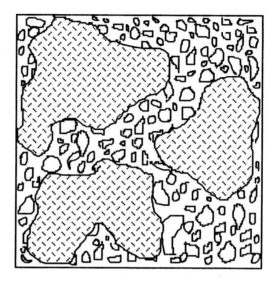

Figure 6.6.1. Composite porous medium with two porous phases.

other elastic constants defined by

$$\frac{1}{K_s^*} = -\frac{1}{V}\left(\frac{\partial V}{\partial p_f}\right)_{p_d=\text{const}}$$

and

$$\frac{1}{K_\phi^*} = -\frac{1}{V_\phi}\left(\frac{\partial V_\phi}{\partial p_f}\right)_{p_d=\text{const}}$$

where V is the total sample volume, V_ϕ is the total pore volume, p_f is the pore pressure, and $p_d = p - p_f$ is the differential pressure with p as the confining pressure. For a single-phase porous medium made up of a single solid mineral constituent with modulus K_0, the two moduli are equal to the mineral modulus: $K_s = K_\phi = K_0$. The relation between K_{sat}^* and K_{dry}^* for a composite porous medium is

$$K_{\text{sat}}^* = K_{\text{dry}}^* + \alpha^* C$$

$$C = \frac{\alpha^*}{\frac{\alpha^*}{K_s^*} + \phi\left(\frac{1}{K_f} - \frac{1}{K_\phi^*}\right)}$$

$$\alpha^* = 1 - \frac{K_{\text{dry}}^*}{K_s^*}$$

where K_f is the fluid bulk modulus. The constants K_s^* and K_ϕ^* can be expressed in terms of the moduli of the two porous constituents making up the composite medium. The key idea leading to the results is that whenever two scalar fields such as p_d and p_f can be varied independently in a linear composite with only two constituents, there exists a special value of the increment ratio $\delta p_d / \delta p$ that corresponds to an overall expansion or contraction of the medium without any relative shape change. This guarantees the existence of a set of consistency relations, allowing K_s^* and K_ϕ^* to be written in terms of the dry frame modulus and the constituent moduli. By linearity, the coefficients for the special value of $\delta p_d / \delta p$ are also the coefficients for any other arbitrary ratio. The relation for K_s^* is

$$\frac{1/K_0^{(1)} - 1/K_0^{(2)}}{1/K_{dry}^{(2)} - 1/K_{dry}^{(1)}} = \frac{1/K_0^{(1)} - 1/K_s^*}{1/K_{dry}^* - 1/K_{dry}^{(1)}} = \frac{1/K_s^* - 1/K_0^{(2)}}{1/K_{dry}^{(2)} - 1/K_{dry}^*}$$

or, equivalently, in terms of α^*

$$\frac{\alpha^* - \alpha^{(1)}}{\alpha^{(2)} - \alpha^{(1)}} = \frac{K_{dry}^* - K_{dry}^{(1)}}{K_{dry}^{(2)} - K_{dry}^{(1)}}$$

where $\alpha^{(1)} = 1 - K_{dry}^{(1)}/K_0^{(1)}$ and $\alpha^{(2)} = 1 - K_{dry}^{(2)}/K_0^{(2)}$. Other equivalent expressions are

$$\frac{1}{K_s^*} = \frac{1}{K_0^{(1)}} - \frac{1/K_0^{(1)} - 1/K_0^{(2)}}{1/K_{dry}^{(2)} - 1/K_{dry}^{(1)}} [1/K_{dry}^* - 1/K_{dry}^{(1)}]$$

and

$$\frac{1}{K_s^*} = \frac{1}{K_0^{(2)}} - \frac{1/K_0^{(1)} - 1/K_0^{(2)}}{1/K_{dry}^{(2)} - 1/K_{dry}^{(1)}} [1/K_{dry}^* - 1/K_{dry}^{(2)}]$$

The relation for K_ϕ^* is given by

$$\frac{\langle \phi \rangle}{K_\phi^*} = \frac{\alpha^*}{K_s^*} - \left\langle \frac{\alpha(x) - \phi(x)}{K_0(x)} \right\rangle - (\langle \alpha(x) \rangle - \alpha^*) \left(\frac{\alpha^{(1)} - \alpha^{(2)}}{K_{dry}^{(1)} - K_{dry}^{(2)}} \right)$$

where $\langle q(x) \rangle = f^{(1)} q^{(1)} + f^{(2)} q^{(2)}$ denotes the volume average of any quantity q. Gassmann's equation for a single porous medium is recovered correctly from the generalized equations when $K_0^{(1)} = K_0^{(2)} = K_0 = K_s^* = K_\phi^*$ and $K_{dry}^{(1)} = K_{dry}^{(2)} = K_{dry}^*$.

USES

The generalized Gassmann's equations can be used to calculate low-frequency saturated velocities from dry velocities for composite porous media made of two porous constituents. Examples include shaley patches embedded within a sand, microporous grains within a rock with macroporosity, or large nonporous inclusions within an otherwise porous rock.

ASSUMPTIONS AND LIMITATIONS

The preceding equations have the following assumptions:

- The rock is isotropic and made of up to two porous constituents.
- All minerals making up the rock are linear elastic.
- Fluid-bearing rock is completely saturated.
- The porosity in each phase is uniform, and the pore structure in each phase is smaller than the size of inclusions of each porous phase.
- The size of the porous inclusions is big enough to have a well-defined dry frame modulus but is much smaller than the wavelength and the macroscopic sample.

EXTENSIONS

For more than two porous constituents the composite may be modeled by dividing it up into regions of two phases. This approach is restrictive and not always possible.

6.7 MAVKO–JIZBA SQUIRT RELATIONS

SYNOPSIS

The *squirt* or *local flow* model suggests that the fluctuating stresses in a rock caused by the passing of a seismic wave induce pore pressure gradients at virtually all scales of pore space heterogeneity – particularly on the scale of individual grains and pores. These gradients impact the viscoelastic behavior of the rock; at high frequencies, when the gradients are unrelaxed, all elastic moduli (including the shear modulus) will be stiffer than at low frequencies, when the gradients are relaxed (the latter case modeled by Gassmann). Mavko and Jizba (1991) derived simple theoretical formulas for predicting the very high frequency moduli of saturated rocks in terms of the pressure dependence of dry rocks. The prediction is made in two steps: first, the squirt effect is incorporated as high-frequency "wet frame moduli" K_{uf} and μ_{uf}, which are derived from the normal dry moduli as

$$\frac{1}{K_{uf}} \approx \frac{1}{K_{dry-hiP}} + \left(\frac{1}{K_{fl}} - \frac{1}{K_0}\right)\phi_{soft}$$

$$\left(\frac{1}{\mu_{uf}} - \frac{1}{\mu_{dry}}\right) = \frac{4}{15}\left(\frac{1}{K_{uf}} - \frac{1}{K_{dry}}\right)$$

where

K_{uf} = effective high-frequency unrelaxed wet frame bulk modulus
K_{dry} = effective bulk modulus of dry rock
$K_{dry-hiP}$ = effective bulk modulus of dry rock at very high pressure
K_0 = bulk modulus of mineral material making up rock
K_{fl} = effective bulk modulus of pore fluid
ϕ_{soft} = soft porosity – the amount of porosity that closes at high pressure. This is often small enough to ignore.
μ_{uf} = effective high-frequency unrelaxed wet frame shear modulus
μ_{dry} = effective shear modulus of dry rock

Then these frame moduli are substituted into Gassmann's or Biot's relations to incorporate the remaining fluid saturation effects. For most crustal rocks the amount of squirt dispersion is comparable to or greater than Biot's dispersion, and thus using Biot's theory alone will lead to poor predictions of high-frequency saturated velocities. (Exceptions: very high permeability materials such as ocean sediments and glass beads; at very high effective pressure when most of the soft, crack-like porosity is closed; or near free boundaries such as borehole walls).

A more detailed analysis of the frequency dependence of the squirt mechanism is presented in Section 6.8.

Although the formulation presented here is independent of any idealized crack shape, the squirt behavior is also implicit in virtually all published formulations for the effective moduli based on elliptical cracks (see Sections 4.7–4.11). In most of those models, the cavities are treated as isolated with respect to flow, thus simulating the high-frequency limit of the squirt model.

USES

The Mavko–Jizba squirt relations can be used to calculate high-frequency saturated rock velocities from dry rock velocities.

ASSUMPTIONS AND LIMITATIONS

The use of the Mavko–Jizba squirt relations requires the following considerations:

- High seismic frequencies that are ideally suited for ultrasonic laboratory measurements are assumed. In situ seismic velocities generally will not have either squirt or Biot dispersion and should be described by using Gassmann's equations. Sonic logging frequencies may or may not be within the range of validity, depending on the rock type and fluid viscosity.
- The rock is isotropic.
- All minerals making up rock have the same bulk and shear moduli.
- Fluid-bearing rock is completely saturated.

EXTENSIONS

The Mavko–Jizba squirt relations can be extended in the following ways:

- For mixed mineralogy, one can usually use an effective average modulus for K_0.
- For clay-filled rocks, it often works best to consider the "soft" clay to be part of the pore-filling phase rather than part of the mineral matrix. Then the pore fluid is "mud," and its modulus can be estimated with an isostress calculation.
- The anisotropic form of these squirt relations has been found by Mukerji and Mavko (1994) and is discussed in Section 6.10.

6.8 EXTENSION OF MAVKO–JIZBA SQUIRT RELATIONS FOR ALL FREQUENCIES

SYNOPSIS

The Mavko and Jizba (1991) squirt relations (see Section 6.7) predict the very high frequency moduli of saturated rocks. At a low frequency, these moduli can be calculated from Gassmann's (1951) equations. Dvorkin, Mavko, and Nur (1995) introduced a model for calculating these moduli, velocities, and attenuations at any intermediate frequency. As input, the model uses such experimentally measurable parameters as the dry rock elastic properties at a given effective pressure, the dry rock bulk modulus at very high effective pressure, and the bulk moduli of the solid and fluid phases, rock density, and porosity. One additional parameter (Z), which determines the frequency scale of the dispersion, is proportional to the **characteristic squirt-flow length**. This parameter can be found by matching the theoretical velocity to that measured experimentally at a given frequency. Then the theory can be used to calculate velocities and attenuation at any frequency and with any pore fluid. The algorithm for calculating velocities and attenuation at a given frequency follows.

STEP 1: Calculate the bulk modulus of the dry modified solid (K_{msd}) from

$$\frac{1}{K_{msd}} = \frac{1}{K_0} - \frac{1}{K_{dry\text{-}hiP}} + \frac{1}{K_{dry}}$$

STEP 2: Calculate the ratio of the induced pore pressure increment to the confining stress increment ($dP/d\sigma$) from

$$\frac{dP}{d\sigma} = -\left[\alpha_0\left(1 + \frac{\phi K_{dry}}{\alpha_0^2 F_0}\right)\right]^{-1}$$

where

$$\frac{1}{F_0} = \frac{1}{K_{fl}} + \frac{1}{\phi Q_0}$$

$$\alpha_0 = 1 - \frac{K_{dry}}{K_0}$$

$$Q_0 = \frac{K_0}{\alpha_0 - \phi}$$

STEP 3: Assume a certain value for the frequency dispersion parameter Z (start with $Z = 0.001$) and calculate the bulk modulus of the saturated modified solid (K_{ms}) from

$$K_{ms} = \frac{K_{msd} + \alpha K_0[1 - f(\xi)]}{1 + \alpha f(\xi)\, dP/d\sigma}$$

where

$$\alpha = 1 - \frac{K_{msd}}{K_0}, \qquad f(\xi) = \frac{2J_1(\xi)}{\xi J_0(\xi)}, \qquad \xi = Z\sqrt{i\omega}$$

ω is angular frequency, and J_0 and J_1 are Bessel functions of zero and first order, respectively.

STEP 4: Calculate the bulk modulus of the modified frame (K_m) from

$$\frac{1}{K_m} = \frac{1}{K_{ms}} + \frac{1}{K_{dry\text{-}hiP}} - \frac{1}{K_0}$$

STEP 5: Calculate the bulk modulus of saturated rock (K_r) from

$$K_r = \frac{K_m}{1 + \alpha_m \, dP/d\sigma}$$

where

$$\alpha_m = 1 - \frac{K_m}{K_{ms}}$$

STEP 6: Calculate the shear modulus of the modified frame (μ_m) from

$$\frac{1}{\mu_{dry}} - \frac{1}{\mu_m} = \frac{4}{15}\left(\frac{1}{K_{dry}} - \frac{1}{K_{md}}\right)$$

where

$$\frac{1}{K_{md}} = \frac{1}{\tilde{K}_{ms}} + \frac{1}{K_{dry\text{-}hiP}} - \frac{1}{K_0}$$

$$\tilde{K}_{ms} = K_{msd} + \alpha K_0[1 - f(\xi)]$$

STEP 7: Finally, calculate velocities V_P and V_S and inverse quality factors Q_P^{-1} and Q_S^{-1} from

$$V_P = \sqrt{\frac{\mathrm{Re}\left(K_r + \frac{4}{3}\mu_m\right)}{\rho}}, \qquad V_S = \sqrt{\frac{\mathrm{Re}(\mu_m)}{\rho}}$$

$$Q_P^{-1} = \frac{\mathrm{Im}\left(K_r + \frac{4}{3}\mu_m\right)}{\mathrm{Re}\left(K_r + \frac{4}{3}\mu_m\right)}, \qquad Q_S^{-1} = \frac{\mathrm{Im}(\mu_m)}{\mathrm{Re}(\mu_m)}$$

STEP 8: The velocities and inverse quality factors have been found for an assumed Z value. To find the true Z value, one has to change it until the theoretical value of one of the four parameters (V_P, V_S, Q_P^{-1}, or Q_S^{-1}) matches the experimentally measured value at a given frequency. It is preferred that V_P be used for this purpose. The Z value thus obtained should be used for calculating the velocities and quality factors at varying frequencies. The Z value can also be used for a different pore fluid. In the latter case, use the following value for Z:

$$Z_{new} = Z\sqrt{\frac{\eta_{new}}{\eta}}$$

where subscript *new* indicates the new pore fluid.
   The notations are

$$\begin{aligned}
K_{dry} &= \text{effective bulk modulus of dry rock}\\
K_{dry\text{-}hiP} &= \text{effective bulk modulus of dry rock at very high pressure}\\
K_0 &= \text{bulk modulus of mineral material making up rock}\\
K_{fl} &= \text{effective bulk modulus of pore fluid}\\
\eta &= \text{viscosity of pore fluid}
\end{aligned}$$

ϕ = porosity

μ_{dry} = effective shear modulus of dry rock

ρ = rock density

ω = angular frequency

USES

The extension of the Mavko–Jizba squirt relations can be used to calculate saturated rock velocities and attenuation at any frequency.

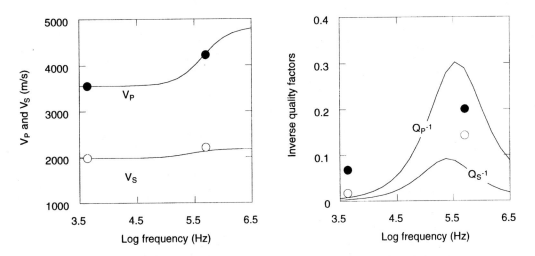

Figure 6.8.1. The V_S and inverse quality factors have been predicted by determining Z from V_P measurements (0.18 porosity limestone). Closed symbols are for P-waves; open symbols are for S-waves.

ASSUMPTIONS AND LIMITATIONS

The following assumptions underlie the extension of the Mavko–Jizba squirt relations:

- The rock is isotropic.
- All minerals making up the rock have the same bulk and shear moduli.
- Fluid-bearing rock is completely saturated.

EXTENSIONS

Additional extensions of the Mavko–Jizba squirt relations include the following:

- For mixed mineralogy, one can usually use an effective average modulus for K_0.
- Murphy, Winkler, and Kleinberg (1984) introduced a micromechanical model to describe the squirt-flow mechanism. They considered a composite grain contact stiffness, which is the parallel combination of the solid–solid contact stiffness and the stiffness of a fluid-filled gap.

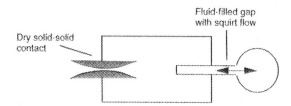

Figure 6.8.2

When modeling the solid–solid contact, surface energy is taken into account. The hydrodynamic contact model takes into account the squirt flow of pore fluid between a thin crack and a stiff, large pore. The results qualitatively match the observed velocity dispersion and attenuation in sandstones.

6.9 BISQ

SYNOPSIS

Dvorkin and Nur (1993) and Dvorkin, Nolen–Hoeksema, and Nur (1994) introduced a unified Biot–squirt (BISQ) model. The model is applicable to rocks at high pressure with compliant cracks closed. The rock is partially saturated or apparently fully saturated (meaning that there are small undetectable amounts of free gas left in pores). The zero-frequency velocity limit, as given by BISQ, is

$$V_{P0} = \sqrt{\frac{M_{dry}}{\rho}}$$

The M_{dry} term is the dry-rock uniaxial-strain modulus ($M_{dry} = \rho_{dry} V_{P\text{-}dry}{}^2$), ρ is rock density (at saturation), and $V_{P\text{-}dry}$ and ρ_{dry} are the dry-rock P-wave velocity and density, respectively. The BISQ high-frequency velocity limit is the same as in the Biot theory.

BISQ gives the following expressions for P-wave velocity (V_P), attenuation coefficient (a_P), and the inverse quality factor (Q_P^{-1}) at **apparently full saturation:**

$$V_P = \frac{1}{\mathrm{Re}(\sqrt{Y})}, \qquad a_P = \omega\,\mathrm{Im}(\sqrt{Y}), \qquad Q_P^{-1} = \frac{2a_P V_P}{\omega}$$

$$Y = -\frac{B}{2A} - \sqrt{\left(\frac{B}{2A}\right)^2 - \frac{C}{A}}, \qquad A = \frac{\phi F_{\mathrm{sq}} M_{\mathrm{dry}}}{\rho_2^2}$$

$$B = \left[F_{\mathrm{sq}}\left(2\gamma - \phi - \phi\frac{\rho_1}{\rho_2}\right) - \left(M_{\mathrm{dry}} + F_{\mathrm{sq}}\frac{\gamma^2}{\phi}\right)\left(1 + \frac{\rho_a}{\rho_2} + i\frac{\omega_c}{\omega}\right) \right] \Big/ \rho_2$$

$$C = \frac{\rho_1}{\rho_2} + \left(1 + \frac{\rho_1}{\rho_2}\right)\left(\frac{\rho_a}{\rho_2} + i\frac{\omega_c}{\omega}\right), \qquad F_{\mathrm{sq}} = F\left[1 - \frac{2 J_1(\lambda R)}{\lambda R J_0(\lambda R)}\right]$$

$$\lambda^2 = \frac{\rho_{\mathrm{fl}}\omega^2}{F}\left(\frac{\phi + \rho_a/\rho_{\mathrm{fl}}}{\phi} + i\frac{\omega_c}{\omega}\right), \qquad \rho_1 = (1 - \phi)\rho_s, \qquad \rho_2 = \phi\rho_{\mathrm{fl}}$$

$$\omega_c = \frac{\eta\phi}{k\rho_{\mathrm{fl}}}, \qquad \gamma = 1 - \frac{K_{\mathrm{dry}}}{K_0}, \qquad \frac{1}{F} = \frac{1}{K_{\mathrm{fl}}} + \frac{1}{\phi K_0}\left(1 - \phi - \frac{K_{\mathrm{dry}}}{K_0}\right)$$

where R is the **characteristic squirt-flow length**, ϕ is porosity, ρ_s and ρ_{fl} are the solid-phase and fluid-phase densities, respectively; $\rho_a = (1 - \alpha)\phi\rho_{\mathrm{fl}}$ is the Biot inertial-coupling density; η is the pore-fluid viscosity; k is rock permeability; K_{dry} is the dry rock bulk modulus; K_0 and K_{fl} are the solid-phase and the fluid-phase bulk moduli, respectively; ω is the angular frequency; and J_0 and J_1 are Bessel functions of zero and first order, respectively. The tortuosity α (sometimes called the structure factor) is a purely geometrical factor independent of the solid or fluid densities and is always greater than 1 (see Section 6.1).

All input parameters, except for the **characteristic squirt-flow length**, are experimentally measurable. The latter has to be either guessed (it should have the same order of magnitude as the average grain size or the average crack length) or adjusted by using an experimental measurement of velocity versus frequency (see Section 6.8).

For **partially-saturated** rock at saturation S,

$$R_s = R\sqrt{S}$$

has to be used instead of R in the preceding formulas. To avoid numerical problems (caused by resonance) at high frequencies,

$$\lambda^2 = i\frac{\rho_{\mathrm{fl}}\omega\omega_c}{F}$$

can be used instead of

$$\lambda^2 = \frac{\rho_{fl}\omega^2}{F}\left(\frac{\phi + \rho_a/\rho_{fl}}{\phi} + i\frac{\omega_c}{\omega}\right)$$

At lower frequencies ($\omega_c/\omega \gg 1$), the following simplified formulas can be used:

$$Y = \frac{(1-\phi)\rho_s + \phi\rho_{fl}}{M_{dry} + F_{sq}\gamma^2/\phi}, \qquad F_{sq} = F\left[1 - \frac{2J_1(\xi)}{\xi J_0(\xi)}\right]$$

$$\xi = \sqrt{i\frac{R^2\omega}{\kappa}}, \qquad \kappa = \frac{kF}{\eta\phi}$$

The BISQ formulas give the Biot theory expressions for the velocity and attenuation if $F_{sq} = F$.

USES

The BISQ formulas can be used to calculate partially saturated rock velocities and attenuation (at high pressure) at any frequency.

ASSUMPTIONS AND LIMITATIONS

The BISQ formulas are based on the following assumptions:

- The rock is isotropic.
- All minerals making up the rock have the same bulk and shear moduli.

EXTENSIONS

- For mixed mineralogy, one can usually use an effective average modulus for K_0.

6.10 ANISOTROPIC SQUIRT

SYNOPSIS

The *squirt* or *local flow* model suggests that the fluctuating stresses in a rock caused by the passing of a seismic wave induce pore-pressure gradients at virtually all scales of pore space heterogeneity – particularly on the scale of individual grains and pores. These gradients impact the viscoelastic behavior of the rock; at high frequencies, when the gradients are unrelaxed, all elastic moduli will be stiffer than at low frequencies, when the gradients are relaxed (the latter case is modeled by the Brown and Korringa relation, which is the anisotropic equivalent of Gassmann). Mukerji and Mavko (1994) derived simple theoretical formulas for predicting the very high frequency compliances of saturated anisotropic rocks in terms of the pressure dependence of dry rocks. The prediction is made in two steps: first, the squirt effect is incorporated as high-frequency "wet frame compliances" $S_{ijkl}^{(\text{wet})}$, which are derived from the dry compliances $S_{ijkl}^{(\text{dry})}$. Then these wet frame compliances are substituted into the Brown and Korringa (see Section 6.5) or Biot relations (see Section 6.1) (in place of the dry compliances) to incorporate the remaining fluid saturation effects. For most crustal rocks the amount of squirt dispersion is comparable with or greater than Biot's dispersion, and thus using Biot's theory alone will lead to poor predictions of high-frequency saturated velocities. Exceptions include very high permeability materials such as ocean sediments and glass beads, rocks at very high effective pressure, when most of the soft, crack-like porosity is closed, and rocks near free boundaries such as borehole walls.

The wet frame compliance is given by (repeated indices imply summation)

$$S_{ijkl}^{(\text{wet})} \approx S_{ijkl}^{(\text{dry})} - \frac{\Delta S_{\alpha\alpha\beta\beta}^{(\text{dry})}}{1 + \phi_{\text{soft}}(\beta_{\text{f}} - \beta_0)/\Delta S_{\alpha\alpha\beta\beta}^{(\text{dry})}} G_{ijkl}$$

where $\Delta S_{ijkl}^{(\text{dry})} = S_{ijkl}^{(\text{dry})} - S_{ijkl}^{(\text{dry high P})}$ is the change in dry compliance between the pressure of interest and very high confining pressure, ϕ_{soft} is the soft porosity that closes under high confining pressure, and β_{f} and β_0 are the fluid and mineral compressibilities, respectively. The soft porosity is often small enough to ignore the second term in the denominator. The tensor G_{ijkl} represents the fraction of the total compliance that is caused by volumetric deformation of crack-like pore space with different orientations for a given externally applied load. The tensor depends on the symmetry of the crack distribution function and is expressed as an integral over all orientations:

$$G_{ijkl} = \int f(\Omega) n_i n_j n_k n_l \, d\Omega$$

where $f(\Omega)$ is the normalized crack orientation distribution function so that its integral over all angles equals unity, and n_i is the unit normal to the crack faces. Elements of G_{ijkl} with any permutation of a given set $ijkl$ are equal. Note that G_{ijkl} has more symmetries than the elastic compliance tensor. For **isotropic** symmetry the elements of G_{ijkl} are given by the following table:

ij	11	22	33	23	13	12
kl						
11	1/5	1/15	1/15	0	0	0
22	1/15	1/5	1/15	0	0	0
33	1/15	1/15	1/5	0	0	0
23	0	0	0	1/15	0	0
13	0	0	0	0	1/15	0
12	0	0	0	0	0	1/15

This table gives exactly the same result as the isotropic equations of Mavko and Jizba (1991) presented in Section 6.7.

When the rock is **transversely isotropic** with the 3-axis as the axis of rotational symmetry, the five independent components of G_{ijkl} are

$$G_{1111} = \Delta \tilde{S}_{1111}^{(dry)} - \frac{4\alpha}{1-4\alpha}\left(\Delta \tilde{S}_{1122}^{(dry)} + \Delta \tilde{S}_{1133}^{(dry)}\right)$$

$$G_{1122} = \Delta \tilde{S}_{1122}^{(dry)}/(1-4\alpha)$$

$$G_{1133} = \Delta \tilde{S}_{1133}^{(dry)}/(1-4\alpha)$$

$$G_{3333} = \Delta \tilde{S}_{3333}^{(dry)} - \frac{8\alpha \Delta \tilde{S}_{1133}^{(dry)}}{1-4\alpha}$$

$$G_{2323} = \frac{\Delta \tilde{S}_{2323}^{(dry)}}{1-4\alpha} - \frac{\Delta \tilde{S}_{1111}^{(dry)} + \Delta \tilde{S}_{3333}^{(dry)}}{4(1-4\alpha)} + \frac{G_{1111} + G_{3333}}{4}$$

The nine independent components of G_{ijkl} for **orthorhombic** symmetry are

$$G_{1111} = \Delta \tilde{S}_{1111}^{(dry)} - \frac{4\alpha}{1-4\alpha}\left(\Delta \tilde{S}_{1122}^{(dry)} + \Delta \tilde{S}_{1133}^{(dry)}\right)$$

$$G_{2222} = \Delta \tilde{S}_{2222}^{(dry)} - \frac{4\alpha}{1-4\alpha}\left(\Delta \tilde{S}_{1122}^{(dry)} + \Delta \tilde{S}_{2233}^{(dry)}\right)$$

$$G_{3333} = \Delta \tilde{S}_{3333}^{(dry)} - \frac{4\alpha}{1-4\alpha}\left(\Delta \tilde{S}_{1133}^{(dry)} + \Delta \tilde{S}_{2233}^{(dry)}\right)$$

$$G_{1122} = \Delta \tilde{S}_{1122}^{(dry)}/(1-4\alpha)$$

$$G_{1133} = \Delta \tilde{S}_{1133}^{(dry)}/(1-4\alpha)$$

$$G_{2233} = \Delta \tilde{S}_{2233}^{(dry)}/(1-4\alpha)$$

$$G_{2323} = \frac{\Delta \tilde{S}_{2323}^{(dry)}}{1-4\alpha} - \frac{\Delta \tilde{S}_{2222}^{(dry)} + \Delta \tilde{S}_{3333}^{(dry)}}{4(1-4\alpha)} + \frac{G_{2222} + G_{3333}}{4}$$

$$G_{1313} = \frac{\Delta \tilde{S}_{1313}^{(dry)}}{1-4\alpha} - \frac{\Delta \tilde{S}_{1111}^{(dry)} + \Delta \tilde{S}_{3333}^{(dry)}}{4(1-4\alpha)} + \frac{G_{1111} + G_{3333}}{4}$$

$$G_{1212} = \frac{\Delta \tilde{S}_{1212}^{(dry)}}{1-4\alpha} - \frac{\Delta \tilde{S}_{1111}^{(dry)} + \Delta \tilde{S}_{2222}^{(dry)}}{4(1-4\alpha)} + \frac{G_{1111} + G_{2222}}{4}$$

where

$$\Delta \tilde{S}_{ijkl}^{(\text{dry})} = \frac{\Delta S_{ijkl}^{(\text{dry})}}{\Delta S_{\alpha\alpha\beta\beta}^{(\text{dry})}}$$

$$\alpha = \left(\Delta \tilde{S}_{\alpha\beta\alpha\beta}^{(\text{dry})} - 1\right)/4$$

Computed from the dry data, α is the ratio of the representative shear-to-normal compliance of a crack set, including all elastic interactions with other cracks. When the orthorhombic anisotropy is due to three mutually perpendicular crack sets superposed on a general orthorhombic background with the crack normals along the three symmetry axes, the wet frame compliances are obtained from

$$S_{ijkl}^{(\text{dry})} - S_{ijkl}^{(\text{wet})} \approx \frac{\Delta S_{1111}^{(\text{dry})}}{1 + \phi_{\text{soft}}^{(1)}(\beta_f - \beta_0)/\Delta S_{1111}^{(\text{dry})}}\delta_{i1}\delta_{j1}\delta_{k1}\delta_{l1}$$

$$+ \frac{\Delta S_{2222}^{(\text{dry})}}{1 + \phi_{\text{soft}}^{(2)}(\beta_f - \beta_0)/\Delta S_{2222}^{(\text{dry})}}\delta_{i2}\delta_{j2}\delta_{k2}\delta_{l2}$$

$$+ \frac{\Delta S_{3333}^{(\text{dry})}}{1 + \phi_{\text{soft}}^{(3)}(\beta_f - \beta_0)/\Delta S_{3333}^{(\text{dry})}}\delta_{i3}\delta_{j3}\delta_{k3}\delta_{l3}$$

where δ_{ij} is the Kronecker delta, and $\phi_{\text{soft}}^{(i)}$ refers to the soft porosity caused by the ith crack set. The above expressions assume that the intrinsic compliance tensor of planar crack-like features is sparse, the largest components being the normal and shear compliances, whereas the other components are approximately zero. This general property of planar crack formulations reflects an approximate decoupling of normal and shear deformation of the crack and decoupling of the in-plane and out-of-plane compressive deformation. In the case of a *single crack set* (with crack normal along the 3-axis) the wet frame compliances can be calculated from the dry compliances for a completely general, nonsparse crack compliance as

$$S_{ijkl}^{(\text{wet})} = S_{ijkl}^{(\text{dry})} - \frac{\Delta S_{\alpha\alpha ij}^{(\text{dry})}\Delta S_{\alpha\alpha kl}^{(\text{dry})}}{\Delta S_{\alpha\alpha\beta\beta}^{(\text{dry})} + \phi_{\text{soft}}(\beta_f - \beta_0)}$$

Little or no change of the $\Delta S_{1111}^{(\text{dry})}$ and $\Delta S_{2222}^{(\text{dry})}$ dry compliances with stress would indicate that all the soft, crack-like porosity is aligned normal to the 3-axis. However, a rotationally symmetric distribution of cracks may often be a better model of crack-induced transversely isotropic rocks than just a single set of aligned cracks. In this case the equations in terms of G_{ijkl} should be used.

The anisotropic squirt formulation presented here does not assume any idealized crack geometries. Because the high-frequency saturated compliances are predicted entirely in terms of the measured dry compliances, the formulation automatically incorporates all elastic pore interactions, and there is no limitation to low crack density.

Although the formulation presented here is independent of any idealized crack shape, the squirt behavior is also implicit in virtually all published formulations for the effective moduli based on elliptical cracks (see Sections 4.7–4.11). In most of those models, the cavities are treated as isolated with respect to flow, thus simulating the high-frequency limit of the squirt model.

USES

The anisotropic squirt formulation can be used to calculate high-frequency saturated rock velocities from dry rock velocities.

ASSUMPTIONS AND LIMITATIONS

The use of the anisotropic squirt formulation requires the following considerations:

- High seismic frequencies ideally suited for ultrasonic laboratory measurements are assumed. In situ seismic velocities will generally not have either squirt or Biot dispersion and should be described by using Brown and Korringa equations. Sonic logging frequencies may or may not be within the range of validity, depending on the rock type and fluid viscosity.
- All minerals making up rock have the same compliances.
- Fluid-bearing rock is completely saturated.

EXTENSIONS

The following extensions of the anisotropic squirt formulation can be made:

- For mixed mineralogy, one can usually use an effective average modulus for β_0.
- For clay-filled rocks, it often works best to consider the "soft" clay to be part of the pore-filling phase rather than part of the mineral matrix. The pore fluid is then "mud," and its modulus can be estimated with an isostress calculation.

6.11 COMMON FEATURES OF FLUID-RELATED VELOCITY DISPERSION MECHANISMS

SYNOPSIS

Many physical mechanisms have been proposed and modeled to explain velocity dispersion and attenuation in rocks: scattering (see Section 3.12), viscous and inertial fluid effects (see Sections 6.1, 6.2, and 6.7–6.13), hysteresis related to surface forces, thermoelastic effects, phase changes, and so forth. Scattering and surface forces appear to dominate in dry or nearly dry conditions (Tutuncu, 1992; Sharma and Tutuncu, 1994). Viscous fluid mechanisms dominate when there is more than a trace of pore fluids such as in the case of the poroelasticity described by Biot (1956) and the local flow or squirt mechanism (Mavko and Nur, 1975; O'Connell and Budiansky, 1977; Stoll and Bryan, 1970; Stoll, 1989; Dvorkin and Nur, 1993). Extensive reviews of these were given by Knopoff (1964), Mavko, Kjartansson, and Winkler (1979), and Bourbié et al. (1987), among others.

This section highlights some features that attenuation–dispersion models have in common. These suggest a simple approach to analyzing dispersion that bypasses some of the complexity of the individual theories, which is often not warranted by available data.

Although the various dispersion mechanisms and their mathematical descriptions are distinct, most can be described by the following three key parameters (see Figure 6.11.1):

1) A **low-frequency** limiting velocity V_0 (or modulus, M_0) often referred to as the "relaxed" state.
2) A **high-frequency** limiting velocity V_∞ (or modulus, M_∞) referred to as the "unrelaxed" state.
3) A **characteristic frequency**, f_c, that separates high-frequency behavior from low-frequency behavior and specifies the range in which velocity changes most rapidly.

HIGH- AND LOW-FREQUENCY LIMITS

Of the three key parameters, usually the low- and high-frequency limits can be estimated most easily. These require the fewest assumptions about the rock microgeometry and are, therefore, the most robust. In rocks, the velocity (or modulus) generally increases with frequency (though not necessarily monotonically in the

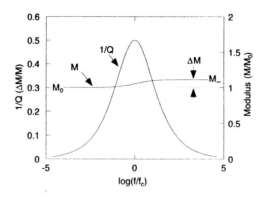

Figure 6.11.1

case of scattering), and thus $M_\infty > M_0$. The total amount of dispersion between very low frequency and very high frequency $(M_\infty - M_0)/M = \Delta M/M$ is referred to as the **Modulus defect** (Zener, 1948), where $M = \sqrt{M_0 M_\infty}$.

One of the first steps in analyzing any dispersion mechanism should be to estimate the modulus defect to see whether the effect is large enough to warrant any additional modeling. In most situations all but one or two mechanisms can be eliminated based on the size of the modulus defect alone.

As an example, consider the local flow or squirt mechanism, which is discussed in Sections 6.7–6.10 and 6.12. The squirt model recognizes that natural heterogeneities in pore stiffnesses, pore orientation, fluid compressibility, and saturation can cause spatial variations in wave-induced pore pressures at the pore scale. At sufficiently low frequencies, there is time for the fluid to flow and eliminate these variations; hence, the very low frequency limiting bulk and shear moduli can be predicted by Gassmann's theory (Section 6.3) or by Brown and Korringa's theory (Section 6.5) if the rock is anisotropic. At high frequencies, the pore pressure variations persist, causing the rock to be stiffer. The high-frequency limiting moduli may be estimated from dry rock data by using the Mavko and Jizba method (Section 6.7). These low- and high-frequency limits are relatively easy to estimate and require minimum assumptions about pore microgeometry. In contrast, calculating the detailed frequency variation between the low- and high-frequency regimes requires estimates of the pore aspect ratio or throat size distributions.

Other simple estimates of the high- and low-frequency limits associated with the squirt mechanism can be made by using ellipsoidal crack models such as the Kuster and Toksöz model (Section 4.7), the self-consistent model (Section 4.8), the DEM model (Section 4.9), or Hudson's model (Section 4.10). In each case, the *dry rock* is modeled by setting the inclusion moduli to zero. The *high-frequency saturated* rock conditions are simulated by assigning fluid moduli to the inclusions. Because each model treats the cavities as isolated with respect to flow, this yields the unrelaxed moduli for squirt. The *low-frequency saturated* moduli are found by taking the model-predicted effective moduli for dry cavities and saturating them with the Gassmann low-frequency relations (see Section 6.3).

The characteristic frequency is also simple to estimate for most models but usually depends more on poorly determined details of grain and pore microgeometry. Hence, the estimated critical frequency is usually less robust. Table 6.11.1 summarizes approaches to estimating the low-frequency moduli, high-frequency moduli, and characteristic frequencies for five important categories of velocity dispersion and attenuation.

Each mechanism shown in the table has an f_c value that depends on poorly determined parameters. The Biot (see Section 6.1) and patchy (see Section 6.12) models depend on the permeability. The squirt (see Sections 6.7–6.10) and viscous shear (Walsh, 1969) models require the crack aspect ratio. Furthermore, the formula for f_c–squirt is only a rough approximation, for it is not clear how reliable is the dependence on α^3. The patchy saturation (see Section 6.12) and scattering (see Sections 3.8–3.12) models depend on the scale of saturation and scattering heterogeneities. Unlike the other parameters, which are determined by grain and pore size, saturation and scattering can involve all scales, ranging from the pore scale to basin scale.

Figure 6.11.2 compares some values for f_c predicted by the various expressions in the table. The following parameters (or ranges of parameters) were used:

$$\phi = 0.2 = \text{porosity}$$

$$\rho_{fl} = 1.0 \text{ g/cm}^3 = \text{fluid density}$$

$$\eta = 1\text{–}10 \text{ cP} = \text{fluid viscosity}$$

$$\kappa = 1\text{–}1000 \text{ mD} = \text{permeability}$$

$$\alpha = 10^{-3}\text{–}10^{-4} = \text{crack aspect ratio}$$

$$V = 3,500 \text{ m/s} = \text{wave velocity}$$

$$a, L = 0.1\text{–}10 \text{ m} = \text{characteristic scale of heterogeneity}$$

$$\mu = 17 \text{ GPa} = \text{rock shear modulus}$$

$$K = 18 \text{ GPa} = \text{rock bulk modulus}$$

CAUTION: Other values for rock and fluid parameters can change f_c considerably.

COMPLETE FREQUENCY DEPENDENCE

Figure 6.11.3 compares the complete *normalized* velocity versus frequency dependence predicted by the Biot, patchy saturation, and scattering models. Although there are differences, each follows roughly the same trend from the low-frequency limits to the high-frequency limits, and the most rapid transition is in the range $f \approx f_c$. This is not always strictly true for the Biot mechanism, where certain combinations of parameters can cause the transition to occur at frequencies far from f_{biot}. All are qualitatively similar to the dispersion predicted by the standard

TABLE 6.11.1. High- and low-frequency limits and characteristic frequency of dispersion mechanisms.

Mechanism[a]	Low-frequency limit[a]	High-frequency limit[a]	Characteristic frequency (f_c)
Biot [6.1]	Gassmann's relations [6.3]	Biot's high-frequency formula [6.1]	$f_{biot} \approx \phi\eta/2\pi\rho_{fl}\kappa$
Squirt [6.7–6.10]			
[b]	Gassman's relations [6.3]	Mavko–Jizba relations [6.7]	$f_{squirt} \approx K_0\alpha^3/\eta$
[c]	Kuster–Toksöz dry relations [4.7] → Gassmann	Kuster–Toksöz saturated relations	,,
[c]	DEM dry relations [4.9] → Gassmann	DEM saturated relations	,,
[c]	Self-consistent dry relations [4.8] → Gassmann	Self-consistent saturated relations	,,
[d]	Hudson's dry relations [4.10] → Brown and Korringa [6.5]	Hudson's saturated relations	,,
Patchy saturation [6.11]			
[e]	Gassman's relations [6.3]	Hill equation [4.5]	$f_{patchy} \approx \kappa/L^2\eta(\beta_p + \beta_{fl})$
[f]	Gassman's relations [6.3]	White high-frequency formula [6.13]	$f_{patchy} \approx \kappa K_s/\pi L^2\eta$
[g]	Generalized Gassman's relations [6.6]	Dutta–Odé high-frequency formula [6.13]	,,
Viscous shear Scattering [3.8–3.12]	[h]	[h]	$f_{visc.crack} \approx \alpha\mu/2\pi\eta$
[i]	Effective medium theory [4.1–4.12]	Ray theory [3.9–3.11]	$f_{scatter} \approx V/2\pi a$
[j]	Backus average [4.12]	Time average [3.9–3.11]	,,

[a] Numbers in brackets [ ] refer to sections in this book.

[b] Inputs are measured dry rock moduli; no idealized pore geometry.

[c] Pore space is modeled as idealized ellipsoidal cracks.

[d] Anisotropic rock modeled as idealized penny-shaped cracks with preferred orientations.

[e] Dry rock is homogeneous; saturation has arbitrarily shaped patches.

[f] Dry rock is homogeneous; saturation is in spherical patches. Same limits as ([e]).

[g] Dry rock can be heterogeneous; saturation is in spherical patches.

[h] Mechanism modeled by Walsh (1969) is related to shearing of penny-shaped cracks with viscous fluids. Interesting only for extremely high viscosity or extremely high frequency.

[i] General heterogeneous three-dimensional medium.

[j] Normal incidence propagation through layered medium.

ϕ = porosity
K_0, μ = bulk and shear moduli of mineral
K_s = saturated rock modulus
β_{fl} = compressibility of the pore fluid
β_p = compressibility of the pore space
a = characteristic size (or correlation length) of scatterers
L = characteristic size (or correlation length) of saturation heterogeneity
ρ_{fl} = density of the pore fluid
η = viscosity of the pore fluid
α = pore aspect ratio
κ = rock permeability
V = wave velocity at f_c

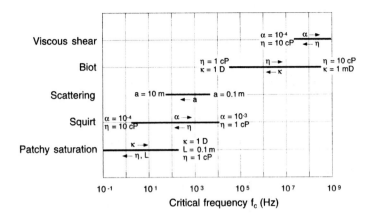

Figure 6.11.2 Comparison of characteristic frequencies for typical rock and fluid parameters. Arrows show the direction of change as the labeled parameter increases.

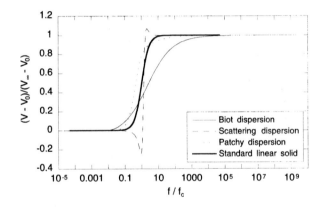

Figure 6.11.3

linear solid (see Section 3.6):

$$\text{Re}(M(\omega)) = \frac{M_0 M_\infty \left[1 + \left(\frac{f}{f_c}\right)^2\right]}{M_\infty + \left(\frac{f}{f_c}\right)^2 M_0}$$

Because the value of f_c and the resulting curves depend on poorly determined parameters, we suggest that a simple and practical way to estimate dispersion curves is simply to use the standard linear solid. The uncertainty in the rock microgeometry, permeability, and heterogeneous scales – as well as the approximations in the theories themselves – often makes a more detailed analysis unwarranted.

ATTENUATION

The Kramers–Kronig relations (see Section 3.7) completely specify the relation between velocity dispersion and attenuation. If velocity is known for all frequencies, then attenuation is determined for all frequencies. The attenuation versus frequency for the standard linear solid is given by

$$\frac{1}{Q} = \frac{M_\infty - M_0}{\sqrt{M_\infty M_0}} \frac{\left(\frac{f}{f_c}\right)}{1 + \left(\frac{f}{f_c}\right)^2}$$

Similarly, it can be argued (see Section 3.6) that, in general, the order of magnitude of attenuation can be determined from the modulus defect:

$$\frac{1}{Q} \approx \frac{M_\infty - M_0}{\sqrt{M_\infty M_0}}$$

USES

The preceding simplified relations can be used to estimate velocity dispersion and attenuation in rocks.

ASSUMPTIONS AND LIMITATIONS

This discussion is based on the premise that difficulties in measuring attenuation and velocity dispersion, and in estimating aspect ratios, permeability, and heterogeneous scales, often make detailed analysis of dispersion unwarranted. Fortunately, much about attenuation–dispersion behavior can be estimated robustly from the high- and low-frequency limits and the characteristic frequency of each physical mechanism.

6.12 PARTIAL AND MULTIPHASE SATURATIONS

SYNOPSIS

One of the most fundamental observations of rock physics is that seismic velocities are sensitive to pore fluids. The first-order low-frequency effects for *single fluid phases* are often described quite well with Gassmann's (1951) relations, which are discussed in Section 6.3. We summarize here variations on those fluid effects that result from *partial or mixed saturations*.

CAVEAT ON VERY DRY ROCKS

Figure 6.12.1 illustrates some key features of the saturation problem. The data are for limestones measured by Cadoret (1993) using the resonant bar technique at 1 kHz. The very dry rock velocity is approximately 2.84 km/s. Upon initial introduction of moisture (water), the velocity drops by about 4 percent. This apparent softening of the rock occurs at tiny volumes of pore fluid equivalent to a few monolayers of liquid if distributed uniformly over the internal surfaces of the pore space. These amounts are hardly sufficient for a fluid dynamic description, as in the Biot–Gassmann theories. Similar behavior has been reported in sandstones by Murphy (1982), Knight and Dvorkin (1992), and Tutuncu (1992).

This velocity drop has been attributed to softening of cements (sometimes called "chemical weakening"), to clay swelling, and to surface effects. In the latter model, very dry surfaces attract each other via cohesive forces, giving a mechanical effect resembling an increase in effective stress. Water or other pore fluids disrupt these forces. A fairly thorough treatment of the subject is found in the articles of Sharma and Tutuncu (Sharma and Tutuncu, 1994; Tutuncu, 1992; Tutuncu and Sharma, 1992; Sharma, Tutuncu, and Podia, 1994).

After the first few percent of saturation, additional fluid effects are primarily elastic and fluid dynamic and are amenable to analysis, for example, by the Biot–Gassmann (Sections 6.1–6.3) and Squirt (Sections 6.7–6.9) models.

Several authors (Cadoret, 1993; Murphy et al., 1991) have pointed out that classical fluid mechanical models such as the Biot–Gassmann theories perform poorly when the measured very dry rock values are used for the "dry rock" or "dry frame." They can be fairly accurate if the extrapolated "moist" rock modulus (see Figure 6.12.1) is used instead. For this reason, and to avoid the artifacts of ultradry rocks, it is often recommended to use samples that are at room conditions or that have been prepared in a constant humidity environment as "dry rock" data. For the rest of this section, it is assumed that the ultradry artifacts have been avoided.

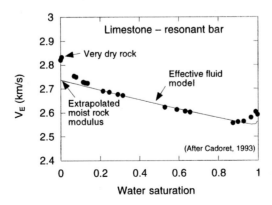

Figure 6.12.1

CAVEAT ON FREQUENCY

It is well known that the Gassmann theory (see Section 6.3) is valid only at sufficiently **low frequencies** so that the induced pore pressures are equilibrated throughout the pore space (i.e., that there is sufficient time for the pore fluid to flow and eliminate wave-induced pore pressure gradients). This limitation to low frequencies explains why Gassmann's relation works best for very-low-frequency in situ seismic data (<100 Hz) and may perform less well as frequencies increase toward sonic logging ($\approx 10^4$ Hz) and laboratory ultrasonic measurements ($\approx 10^6$ Hz). Knight and Nolen–Hoeksema (1990) studied the effects of pore scale fluid distributions on ultrasonic measurements.

EFFECTIVE FLUID MODEL

The most common approach to modeling partial saturation (air/water or gas/water) or mixed fluid saturations (gas/water/oil) is to replace the collection of phases with a single "effective fluid."

When a rock is stressed by a passing wave, pores are always elastically compressed more than the solid grains. This pore compression tends to induce increments of pore fluid pressure, which resist the compression; hence, pore phases with the largest bulk modulus K_{fl} stiffen the rock most. For single fluid phases the effect is described quite elegantly by Gassmann's (1951) relation (see Section 6.3):

$$\frac{K_{sat}}{K_0 - K_{sat}} = \frac{K_{dry}}{K_0 - K_{dry}} + \frac{K_{fl}}{\phi(K_0 - K_{fl})}, \qquad \mu_{sat} = \mu_{dry}$$

where

$$
\begin{aligned}
K_{dry} &= \text{effective bulk modulus of dry rock} \\
K_{sat} &= \text{effective bulk modulus of the rock with pore fluid} \\
K_0 &= \text{bulk modulus of mineral material making up rock} \\
K_{fl} &= \text{effective bulk modulus of pore fluid} \\
\phi &= \text{porosity} \\
\mu_{dry} &= \text{effective shear modulus of dry rock} \\
\mu_{sat} &= \text{effective shear modulus of rock with pore fluid}
\end{aligned}
$$

Implicit in Gassmann's relation is the stress-induced pore pressure

$$\frac{dP}{d\sigma} = \frac{1}{1 + K_\phi\left(\frac{1}{K_{fl}} - \frac{1}{K_0}\right)} = \frac{1}{1 + \phi\left(\frac{1}{K_{fl}} - \frac{1}{K_0}\right)\left(\frac{1}{K_{dry}} - \frac{1}{K_0}\right)^{-1}}$$

where P is the increment of pore pressure, and σ is the applied hydrostatic stress.

If there are multiple pore fluid phases with different fluid bulk moduli, then there is a tendency for each to have a different induced pore pressure. However, when the phases are intimately mixed at the finest scales, these pore pressure increments can equilibrate with each other to a single average value. This is an *isostress* situation, and therefore the effective bulk modulus of the mixture of fluids is described well by the Reuss average (see Section 4.2) as follows:

$$\frac{1}{K_{fl}} = \sum \frac{S_i}{K_i}$$

where K_{fl} is the effective bulk modulus of the fluid mixture, K_i denotes the bulk moduli of the individual gas and fluid phases, and S_i represents their saturations. The rock moduli can often be predicted quite accurately by inserting this effective fluid modulus into Gassmann's relation. The effective fluid (air + water) prediction is superimposed on the data in Figure 6.12.1 and reproduces the overall trend quite well. This approach has been discussed, for example, by Domenico (1976), Murphy (1984), Mavko and Nolen–Hoeksema (1994), Cadoret (1993), and many others.

CAUTION: It is thought that the effective fluid model is valid only when all of the fluid phases are mixed at the finest scale.

CRITICAL RELAXATION SCALE

A critical assumption in the effective fluid model represented by the Reuss average is that differences in wave-induced pore pressure have time to flow and equilibrate among the various phases. As discussed in Section 8.1, the characteristic relaxation time or diffusion time for heterogeneous pore pressures of scale L is

$$\tau \approx \frac{L^2}{D}$$

where $D = kK_{fl}/\eta$ is the diffusivity, k is the permeability, K_{fl} is the fluid bulk modulus, and η is the viscosity. Therefore, at a seismic frequency $f = 1/\tau$, pore pressure heterogeneities caused by saturation heterogeneities will have time to relax and reach a local isostress state over scales smaller than

$$L_c \approx \sqrt{\tau D} = \sqrt{D/f}$$

and will be described *locally* by the effective fluid model mentioned in the previous discussion. Spatial fluctuations on scales larger than L_c will tend to persist and will not be described well by the effective fluid model.

PATCHY SATURATION

Consider the situation of a homogeneous rock type with spatially variable saturation $S_i(x, y, z)$. Each "patch" or pixel at scale $\approx L_c$ will have fluid phases equilibrated within the patch at scales smaller than L_c, but neighboring patches at scales
$> L_c$ will not be equilibrated with each other. Each patch will have a different effective fluid described approximately by the Reuss average. Consequently, the rock in
each patch will have a different bulk modulus describable locally with Gassmann's
relations. Yet, the shear modulus will remain unchanged and spatially uniform.

The effective moduli of the rock with spatially varying bulk modulus but uniform shear modulus is described exactly by the equation of Hill (1963) discussed
in Section 4.5:

$$K_{\text{eff}} = \left\langle \frac{1}{K + \frac{4}{3}\mu} \right\rangle^{-1} - \frac{4}{3}\mu$$

This striking result states that the effective moduli of a composite with uniform
shear modulus can be found *exactly* by knowing only the volume fractions of the
constituents independent of the constituent geometries. There is no dependence,
for example, on ellipsoids, spheres, or other idealized shapes.

Figure 6.12.2 shows the effective bulk modulus versus water saturation for
another limestone (data from Cadoret, 1993). Unlike the effective fluid behavior,
which shows a small decrease in velocity with increasing saturation and then an
abrupt increase as S_W approaches unity, the patchy model predicts a monotonic,
almost linear increase in velocity from the dry to saturated values. The deviation
of the data from the effective fluid curve at saturations greater than 0.8 is an
indication of patchy saturation (Cadoret, 1993).

The velocity versus saturation curves shown in Figure 6.12.2 for the effective
fluid model and patchy saturation model represent approximate lower and upper
bounds, respectively, at low frequencies. The lower effective fluid curve is achieved
when the fluid phases are mixed at the finest scales. The upper patchy saturation

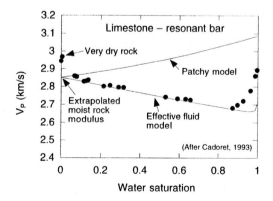

Figure 6.12.2

curve is achieved when there is the greatest separation of phases: when each patch of size $>L_c$ has only a single phase. Any velocity between these "bounds" can occur when there is a range of saturation scales.

SATURATION-RELATED VELOCITY DISPERSION

The difference between effective fluid behavior and patchy saturation behavior is largely a matter of the scales at which the various phases are mixed. The critical relaxation scale L_c, which separates the two domains, is related to the seismic frequency. Hence, spatially varying saturations can be a source of velocity dispersion. Attempts to quantify this velocity dispersion have been made by White (1975) and by Dutta and Odé (1979). See Sections 6.11 and 6.13.

VOIGT AVERAGE APPROXIMATION

It can be shown that an approximation to the patchy saturation "upper bound" can be found by first computing the Voigt average (see Section 4.2) of the fluid modulus:

$$K_{fl} = \sum S_i K_i$$

where K_{fl} is the effective bulk modulus of the fluid mixture, K_i denotes the bulk moduli of the individual gas and fluid phases, and S_i represents their saturations. Next, this average fluid is put into Gassmann's equations to predict the overall rock moduli. This is in contrast to the effective fluid model discussed earlier in which the Reuss average of the fluid moduli was put into the Gassmann equations.

6.13 PARTIAL SATURATION: WHITE AND DUTTA–ODÉ MODEL FOR VELOCITY DISPERSION AND ATTENUATION

SYNOPSIS

Consider the situation of a reservoir rock with spatially variable saturation $S_i(x, y, z)$. Each "patch" at scale $\approx L_c$ (where $L_c \approx \sqrt{\tau D} = \sqrt{D/f}$ is the critical fluid diffusion relaxation scale, $D = kK_{fl}/\eta$ is the diffusivity, k is the permeability, K_{fl} is the fluid bulk modulus, and η is the viscosity) will have fluid phases equilibrated within the patch at scales smaller than L_c, but neighboring patches at scales $>L_c$ will not be equilibrated with each other. Fluid flow resulting from

unequilibrated pore pressures between patches of different fluids will cause attenuation and dispersion of seismic waves traveling through the rock. White (1975)
modeled the seismic effects of patchy saturation by considering porous rocks saturated with brine but containing spherical gas-filled regions. Dutta and Odé (1979)
gave a more rigorous solution for White's model by using Biot's equations for
poroelasticity. The patches (spheres) of heterogeneous saturation are much larger
than the grain scale but are much smaller than the wavelength. The idealized geometry of a unit cell consists of a gas-filled sphere of radius a placed at the center
of a brine-saturated spherical shell of outer radius b ($b > a$). Adjacent unit cells
do not interact with each other. The gas saturation is $S_g = a^3/b^3$. The inner region
will be denoted by subscript 1, and the outer shell by subscript 2. In the more rigorous Dutta and Odé formulation, the dry frame properties (denoted by subscript
"dry" in the following equations) in the two regions may be different. However,
in White's approximate formulation the dry frame properties are assumed to be
the same in regions 1 and 2. White's equations as given below (incorporating a
correction pointed out by Dutta and Seriff, 1979) yield results that agree very
well with the Dutta–Odé results. The complex bulk modulus, K^*, for the partially
saturated porous rock as a function of angular frequency ω is given by

$$K^* = [K_\infty/(1 - K_\infty W)] = K_r^* + i K_i^*$$

where

$$W = \frac{3a^2(R_1 - R_2)(-Q_1 + Q_2)}{b^3 i\omega(Z_1 + Z_2)}$$

$$R_1 = \frac{K_1 - K_{dry_1}}{1 - K_{dry_1}/K_{0_1}} \frac{3K_2 + 4\mu_2}{K_2(3K_1 + 4\mu_2) + 4\mu_2(K_1 - K_2)S_g}$$

$$R_2 = \frac{K_2 - K_{dry_2}}{1 - K_{dry_2}/K_{0_2}} \frac{3K_1 + 4\mu_1}{K_2(3K_1 + 4\mu_2) + 4\mu_2(K_1 - K_2)S_g}$$

$$Z_1 = \frac{\eta_1 a}{\kappa_1}\left[\frac{(1 - e^{-2\alpha_1 a})}{(\alpha_1 a - 1) + (\alpha_1 a + 1)e^{-2\alpha_1 a}}\right]$$

$$Z_2 = -\frac{\eta_2 a}{\kappa_2}\left[\frac{(\alpha_2 b + 1) + (\alpha_2 b - 1)e^{2\alpha_2(b-a)}}{(\alpha_2 b + 1)(\alpha_2 a - 1) - (\alpha_2 b - 1)(\alpha_2 a + 1)e^{2\alpha_2(b-a)}}\right]$$

$$\alpha_j = (i\omega\eta_j/\kappa_j K_{Ej})^{1/2}$$

$$K_{Ej} = \left[1 - \frac{K_{fj}(1 - K_j/K_{0j})(1 - K_{dry j}/K_{0j})}{\phi K_j(1 - K_{fj}/K_{0j})}\right]K_{Aj}$$

$$K_{Aj} = \left(\frac{\phi}{K_{fj}} + \frac{1 - \phi}{K_{0j}} - \frac{K_{dry j}}{K_{0j}^2}\right)^{-1}$$

$$Q_j = \frac{(1 - K_{dry j}/K_{0j})K_{Aj}}{K_j}$$

Here, $j = 1$ or 2 denotes quantities corresponding to the two different regions; η and K_f are the fluid viscosity and bulk modulus, respectively; ϕ is the porosity; κ is the permeability; and K_0 is the bulk modulus of the solid mineral grains. The saturated bulk and shear moduli, K_j and μ_j, respectively, of region j are obtained from Gassmann's equation using $K_{\text{dry}j}$, $\mu_{\text{dry}j}$, and K_{fj}. When the dry frame moduli are the same in both regions, $\mu_1 = \mu_2 = \mu_{\text{dry}}$ because Gassmann predicts no change in the shear modulus upon fluid substitution. At the high-frequency limit, when there is no fluid flow across the fluid interface between regions 1 and 2, the bulk modulus is given by

$$K_\infty(\text{no-flow}) = \frac{K_2(3K_1 + 4\mu_2) + 4\mu_2(K_1 - K_2)S_g}{(3K_1 + 4\mu_2) - 3(K_1 - K_2)S_g}$$

This assumes that the dry frame properties are the same in the two regions. The low-frequency limiting bulk modulus is given by Gassmann's relation with an effective fluid modulus equal to the Reuss average of the fluid moduli. In this limit the fluid pressure is constant and uniform throughout the medium.

$$K(\text{low frequency}) = \frac{K_2(K_1 - K_{\text{dry}}) + S_g K_{\text{dry}}(K_2 - K_1)}{(K_1 - K_{\text{dry}}) + S_g(K_2 - K_1)}$$

Note that in White (1975) the expressions for K_E and Q had the P-wave modulus $M_j = K_j + 4\mu_j/3$ in the denominator instead of K_j. As pointed out by Dutta and Seriff (1979), this form does not give the right low-frequency limit. An estimate of the transition frequency separating the relaxed and unrelaxed (no-flow) states is given by

$$f_c \approx \frac{\kappa_2 K_{E2}}{\pi \eta_2 b^2}$$

which has the length-squared dependence characteristic of diffusive phenomena. When the central sphere is saturated with a very compressible gas (caution: this may not hold at reservoir pressures) $R_1, Q_1, Z_1 \approx 0$. The expression for the effective complex bulk modulus then reduces to

$$K^* = \frac{K_\infty}{1 + 3a^2 R_2 Q_2 K_\infty / b^3 i\omega Z_2}$$

Dutta and Odé obtained more rigorous solutions for the same spherical geometry by solving a boundary value problem involving Biot's poroelastic field equations. They considered steady-state, time-harmonic solutions for u and w, the displacement of the porous frame and the displacement of the pore fluid relative to the frame, respectively. The solutions in the two regions are given in terms of spherical Bessel and Neumann functions, j_1 and n_1, of order 1. The general solution for w in region 1 for purely radial motion is

$$w(r) = w_c(r) + w_d(r)$$

$$w_c(r) = C_1 j_1(k_c r) + C_2 n_1(k_c r)$$

$$w_d(r) = C_3 j_1(k_d r) + C_4 n_1(k_d r)$$

where C_1, C_2, C_3, C_4 are integration constants to be determined from the boundary conditions at $r = 0, r = a$, and $r = b$. A similar solution holds for region 2 but with integration constants C_5, C_6, C_7, C_8. The wavenumbers k_c and k_d in each region are given in terms of the moduli and density corresponding to that region:

$$\frac{k_c^2}{\omega^2} = \frac{\rho_f(M\sigma_c + 2\gamma D) - \rho(2\gamma D\sigma_c + 2D)}{(4\gamma^2 D^2 - 2DM)}$$

$$\frac{k_d^2}{\omega^2} = \frac{\rho_f(M\sigma_d + 2\gamma D) - \rho(2\gamma D\sigma_d + 2D)}{(4\gamma^2 D^2 - 2DM)}$$

where

$$\rho = (1 - \phi)\rho_0 + \phi\rho_f$$

$$\gamma = 1 - \frac{K_{dry}}{K_0}$$

$$M = K + \frac{4}{3}\mu$$

$$D = \frac{K_0}{2}\left[\gamma + \frac{\phi}{K_f}(K_0 - K_f)\right]^{-1}$$

and σ_c and σ_d are the two complex roots of the quadratic equation

$$(\rho_f M - 2\rho\gamma D)\sigma^2 + \left(Mm - 2\rho D - \frac{i\eta M}{\kappa\omega}\right)\sigma + \left(2m\gamma D - 2\rho_f D - 2i\frac{\eta\gamma D}{\kappa\omega}\right) = 0$$

where $m = s\rho_f/\phi$. The tortuosity parameter s (sometimes called the structure factor) is a purely geometrical factor independent of the solid or fluid densities and is never less than 1 (see Section 6.1 on Biot theory). For idealized geometries and uniform flow, s usually lies between 1 and 5. Berryman (1981) obtained the relation:

$$s = 1 - \xi(1 - 1/\phi)$$

where $\xi = 1/2$ for spheres and lies between 0 and 1 for other ellipsoids. For uniform cylindrical pores with axes parallel to the pore pressure gradient, s equals 1 (the minimum possible value), whereas for a random system of pores with all possible orientations, $s = 3$ (Stoll, 1977). The solutions for u are given as

$$u(r) = u_c(r) + u_d(r)$$

$$u_c(r) = \sigma_d w_c(r); \quad u_d(r) = \sigma_c w_d(r)$$

The integration constants are obtained from the following boundary conditions:

(1) $r \to 0, w_1(r) \to 0,$

(2) $r \to 0, u_1(r) \to 0,$

(3) $r = a, u_1(r) = u_2(r),$

(4) $r = a, w_1(r) = w_2(r),$

(5) $r = a, \tau_1(r) = \tau_2(r),$

(6) $r = a, p_1(r) = p_2(r),$

(7) $r = b, \tau_2(r) = -\tau_0,$

(8) $r = b, w_2(r) = 0$

The first two conditions imply no displacements at the center of the sphere because of purely radial flow and finite fluid pressure at the origin. These require $C_2 = C_4 = 0$. Conditions (3)–(6) come from continuity of displacements and stresses at the interface. The bulk radial stress is denoted by τ, and p denotes the fluid pressure. These parameters are obtained from u and w by

$$\tau = M\frac{\partial u}{\partial r} + 2(M - 2\mu)\frac{u}{r} + 2\gamma D\left(\frac{\partial w}{\partial r} + \frac{2}{r}w\right)$$

$$p = -2\gamma D\left(\frac{\partial u}{\partial r} + \frac{2}{r}u\right) - 2D\left(\frac{\partial w}{\partial r} + \frac{2}{r}w\right)$$

Condition (7) gives the amplitude τ_0 of the applied stress at the outer boundary, and condition (8) implies that the outer boundary is jacketed. These boundary conditions are not unique and could be replaced by others if they happen to be appropriate for the situation under consideration. The jacketed outer boundary is consistent with noninteracting unit cells. The remaining six integration constants are obtained by solving the linear system of equation given by the boundary conditions. Solving the linear system requires considerable care (Dutta and Odé, 1979). The equations may become ill-conditioned because of the wide range of the arguments of the spherical Bessel and Neumann functions. Once the complete solution for $u(r)$ is obtained, the effective complex bulk modulus for the partially saturated medium is given by

$$K^*(\omega) = -\frac{\tau_0}{\Delta V / V} = -\frac{\tau_0 b}{3u(b)}$$

The P-wave velocity V_P^* and attenuation coefficient α_P^* are given in terms of the complex P-wave modulus M^* and the effective density ρ^*

$$\rho^* = S_g[(1 - \phi)\rho_0 + \phi\rho_{f1}] + (1 - S_g)[(1 - \phi)\rho_0 + \phi\rho_{f2}]$$
$$M^* = (M_r^* + iM_i^*) = (K_r^* + 4\mu^*/3 + iK_i)$$
$$\mu^* = \mu_1 = \mu_2 = \mu_{dry}$$
$$V_P^* = (|M^*|/\rho^*)^{1/2}/\cos(\theta_P^*/2)$$
$$\alpha_P^* = \omega \tan(\theta_P^*/2)/V_P^*$$
$$\theta_P^* = \tan^{-1}(M_i^*/M_r^*)$$

USES

The White and Dutta–Odé models can be used to calculate velocity dispersion and attenuation in porous media with patchy partial saturation.

ASSUMPTIONS AND LIMITATIONS

The White and Dutta–Odé models are based on the following assumptions:

- The rock is isotropic.
- All minerals making up the rock are isotropic, linear elastic.
- The patchy saturation has an idealized geometry consisting of a sphere saturated with one fluid within a spherical shell saturated with another fluid.
- The porosity in each saturated region is uniform, and the pore structure is smaller than the size of the spheres.
- The patches (spheres) of heterogeneous saturation are much larger than the grain scale but are much smaller than the wavelength.

6.14 WAVES IN PURE VISCOUS FLUID

SYNOPSIS

Acoustic waves in pure viscous fluid are dispersive and attenuate. This occurs owing to the shear component in the wave-induced deformation of an elementary fluid volume. The linearized wave equation in a viscous fluid can be derived from the Navier–Stokes equation (Schlichting, 1951) as

$$\rho\frac{\partial^2 u}{\partial t^2} = c_0^2\frac{\partial^2 u}{\partial x^2} + \frac{4}{3}\eta\frac{\partial^3 u}{\partial x^2 \partial t}, \qquad c_0^2 = \frac{K_f}{\rho}$$

Then the wave-number-to-angular-frequency ratio is

$$\frac{k}{\omega} = \sqrt{\frac{c_0^2 + i\gamma\omega}{c_0^4 + \gamma^2\omega^2}} = \frac{e^{i\frac{\arctan z}{2}}}{c_0\sqrt[4]{1 + z^2}}$$

$$= \frac{1}{c_0\sqrt[4]{1 + z^2}}\left(\sqrt{\frac{\sqrt{1 + z^2} + 1}{2\sqrt{1 + z^2}}} + i\sqrt{\frac{\sqrt{1 + z^2} - 1}{2\sqrt{1 + z^2}}}\right)$$

$$\gamma = \frac{4\eta}{3\rho}, \qquad z = \frac{\gamma\omega}{c_0^2}$$

and the phase velocity, attenuation coefficient, and inverse quality factor are

$$V = c_0\sqrt{\frac{2(1 + z^2)}{\sqrt{1 + z^2} + 1}}$$

$$a = \frac{\omega}{c_0}\sqrt{\frac{\sqrt{1 + z^2} - 1}{2(1 + z^2)}}$$

$$Q^{-1} = 2\sqrt{\frac{\sqrt{1 + z^2} - 1}{\sqrt{1 + z^2} + 1}}$$

If $z \ll 1$ these expressions can be simplified to

$$V = c_0\left[1 + 2\left(\frac{4\pi\eta f}{3\rho c_0^2}\right)^2\right], \qquad a = \frac{8\pi^2\eta f^2}{3\rho c_0^3}, \qquad Q^{-1} = \frac{8\pi\eta f}{3\rho c_0^2}$$

where

$$u = \text{displacement}$$
$$\rho = \text{density}$$
$$\eta = \text{viscosity}$$
$$K_f = \text{bulk modulus of the fluid}$$
$$x = \text{coordinate}$$
$$t = \text{time}$$
$$k = \text{wave-number}$$
$$\omega = \text{angular frequency}$$
$$V = \text{phase velocity}$$
$$a = \text{attenuation coefficient}$$
$$Q = \text{quality factor}$$
$$f = \text{frequency}$$

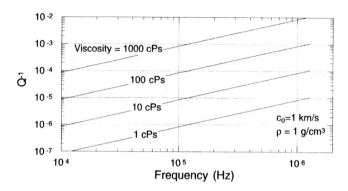

Figure 6.14.1

Attenuation is very small at low frequency and for low-viscosity fluid; however, it may become noticeable in high-viscosity fluids.

USES

The equations presented in this section can be used for estimating the frequency dependence of acoustic velocity in viscous fluids.

ASSUMPTIONS AND LIMITATIONS

The preceding equations assume that the fluid is Newtonian. Many high-viscosity oils are non-Newtonian (i.e., their flow cannot be accurately described by the Navier–Stokes equation).

6.15 PHYSICAL PROPERTIES OF GASES AND FLUIDS

SYNOPSIS

The **bulk modulus** (K) of a fluid or a gas is defined as

$$K = \frac{1}{\beta} = -\frac{dP}{(dV/V)} = \rho \frac{dP}{d\rho}$$

where β is compressibility, P is pressure, and V is volume. For small pressure variations (typical for wave propagation), the pressure variation is related to the density variation through the acoustic velocity c_0 (which is 1500 m/s for water at room conditions) as follows:

$$dP = c_0{}^2 \, d\rho.$$

Therefore,

$$K = \rho c_0{}^2.$$

Batzle and Wang (1992) have summarized some important properties of reservoir fluids.

BRINE

The **density of brine** ρ_B of salinity S of sodium chloride is

$$\rho_B = \rho_w + S\{0.668 + 0.44S + 10^{-6}[300P - 2400PS$$
$$+ T(80 + 3T - 3300S - 13P + 47PS)]\}$$

where the **density of pure water** (ρ_w) is

$$\rho_w = 1 + 10^{-6}(-80T - 3.3T^2 + 0.00175T^3$$
$$+ 489P - 2TP + 0.016T^2P - 1.3 \times 10^{-5}T^3P$$
$$- 0.333P^2 - 0.002TP^2)$$

In these formulas pressure P is in MPa, temperature T is in degrees Celsius, salinity S is in fractions of one (parts per million divided by 10^6), and density (ρ_B and ρ_w) is in g/cm^3.

The **acoustic velocity in brine** V_B in m/s is

$$V_B = V_w + S(1170 - 9.6T + 0.055T^2 - 8.5 \times 10^{-5}T^3$$
$$+ 2.6P - 0.0029TP - 0.0476P^2)$$
$$+ S^{1.5}(780 - 10P + 0.16P^2) - 1820S^2$$

where the **acoustic velocity in pure water** V_w in m/s is

$$V_w = \sum_{i=0}^{4} \sum_{j=0}^{3} w_{ij} T^i P^j$$

and coefficients w_{ij} are

$$
\begin{aligned}
&w_{00} = 1402.85 &\quad &w_{02} = 3.437 \times 10^{-3} \\
&w_{10} = 4.871 &\quad &w_{12} = 1.739 \times 10^{-4} \\
&w_{20} = -0.04783 &\quad &w_{22} = -2.135 \times 10^{-6} \\
&w_{30} = 1.487 \times 10^{-4} &\quad &w_{32} = -1.455 \times 10^{-8} \\
&w_{40} = -2.197 \times 10^{-7} &\quad &w_{42} = 5.230 \times 10^{-11} \\
&w_{01} = 1.524 &\quad &w_{03} = -1.197 \times 10^{-5} \\
&w_{11} = -0.0111 &\quad &w_{13} = -1.628 \times 10^{-6} \\
&w_{21} = 2.747 \times 10^{-4} &\quad &w_{23} = 1.237 \times 10^{-8} \\
&w_{31} = -6.503 \times 10^{-7} &\quad &w_{33} = 1.327 \times 10^{-10} \\
&w_{41} = 7.987 \times 10^{-10} &\quad &w_{43} = -4.614 \times 10^{-13}
\end{aligned}
$$

We define the **gas–water ratio** R_G as the ratio of the volume of dissolved gas at standard conditions to the volume of brine. Then, for temperatures below 250°C, the maximum amount of methane that can go into solution in brine is

$$
\log_{10}(R_G) = \log_{10}(0.712P|T - 76.71|^{1.5} + 3676P^{0.64}) \\
- 4 - 7.786S(T + 17.78)^{-0.306}
$$

If K_B is the bulk modulus of the gas-free brine, and K_G is that of brine with gas–water ratio R_G, then

$$
\frac{K_B}{K_G} = 1 + 0.0494 R_G
$$

(i.e., bulk modulus decreases linearly with increasing gas content). As far as the density of the brine is concerned, experimental data are sparse, but the consensus is that the density is almost independent of the amount of dissolved gas.

The **viscosity of brine** η in cPs for temperatures below 250°C is

$$
\eta = 0.1 + 0.333S + (1.65 + 91.9S^3)e^{-[0.42(S^{0.8} - 0.17)^2 + 0.045]T^{0.8}}
$$

GAS

Natural gas is characterized by its gravity G, which is the ratio of gas density to air density at 15.6°C and atmospheric pressure. The gravity of methane is 0.56. It may be as large as 1.8 for heavier natural gases. Algorithms for calculating **gas density** and **bulk modulus** follow.

STEP 1: Calculate absolute temperature T_a as

$$
T_a = T + 273.15
$$

where T is in degrees Celsius.

STEP 2: Calculate pseudopressure P_r and pseudotemperature T_r as

$$
P_r = \frac{P}{4.892 - 0.4048G}, \qquad T_r = \frac{T_a}{94.72 + 170.75G}
$$

where pressure is in MPa.

STEP 3: Calculate **density** ρ_G in g/cm³ as

$$\rho_G \approx \frac{28.8GP}{ZRT_a}$$

$$Z = aP_r + b + E, \quad E = cd$$

$$d = \exp\left\{-\left[0.45 + 8\left(0.56 - \frac{1}{T_r}\right)^2\right]\frac{P_r^{1.2}}{T_r}\right\}$$

$$c = 0.109(3.85 - T_r)^2, \quad b = 0.642T_r - 0.007T_r^4 - 0.52$$

$$a = 0.03 + 0.00527(3.5 - T_r)^3$$

$$R = 8.31441 \text{ J/g-mole deg (gas constant)}$$

STEP 4: Calculate the adiabatic **bulk modulus** K_G in MPa as

$$K_G \approx \frac{P\gamma}{1 - \frac{P_r}{Z}f}, \quad \gamma = 0.85 + \frac{5.6}{P_r + 2} + \frac{27.1}{(P_r + 3.5)^2} - 8.7e^{-0.65(P_r+1)}$$

$$f = cdm + a, \quad m = 1.2\left\{-\left[0.45 + 8\left(0.56 - \frac{1}{T_r}\right)^2\right]\frac{P_r^{0.2}}{T_r}\right\}$$

The preceding approximate expressions for ρ_G and K_G are valid as long as P_r and T_r are not both within 0.1 of unity.

OIL

Oil density under room conditions may vary from under 0.5 to 1 g/cm³, and most produced oils are in the 0.7 to 0.8 g/cm³ range. A reference (standard) density that can be used to characterize an oil ρ_0 is measured at 15.6°C and atmospheric pressure. A widely used classification of crude oil is the American Petroleum Institute oil gravity (API gravity). It is defined as

$$\text{API} = \frac{141.5}{\rho_0} - 131.5$$

where density is in g/cm³. API gravity may be about 5 for very heavy oils and about 100 for light condensates.

Acoustic velocity in oil V_P may generally vary with temperature T and **molecular weight** M:

$$V_P(T, M) = V_0 - b\Delta T - a_m\left(\frac{1}{M} - \frac{1}{M_0}\right)$$

$$b = 0.306 - 7.6/M$$

In this formula, V_0 is the velocity of oil of molecular weight M_0 at temperature

T_0; a_m is a positive function of temperature, and thus oil velocity increases with molecular weight. When components are mixed, velocity can be approximately calculated as a fractional average of the end components.

For **dead oil** (oil with no dissolved gas), the effects of pressure and temperature on **density** are largely independent. The pressure dependence is

$$\rho_P = \rho_0 + (0.00277P - 1.71 \times 10^{-7}P^3)(\rho_0 - 1.15)^2 + 3.49 \times 10^{-4}P$$

where ρ_P is the density in g/cm^3 at pressure P in MPa. Temperature dependence of density at given pressure P is

$$\rho = \rho_P/[0.972 + 3.81 \times 10^{-4}(T + 17.78)^{1.175}]$$

where temperature is in degrees Celsius.

Acoustic velocity in **dead oil** depends on pressure and temperature as

$$V_P(\text{m/s}) = 2,096\left(\frac{\rho_0}{2.6 - \rho_0}\right)^{1/2} - 3.7T + 4.64P$$
$$+ 0.0115\left[4.12\left(1.08\rho_0^{-1} - 1\right)^{1/2} - 1\right]TP$$

or, in terms of API gravity as

$$V_P(\text{ft/s}) = 15,450(77.1 + \text{API})^{-1/2} - 3.7T + 4.64P$$
$$+ 0.0115[0.36\text{API}^{1/2} - 1]TP$$

LIVE OIL

Large amounts of **gas can be dissolved in an oil**. The original fluid in situ is usually characterized by R_G, the volume ratio of liberated gas to remaining oil at atmospheric pressure and 15.6°C. The maximum amount of gas that can be dissolved in an oil is a function of pressure, temperature, and the composition of both the gas and the oil:

$$R_G^{(\text{max})} = 0.02123\,G\left[P\exp\left(\frac{4.072}{\rho_0} - 0.00377T\right)\right]^{1.205}$$

or, in terms of API gravity:

$$R_G^{(\text{max})} = 2.03G[P\exp(0.02878\text{API} - 0.00377T)]^{1.205}$$

where R_G is in liters/liter (1 L/L = 5.615 ft^3/bbl) and G is the gas gravity. Temperature is in degrees Celsius, and pressure is in MPa.

Velocities in oils with dissolved gas can still be calculated versus pressure and temperature as follows by using the preceding formulas with a pseudodensity

ρ' used instead of ρ_0:

$$\rho' = \frac{\rho_0}{B_0}(1 + 0.001 R_G)^{-1}$$

$$B_0 = 0.972 + 0.00038\left[2.4 R_G\left(\frac{G}{\rho_0}\right)^{1/2} + T + 1.78\right]^{1.175}$$

The true **density of oil with gas** (in g/cm^3) can be calculated as

$$\rho_G = (\rho_0 + 0.0012 G R_G)/B_0$$

EXAMPLE

Calculate the density and acoustic velocity of live oil of 30 API gravity at 80 °C and 20 MPa. The gas–oil ratio R_G is 100, and the gas gravity is 0.6.

Calculate ρ_0 from API as

$$\rho_0 = \frac{141.5}{API + 131.5} = \frac{141.5}{30 + 131.5} = 0.876 \text{ g/cm}^3$$

Calculate B_0 as

$$B_0 = 0.972 + 0.00038\left[2.4 R_G\left(\frac{G}{\rho_0}\right)^{1/2} + T + 1.78\right]^{1.175}$$

$$= 0.972 + 0.00038\left[2.4 \times 100 \times \left(\frac{0.6}{0.876}\right)^{1/2} + 80 + 1.78\right]^{1.175}$$

$$= 1.2577$$

Calculate pseudodensity ρ' as

$$\rho' = \frac{\rho_0}{B_0}(1 + 0.001 R_G)^{-1} = 0.633 \text{ g/cm}^3$$

Calculate the density of oil with gas ρ_G as

$$\rho_G = (\rho_0 + 0.0012 G R_G)/B_0 = 0.7538 \text{ g/cm}^3$$

Correct this density for pressure and find the actual density ρ_P as

$$\rho_P = \rho_G + (0.00277 P - 1.71 \times 10^{-7} P^3)(\rho_G - 1.15)^2 + 3.49 \times 10^{-4} P$$

$$= 0.754 + (0.00277 \times 20 - 1.71 \times 10^{-7} 20^3)(0.754 - 1.15)^2$$

$$+ 3.49 \times 10^{-4} \times 20 = 0.769 \text{ g/cm}^3$$

Calculate velocity in oil with gas V_p as

$$V_p = 2096\left(\frac{\rho'}{2.6 - \rho'}\right)^{1/2} - 3.7T + 4.64P$$

$$+ 0.0115[4.12(1.08/\rho' - 1)^{1/2} - 1]TP$$

$$= 2096\left(\frac{0.633}{2.6 - 0.633}\right)^{1/2} - 3.7 \times 80 + 4.64 \times 20$$

$$+ 0.0115[4.12(1.08/0.633 - 1)^{1/2} - 1] \times 80 \times 20$$

$$= 1.031 \text{ m/s}$$

Viscosity of dead oil (η) decreases rapidly with increasing temperature. At room pressure for a gas-free oil we have

$$\log_{10}(\eta + 1) = 0.505y(17.8 + T)^{-1.163}$$

$$\log_{10}(y) = 5.693 - 2.863/\rho_0$$

Pressure has a smaller influence on viscosity and can be estimated independently of the temperature influence. If oil viscosity is η_0 at a given temperature and room pressure, its viscosity at pressure P and the same temperature is

$$\eta = \eta_0 + 0.145PI$$

$$\log_{10}(I) = 18.6[0.1\log_{10}(\eta_0) + (\log_{10}(\eta_0) + 2)^{-0.1} - 0.985]$$

Viscosity in these formulas is in centipoise, temperature is in degrees Celsius, and pressure is in MPa.

USES

The equations presented in this section are used to estimate acoustic velocities and densities of pore fluids.

ASSUMPTIONS AND LIMITATIONS

The formulas are mostly based on empirical measurements summarized by Batzle and Wang (1992).

PART 7

EMPIRICAL RELATIONS

7.1 VELOCITY–POROSITY MODELS: CRITICAL POROSITY AND NUR'S MODIFIED VOIGT AVERAGE

SYNOPSIS

Nur et al. (1991, 1995) and other workers have championed the simple, if not obvious, idea that the P and S velocities of rocks should trend between the velocities of the mineral grains in the limit of low porosity and the values for a mineral–pore fluid suspension in the limit of high porosity. This idea is based on the observation that for most porous materials there is a **critical porosity**, ϕ_c, that separates their mechanical and acoustic behavior into two distinct domains. For porosities lower than ϕ_c the mineral grains are load-bearing, whereas for porosities greater than ϕ_c the rock simply "falls apart" and becomes a suspension in which the fluid phase is load-bearing. The transition from solid to suspension is implicit in the well-known empirical velocity–porosity relations of Raymer, Hunt, and Gardner (1980) discussed in Section 7.4.

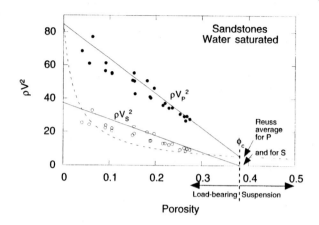

Figure 7.1.1. Critical porosity behavior: ρV_P^2 **(P-wave modulus) versus porosity and** ρV_S^2 **(shear modulus) versus porosity, both trending between the mineral value at zero porosity and the Reuss average at critical porosity.**

In the **suspension domain,** $\phi > \phi_c$, the effective bulk and shear moduli can be estimated quite accurately using by the Reuss (isostress) average:

$$K_R^{-1} = (1 - \phi)K_0^{-1} + \phi K_{fl}^{-1}; \quad \mu_R = 0$$

where K_0 and K_{fl} are the bulk moduli of the mineral material and the fluid, respectively. The effective shear modulus of the suspension is zero because the shear modulus of the fluid is zero.

In the **load-bearing domain,** $\phi < \phi_c$, the moduli decrease rapidly from the mineral values at zero porosity to the suspension values at the critical porosity. Nur found that this dependence can often be approximated with a straight line when expressed as ρV^2 versus porosity. Figure 7.1.1 illustrates this behavior with laboratory ultrasonic sandstone data from Han (1986) for samples at 40 MPa effective pressure and clay content $<10\%$ by volume. In the figure, ρV_p^2 is the P-wave modulus $[K + (4/3)\mu]$, and ρV_S^2 is the shear modulus.

A geometric interpretation of the mineral-to-critical-porosity trend is simply that if we make the porosity large enough, the grains must lose contact and their rigidity. The geologic interpretation is that, at least for clastics, the weak suspension state at critical porosity, ϕ_c, describes the sediment when it is first deposited before compaction and diagenesis. The value of ϕ_c is determined by the grain sorting and angularity at deposition. Subsequent compaction and diagenesis move the sample along an upward trajectory as the porosity is reduced and the elastic stiffness is increased.

Although there is nothing special about a linear trend of ρV^2 versus ϕ, it does describe sandstones fairly well, and it leads to convenient mathematical properties. For dry rocks, the bulk and shear moduli can be expressed as the linear functions

$$K_{\text{dry}} = K_0\left(1 - \frac{\phi}{\phi_c}\right)$$

$$\mu_{\text{dry}} = \mu_0\left(1 - \frac{\phi}{\phi_c}\right)$$

where K_0 and μ_0 are the mineral bulk and shear moduli. Thus, the dry rock bulk and shear moduli trend linearly between K_0, μ_0 at $\phi = 0$ and $K_{\text{dry}} = \mu_{\text{dry}} = 0$ at $\phi = \phi_c$.

This linear dependence can be thought of as a **modified Voigt average** (see Section 4.2), where one end member is the mineral and the other end member is the suspension at the critical porosity, which can be measured or estimated by using the Reuss average. Then we can write

$$M_{\text{MV}} = (1 - \phi')M_0 + \phi'M_c$$

where M_0 and M_c are the moduli (bulk or shear) of the mineral material at zero porosity and of the suspension at the critical porosity. The porosity is scaled by the critical porosity, $\phi' = \phi/\phi_c$, and thus ϕ' ranges from 0 to 1 as ϕ ranges from 0 to ϕ_c. Note that using the suspension modulus M_c in this form automatically incorporates the effect of pore fluids on the modified Voigt average, which is equivalent to applying Gassmann's relations to the dry rock moduli–porosity relations (see Section 6.3 on Gassmann's relations).

The critical porosity value depends on the internal structure of the rock. It may be medium for granular rocks, very small for cracked rocks, and large for foam-like rocks. Examples of critical porosity behavior in sandstones, dolomites, pumice, and cracked igneous rocks are shown in Figure 7.1.2 and Table 7.1.1.

USES

The equations presented in this section can be used for relating velocity and porosity.

ASSUMPTIONS AND LIMITATIONS

The model discussed in this section has the following limitations:

- The critical porosity result is empirical.

TABLE 7.1.1. Typical values of critical porosity.

Material	Critical porosity
Natural rocks	
Sandstones	40%
Limestones	60%
Dolomites	40%
Pumice	80%
Chalks	65%
Rock salt	40%
Cracked igneous rocks	5%
Oceanic basalts	20%
Artificial rocks	
Sintered glass beads	40%
Glass foam	90%

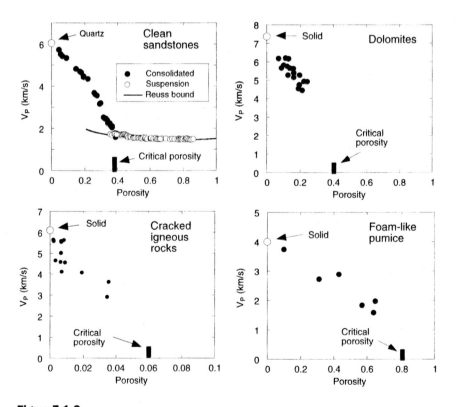

Figure 7.1.2

- Because only the variation with porosity is described, one must apply other corrections to account for parameters such as clay content.

EXTENSIONS

Similar descriptions of the failure strength of porous media can be quantified in terms of the critical porosity.

7.2 VELOCITY–POROSITY MODELS: GEERTSMA'S EMPIRICAL RELATIONS FOR COMPRESSIBILITY

SYNOPSIS

Geertsma (1961) suggested the following empirical estimate of bulk modulus in dry rocks with porosities $0 < \phi < 0.3$:

$$\frac{1}{K_{dry}} = \frac{1}{K_0}(1 + 50\phi)$$

where

$$K_{dry} = \text{dry rock bulk modulus}$$
$$K_0 = \text{mineral bulk modulus}$$

USES

This equation can be used to relate bulk modulus to porosity in rocks empirically.

ASSUMPTIONS AND LIMITATIONS

This relation is empirical.

7.3 VELOCITY–POROSITY MODELS: WYLLIE'S TIME AVERAGE EQUATION

SYNOPSIS

Measurements by Wyllie et al. (1956, 1958, 1963) revealed that a relatively simple monotonic relation often can be found between velocity and porosity in sedimentary rocks when (1) they have relatively uniform mineralogy, (2) they are fluid-saturated, and (3) they are at high effective pressure. Wyllie et al. approximated these relations with the expression

$$\frac{1}{V_P} = \frac{\phi}{V_{P\text{-fl}}} + \frac{1 - \phi}{V_{P\text{-0}}}$$

where V_P, $V_{P\text{-0}}$, and $V_{P\text{-fl}}$ are the P-wave velocities of the saturated rocks, of the mineral material making up the rocks, and of the pore fluid, respectively. Some useful values for $V_{P\text{-0}}$ are shown in Table 7.3.1.

The interpretation of this expression is that the total transit time is the sum of the transit time in the mineral plus the transit time in the pore fluid. Hence, it is often called the **time average equation**.

CAUTION: The time average equation is heuristic and cannot be justified theoretically. The argument that the total transit time can be written as the sum of the transit time in each of the phases is a seismic ray theory assumption and can be correct only if (1) the wavelength is small compared with the typical pore size and grain size, and (2) the pores and grains are arranged as homogeneous layers perpendicular to the ray path. Because neither of these assumptions is even remotely true, the agreement with observations is only fortuitous. Attempts to overinterpret observations in terms of the mineralogy and fluid properties can lead to errors. An illustration of this point is that a form of the time average equation is sometimes used to interpret shear velocities. To do this, a finite value of shear velocity in the fluid is used, which is clearly nonsense.

TABLE 7.3.1. Typical mineral P-wave velocities.

	$V_{P\text{-0}}$ (m/s)
Sandstones	5,480 to 5,950
Limestones	6,400 to 7,000
Dolomites	7,000 to 7,925

USES

Wyllie's time average equation can be used for the following purposes:

- To estimate the expected seismic velocities of rocks with a given mineralogy and pore fluid.
- To estimate the porosity from measurements of seismic velocity and knowledge of the rock type and pore-fluid content.

ASSUMPTIONS AND LIMITATIONS

The use of the time average equation requires the following serious considerations:

- The rock is isotropic.
- The rock must be fluid saturated.
- The time average equation works best if rocks are at high enough effective pressure to be at the "terminal velocity," which is usually on the order of 30 MPa. Most rocks show an increase of velocity with pressure owing to the progressive closing of compliant crack-like parts of the pore space, including microcracks, compliant grain boundaries, and narrow tips of otherwise equant-shaped pores. Usually the velocity appears to level off at high pressure, approaching a limiting "terminal" velocity when, presumably, all the crack-like pore space is closed. Because the compliant fraction of the pore space can have a very small porosity and yet have a very large effect on velocity, its presence, at low pressures, can cause a very poor correlation between porosity and velocity; hence, the requirement for high effective pressure. At low pressures or in uncompacted situations, the time average equation tends to overpredict the velocity and porosity. Log analysts sometimes use a compaction correction, which is an empirical attempt to correct for the effect of compliant porosity. The time average equation underpredicts velocities in consolidated low-to-medium-porosity rocks, and in high-porosity cemented rocks, as shown in Figure 7.3.1.
- The time average relation should not be used to relate velocity to porosity in unconsolidated uncemented rocks (see example in Section 7.4).
- The time average relation works best with primary porosity. Secondary or vuggy porosity tends to be "stiffer" than primary porosity and therefore lowers the velocity less. In these situations, the time average equations tend to underpredict the velocity and the porosity. Empirical corrections can be attempted to adjust for this.
- The time average relation assumes a single homogeneous mineralogy. Empirical corrections for mixed mineralogy, such as shaliness, can be attempted to adjust for this.
- The time average equation usually works best for intermediate porosities. (See the Raymer equations in Section 7.4.)

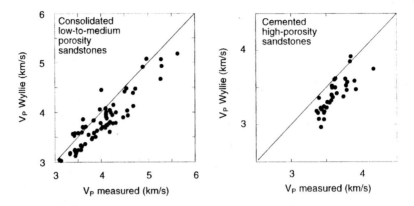

Figure 7.3.1. Comparison of predicted and measured velocity in water-saturated medium-to-low-porosity shaley sandstones (40 MPa effective pressure). The velocity in pure quartz is taken at 6.038 km/s, which follows from the bulk modulus, shear modulus, and density being 38 GPa, 44 GPa, and 2.65 g/cm³, respectively. The velocity in clay is 3.41 km/s, which follows from the bulk modulus, shear modulus, and density being 21 GPa, 7 GPa, and 2.58 g/cm³, respectively.

EXTENSION

The time average equation can be extended in the following ways:

- For mixed mineralogy one can often use an effective average velocity for the mineral material.
- Empirical corrections can sometimes be found for shaliness, compaction, and secondary porosity, but they should be calibrated when possible.

7.4 VELOCITY–POROSITY MODELS: RAYMER–HUNT–GARDNER RELATIONS

SYNOPSIS

Raymer et al. (1980) suggested improvements to Wyllie's empirical velocity-to-travel time relations as follows:

$$V = (1 - \phi)^2 V_0 + \phi V_{\text{fl}}, \quad \phi < 37\%$$

$$\frac{1}{\rho V^2} = \frac{\phi}{\rho_{\text{fl}} V_{\text{fl}}^2} + \frac{1 - \phi}{\rho_0 V_0^2}, \quad \phi > 47\%$$

where V, V_{fl}, and V_0 are the velocities in the rock, the pore fluid, and the minerals, respectively. The terms ρ, ρ_{fl}, and ρ_0 are the densities of the rock, the pore fluid, and the minerals, respectively. Note that the second relation is the same as the isostress or Reuss average (see Section 4.2) of the P-wave moduli. A third expression for intermediate porosities is derived as a simple interpolation of these two:

$$\frac{1}{V} = \frac{0.47 - \phi}{0.10} \frac{1}{V_{37}} + \frac{\phi - 0.37}{0.10} \frac{1}{V_{47}}$$

where V_{37} is calculated from the low-porosity formula at $\phi = 0.37$, and V_{47} is calculated from the high-porosity formula at $\phi = 0.47$.

USES

The Raymer–Hunt–Gardner relations have the following uses:

- To estimate the seismic velocities of rocks with a given mineralogy and pore fluid.
- To estimate the porosity from measurements of seismic velocity and knowledge of the rock type and pore fluid content.

ASSUMPTIONS AND LIMITATIONS

The use of the Raymer–Hunt–Gardner relations requires the following considerations:

- The rock is isotropic.
- All minerals making up the rock have the same velocities.
- The rock is fluid-saturated.
- The method is empirical. See also the discussion and limitations of Wyllie's time average equation (Section 7.3).
- These relations should work best at high enough effective pressure to be at the "terminal velocity," usually on the order of 30 MPa. Most rocks show an increase of velocity with pressure owing to the progressive closing of compliant crack-like parts of the pore space, including microcracks, compliant grain boundaries, and narrow tips of otherwise equant-shaped pores. Usually the velocity appears to level off at high pressure, approaching a limiting "terminal" velocity, when, presumably, all the crack-like pore space is closed. Because the compliant fraction of the pore space can have very small porosity and yet have a very large effect on velocity, its presence, at low pressures, can cause a very poor correlation between porosity and velocity; hence, the requirement for high effective pressure. These relations work well for consolidated low-to-medium porosity and high-porosity cemented sandstones.
- These relations should not be used for unconsolidated uncemented rocks.

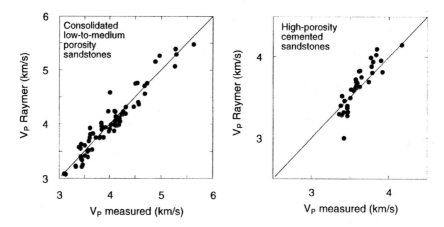

Figure 7.4.1. Comparison of predicted and measured velocities in water-saturated shaley sandstones (40 MPa effective pressure). The velocity in pure quartz is taken at 6.038 km/s, which follows from the bulk modulus, shear modulus, and density being 38 GPa, 44 GPa, and 2.65 g/cm³, respectively. The velocity in clay is 3.41 km/s, which follows from the bulk modulus, shear modulus, and density being 21 GPa, 7 GPa, and 2.58 g/cm³, respectively.

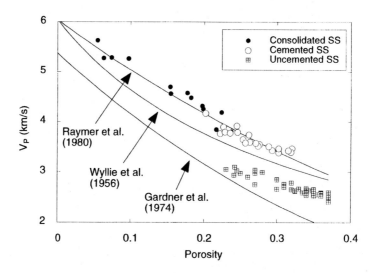

Figure 7.4.2. Velocity versus porosity in water-saturated clay-free sandstones. The Wyllie et al. (1956) equation underestimates the consolidated rock values. The Gardner et al. (1974) equation underpredicts all of the measured values. None of the equations adequately models the uncemented sands. For the predictions, the mineral is taken as quartz.

7.5 VELOCITY–POROSITY–CLAY MODELS: HAN'S EMPIRICAL RELATIONS FOR SHALEY SANDSTONES

SYNOPSIS

Han (1986) found empirical regressions relating ultrasonic (laboratory) velocities to porosity and clay content. These were determined from a set of eighty *well-consolidated* Gulf Coast sandstones with porosities, ϕ, ranging from 3 to 30 percent and clay volume fractions, C, ranging from 0 to 55 percent. The study found that clean sandstone velocities can be related empirically to porosity alone with very high accuracy. When clay is present, the correlation with porosity is relatively poor but becomes very accurate if clay volume is also included in the regression.

TABLE 7.5.1. Han's empirical relations between ultrasonic V_P and V_S in km/s with porosity and clay volume fractions.

Clean sandstones (determined from ten samples)

Water-saturated

40 MPa	$V_P = 6.08 - 8.06\phi$	$V_S = 4.06 - 6.28\phi$

Shaley sandstones (determined from seventy samples)

Water-saturated

40 MPa	$V_P = 5.59 - 6.93\phi - 2.18C$	$V_S = 3.52 - 4.91\phi - 1.89C$
30 MPa	$V_P = 5.55 - 6.96\phi - 2.18C$	$V_S = 3.47 - 4.84\phi - 1.87C$
20 MPa	$V_P = 5.49 - 6.94\phi - 2.17C$	$V_S = 3.39 - 4.73\phi - 1.81C$
10 MPa	$V_P = 5.39 - 7.08\phi - 2.13C$	$V_S = 3.29 - 4.73\phi - 1.74C$
5 MPa	$V_P = 5.26 - 7.08\phi - 2.02C$	$V_S = 3.16 - 4.77\phi - 1.64C$

Dry

40 MPa	$V_P = 5.41 - 6.35\phi - 2.87C$	$V_S = 3.57 - 4.57\phi - 1.83C$

Eberhart–Phillips (1989) used a multivariate analysis to investigate the combined influences of effective pressure, porosity, and clay content on Han's measurements of velocities in water-saturated shaley sandstones. She found that the water-saturated P- and S-wave ultrasonic velocities (in km/s) could be described

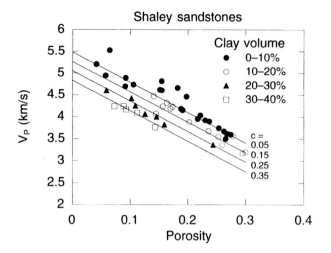

Figure 7.5.1. Han's water-saturated ultrasonic velocity data at 40 MPa compared with his empirical relations evaluated at four different clay fractions.

empirically by

$$V_P = 5.77 - 6.94\phi - 1.73\sqrt{C} + 0.446(P_e - 1.0e^{-16.7P_e})$$
$$V_S = 3.70 - 4.94\phi - 1.57\sqrt{C} + 0.361(P_e - 1.0e^{-16.7P_e})$$

where P_e is the effective pressure in kilobars. The model accounts for 95 percent of the variance and has an rms error of 0.1 km/s.

USES

These relations can be used to relate velocity, porosity, and clay content empirically in shaley sandstones.

ASSUMPTIONS AND LIMITATIONS

The preceding relations have the following limitations:

- These relations are empirical and thus strictly speaking they apply only to the set of rocks studied. However, the result should extend in general to many consolidated sandstones. In any case, the key result is that clay content is an important parameter for quantifying velocity; if possible, the regression coefficients should be recalibrated from cores or logs at the site being studied, but be sure to include clay.

- Han's linear regression coefficients change slightly with confining pressure. They are fairly stable above about 10 MPa; below this, they vary more, and the correlation coefficients degrade.
- A common mistake is to try to overinterpret the empirical coefficients by comparing the equations, for example, to Wyllie's time-average equation (see Section 7.3). This can lead to nonsensical interpreted values for the velocities of water and clay. This is not surprising, for Wyllie's equations are heuristic only.
- It is dangerous to extrapolate the results to values of porosity or clay content outside the range of the experiments. Note, for example, that the intercepts of the various equations corresponding to no porosity and no clay do not agree with each other and generally do not agree with the velocities in pure quartz.

7.6 VELOCITY–POROSITY–CLAY MODELS: TOSAYA'S EMPIRICAL RELATIONS FOR SHALEY SANDSTONES

SYNOPSIS

On the basis of their measurements, Tosaya and Nur (1982) determined empirical regressions relating ultrasonic (laboratory) P- and S-wave velocities to porosity and clay content. For water-saturated rocks at an effective pressure of 40 MPa, they found

$$V_P(\text{km/s}) = 5.8 - 8.6\phi - 2.4C$$
$$V_S(\text{km/s}) = 3.7 - 6.3\phi - 2.1C$$

where ϕ is the porosity, and C is the clay content by volume. See also Han's relation in Section 7.5.

USES

Tosaya's relations can be used to relate velocity, porosity, and clay content empirically in shaley sandstones.

ASSUMPTIONS AND LIMITATIONS

Tosaya's relations have the following limitations:

- These relations are empirical, and thus strictly speaking they apply only to the set of rocks studied. However, the result should extend in general to many consolidated sandstones. In any case, the key result is that clay content is an important parameter for quantifying velocity; if possible, the regression coefficients should be recalibrated from cores or logs at the site being studied, but be sure to include clay.
- The relation given above holds only for the high effective pressure value of 40 MPa.
- A common mistake is to try to overinterpret the empirical coefficients by comparing the equations, for example, to Wyllie's time average equation (see Section 7.3). This can lead to nonsensical interpreted values for the velocities of water and clay. This is not surprising, because Wyllie's equations are heuristic only.
- It is dangerous to extrapolate the results to values of porosity or clay content outside the range of the experiments.

7.7 VELOCITY–POROSITY–CLAY MODELS: CASTAGNA'S EMPIRICAL RELATIONS FOR VELOCITIES

SYNOPSIS

On the basis of log measurements, Castagna et al. (1985) determined empirical regressions relating velocities with porosity and clay content under water-saturated conditions. See also Section 7.8 on V_P–V_S relations.

For mudrock (clastic silicate rock composed primarily of clay and silt-sized particles), they found the relation between V_P and V_S (in km/s) to be

$$V_P \text{ (km/s)} = 1.36 + 1.16 V_S$$

where V_P and V_S are the P- and S-wave velocities.

For shaley sands of the Frio formation they found

$$V_P \text{ (km/s)} = 5.81 - 9.42\phi - 2.21C$$

$$V_S \text{ (km/s)} = 3.89 - 7.07\phi - 2.04C$$

where ϕ is porosity and C is clay volume fraction.

USES

Castagna's relations for velocities can be used to relate velocity, porosity, and clay content empirically in shaley sandstones.

ASSUMPTIONS AND LIMITATIONS

Castagna's empirical relations have the following limitations:

- These relations are empirical, and thus strictly speaking they apply only to the set of rocks studied.
- A common mistake is to try to overinterpret the empirical coefficients by comparing the equations, for example, to Wyllie's time average equation (see Section 7.3). This can lead to nonsensical interpreted values for the velocities of water and clay. This is not surprising because Wyllie's equations are heuristic only; there is no theoretical justification for them, and they do not represent an empirical best fit to any data.

7.8 V_P–V_S RELATIONS

SYNOPSIS

V_P–V_S relations are key to the determination of lithology from seismic or sonic log data as well as for direct seismic identification of pore fluids using, for example, AVO analysis. Castagna et al. (1993) give an excellent review of the subject.

There is a wide and sometimes confusing variety of published V_P–V_S relations and V_S prediction techniques, which at first appear to be quite distinct. However, most reduce to the same two simple steps:

1) Establish empirical relations among V_P, V_S, and porosity, ϕ, for one reference pore fluid – most often water-saturated or dry.
2) Use Gassmann's (1951) relations to map these empirical relations to other pore-fluid states (see Section 6.3).

Although some of the effective medium models summarized in Chapter 4 predict both P- and S-velocities on the basis of idealized pore geometries, the fact remains that the most reliable and most often used V_P–V_S relations are empirical fits to laboratory or log data, or both. The most useful role of theoretical methods is extending these empirical relations to different pore fluids or measurement frequencies, which accounts for the two steps listed above.

We summarize here a few of the popular V_P–V_S relations compared with laboratory and log data sets and illustrate some of the variations that can result from lithology, pore fluids, and measurement frequency.

SOME EMPIRICAL RELATIONS

LIMESTONES

Figure 7.8.1 shows laboratory ultrasonic V_P–V_S data for water-saturated limestones from Pickett (1963), Milholland et al. (1980), and Castagna et al. (1993), as compiled by Castagna et al. (1993). Superimposed, for comparison, are Pickett's (1963) empirical limestone relation derived from laboratory core data:

$$V_S = V_P/1.9 \quad \text{(km/s)}$$

and a least-squares polynomial fit to the data derived by Castagna et al. (1993):

$$V_S = -0.055 V_P^2 + 1.017 V_P - 1.031 \quad \text{(km/s)}$$

At higher velocities, Pickett's straight line fits the data better, although at lower velocities (higher porosities), the data deviate from a straight line and trend toward the water point, $V_P = 1.5$ km/s, $V_S = 0$. In fact, this limit is more accurately described as a suspension of grains in water at the critical porosity (see discussion below), at which the grains lose contact and the shear velocity vanishes.

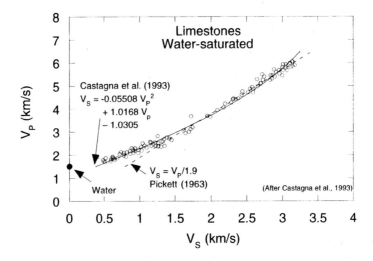

Figure 7.8.1

DOLOMITE

Figure 7.8.2 shows laboratory V_P–V_S data for water-saturated dolomites from Castagna et al. (1993). Superimposed, for comparison, are Pickett's (1963) dolomite (laboratory) relation

$$V_S = V_P/1.8 \quad \text{(km/s)}$$

and a least-squares linear fit (Castagna et al., 1993)

$$V_S = 0.583 V_P - 0.078 \quad \text{(km/s)}$$

For the data shown, the two relations are essentially equivalent. The data range is too limited to speculate about behavior at much lower velocity (higher porosity).

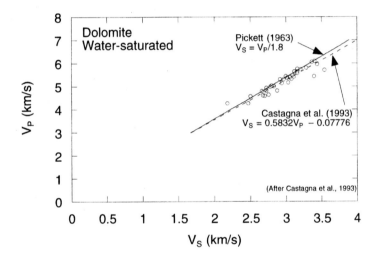

Figure 7.8.2

SANDSTONES AND SHALES

Figures 7.8.3 and 7.8.4 show laboratory V_P–V_S data for water-saturated sandstones and shales from Castagna et al. (1985, 1993) and Thomsen (1986), as compiled by Castagna et al. (1993). Superimposed, for comparison, are a least-squares linear fit to these data offered by Castagna et al. (1993),

$$V_S = 0.804 V_P - 0.856 \quad \text{(km/s)}$$

the famous "mudrock line" of Castagna et al. (1985), which was derived from in situ data,

$$V_S = 0.862 V_P - 1.172 \quad \text{(km/s)}$$

and the following empirical relation of Han (1986), which is based on laboratory

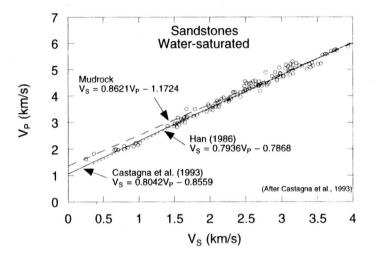

Figure 7.8.3

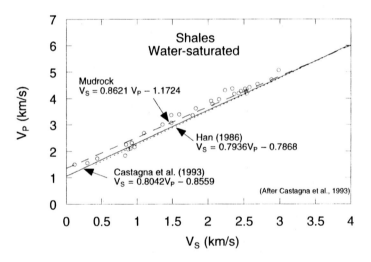

Figure 7.8.4

ultrasonic data:

$$V_S = 0.794 V_P - 0.787 \quad \text{(km/s)}$$

Of these three relations, those by Han and Castagna et al. are essentially the same and give the best overall fit to the sandstones. The mudrock line predicts systematically lower V_S because it is best suited to the most shaley samples, as seen in Figure 7.8.4. Castagna et al. (1993) suggest that if the lithology is well known, one can fine-tune these relations to slightly lower V_S/V_P for high shale content and higher V_S/V_P in cleaner sands. When the lithology is not well constrained, the Han and Castagna et al. lines give a reasonable average.

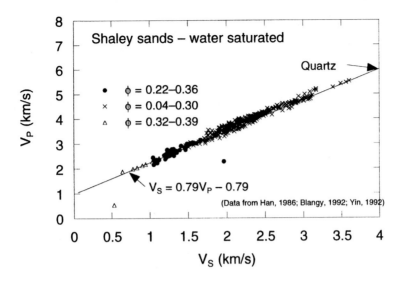

Figure 7.8.5

Figure 7.8.5 compares laboratory ultrasonic data for a larger set of water-saturated sands. The lowest porosity samples ($\phi = 0.04$–0.30) are from a set of consolidated shaley Gulf Coast sandstones studied by Han (1986). The medium porosities ($\phi = 0.22$–0.36) are poorly consolidated North Sea samples studied by Blangy (1992). The very high porosity samples ($\phi = 0.32$–0.39) are unconsolidated clean Ottawa sand studied by Yin (1992). The samples span clay volume fractions from 0 to 55 percent, porosities from 0.04 to 0.39, and confining pressures from 0 to 40 MPa. In spite of this, there is a remarkably systematic trend well represented by Han's relation as follows:

$$V_S = 0.79 V_P - 0.79 \quad \text{(km/s)}$$

SANDSTONES: MORE ON THE EFFECTS OF CLAY

Figure 7.8.6 shows again the ultrasonic laboratory data for seventy water-saturated shaley sandstone samples from Han (1986). The data are separated by clay volume fractions greater than 25 percent and less than 25 percent. Regressions to each part of the data set are shown as follows:

$$V_S = 0.842 V_P - 1.099 \quad \text{clay} > 25\%$$
$$V_S = 0.754 V_P - 0.657 \quad \text{clay} < 25\%$$

The mudrock line (heavier line) is a reasonable fit to the trend but is skewed toward higher clay and lies almost on top of the regression for clay >25 percent.

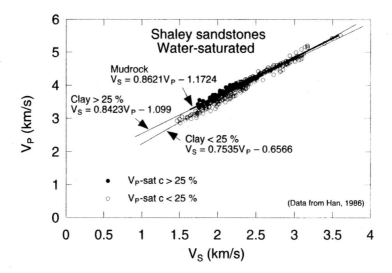

Figure 7.8.6

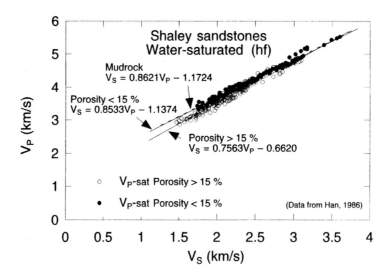

Figure 7.8.7

SANDSTONES: EFFECTS OF POROSITY

Figure 7.8.7 shows the laboratory ultrasonic data for water-saturated shaley sandstones from Han (1986) separated into porosity greater than 15 percent and less than 15 percent. Regressions to each part of the data set are shown as follows:

$$V_S = 0.756V_P - 0.662 \quad \text{porosity} > 15\%$$
$$V_S = 0.853V_P - 1.137 \quad \text{porosity} < 15\%$$

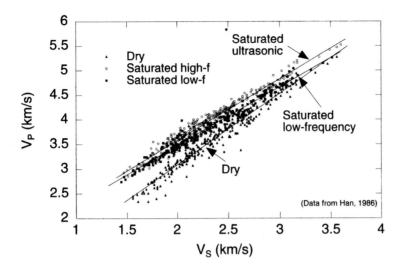

Figure 7.8.8

Note that the low-porosity line is very close to the mudrock line, which as we saw above, fits the high clay values, whereas the high-porosity line is similar to the clean sand (low clay) regression in Figure 7.8.6.

SANDSTONES: EFFECTS OF FLUIDS AND FREQUENCY

Figure 7.8.8 compares V_P–V_S at several conditions based on the shaley sandstone data of Han (1986). The "dry" and "saturated ultrasonic" points are the measured ultrasonic data. The "saturated low-frequency" points are estimates of low-frequency saturated data computed from the dry measurements using the low-frequency Gassmann's relations (see Section 6.3). It is no surprise that the water-saturated samples have higher V_P/V_S because of the well-known larger effects of pore fluids on P-velocities than on S-velocities. Less often recognized is that the velocity dispersion that almost always occurs in ultrasonic measurements appears to increase V_P/V_S systematically.

CRITICAL POROSITY MODEL

The P and S velocities of rocks (as well as their V_P/V_S ratio) generally trend between the velocities of the mineral grains in the limit of low porosity and the values for a mineral–pore fluid suspension in the limit of high porosity. For most porous materials there is a **critical porosity**, ϕ_c, that separates their mechanical and acoustic behavior into two distinct domains. For porosities lower than ϕ_c the mineral grains are load-bearing, whereas for porosities greater than ϕ_c the rock simply "falls apart" and becomes a suspension in which the fluid phase is load-bearing (see Section 7.1 on critical porosity). The transition from solid to suspension is implicit in the Raymer et al. (1980) empirical velocity–porosity relation (see Section 7.4) and the work of Krief et al. (1990), which is discussed below.

A geometric interpretation of the mineral-to-critical-porosity trend is simply that if we make the porosity large enough, the grains must lose contact and their rigidity. The geologic interpretation is that, at least for clastics, the weak suspension state at critical porosity, ϕ_c, describes the sediment when it is first deposited before compaction and diagenesis. The value of ϕ_c is determined by the grain sorting and angularity at deposition. Subsequent compaction and diagenesis move the sample along an upward trajectory as the porosity is reduced and the elastic stiffness is increased.

The value of ϕ_c depends on the rock type. For example $\phi_c \approx 0.4$ for sandstones; $\phi_c \approx 0.7$ for chalks; $\phi_c \approx 0.9$ for pumice and porous glass; and $\phi_c \approx 0.02$–0.03 for granites.

In the **suspension domain**, the effective bulk and shear moduli of the rock K and μ can be estimated quite accurately by using the Reuss (isostress) average (see Section 4.2 on the Voigt–Reuss average and Section 7.1 on critical porosity) as follows:

$$\frac{1}{K} = \frac{\phi}{K_f} + \frac{1-\phi}{K_0}, \quad \mu = 0$$

where K_f and K_0 are the bulk moduli of the fluid and mineral and ϕ is the porosity.

In the **load-bearing domain**, $\phi < \phi_c$, the moduli decrease rapidly from the mineral values at zero porosity to the suspension values at the critical porosity. Nur et al. (1995) found that this dependence can often be approximated with a straight line when expressed as modulus versus porosity. Although there is nothing special about a linear trend of modulus versus ϕ, it does describe sandstones fairly well, and it leads to convenient mathematical properties. For dry rocks, the bulk and shear moduli can be expressed as the linear functions

$$K_{dry} = K_0\left(1 - \frac{\phi}{\phi_c}\right)$$

$$\mu_{dry} = \mu_0\left(1 - \frac{\phi}{\phi_c}\right)$$

where K_0 and μ_0 are the mineral bulk and shear moduli, respectively. Thus, the dry rock bulk and shear moduli trend linearly between K_0, μ_0 at $\phi = 0$, and $K_{dry} = \mu_{dry} = 0$ at $\phi = \phi_c$. At low frequency, changes of pore fluids have little or no effect on the shear modulus. However, it can be shown (see Section 6.3 on Gassmann) that with a change of pore fluids the straight line in the K–ϕ plane remains a straight line, trending between K_0 at $\phi = 0$ and the Reuss average bulk modulus at $\phi = \phi_c$. Thus, the effect of pore fluids on K or $\rho V^2 = K + (4/3)\mu$ is automatically incorporated by the change of the Reuss average at $\phi = \phi_c$.

The relevance of the critical porosity model to V_P–V_S relations is simply that V_S/V_P should generally trend toward the value for the solid mineral material

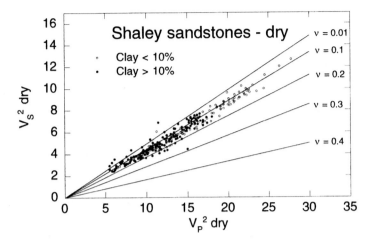

Figure 7.8.9. Velocity data from Han (1986) illustrating that Poisson's ratio is approximately constant for dry sandstones.

in the limit of low porosity and toward the value for a fluid suspension as the porosity approaches ϕ_c (Castagna et al., 1993). Furthermore, if the modulus–porosity relations are linear (or nearly so), then it follows that V_S/V_P for a dry rock at any porosity ($0 < \phi < \phi_c$) will equal the V_S/V_P of the mineral. The same is true if K_{dry} and μ_{dry} are any other functions of porosity but are proportional to each other [$K_{dry}(\phi) \propto \mu_{dry}(\phi)$]. Equivalently, the Poisson's ratio ν for the dry rock will equal the Poisson's ratio of the mineral grains, as is often observed (Pickett, 1963; Krief et al., 1990).

$$\left(\frac{V_S}{V_P}\right)_{dry\ rock} \approx \left(\frac{V_S}{V_P}\right)_{mineral}$$

$$\nu_{dry\ rock} \approx \nu_{mineral}$$

Figure 7.8.9 illustrates the approximately constant dry rock Poisson's ratio observed for a large set of ultrasonic sandstone velocities (from Han, 1986) over a large range of effective pressures ($5 < P_{eff} < 40$ MPa) and clay contents ($0 < C < 55\%$ by volume).

To summarize, the critical porosity model suggests that P- and S-wave velocities trend systematically between their mineral values at zero porosity to fluid suspension values ($V_S = 0$, $V_P = V_{suspension} \approx V_{fluid}$) at the critical porosity ϕ_c, which is a characteristic of each class of rocks. Expressed in the modulus versus porosity domain, if dry rock ρV_P^2 versus ϕ is proportional to ρV_S^2 versus ϕ (for example, both ρV_P^2 and ρV_S^2 are linear in ϕ), then V_S/V_P of the dry rock will be equal to V_S/V_P of the mineral.

The V_P–V_S relation for different pore fluids is found using Gassmann's relation, which is applied automatically if the trend terminates on the Reuss average at ϕ_c (see a discussion of this in Section 6.3 on Gassmann's relation).

KRIEF'S RELATION

Krief et al. (1990) suggested a V_P–V_S prediction technique that very much resembles the critical porosity model. The model again combines the same two elements:

1) An empirical V_P–V_S–ϕ relation for water-saturated rocks, which we will show is approximately the same as predicted by the simple critical porosity model.
2) Gassmann's relation to extend the empirical relation to other pore fluids.

DRY ROCK V_P–V_S–ϕ RELATION

If we model the dry rock as a porous elastic solid, then with great generality we can write the dry rock bulk modulus as

$$K_{dry} = K_0(1 - \beta)$$

where K_{dry} and K_0 are the bulk moduli of the dry rock and mineral and β is Biot's coefficient (see Section 4.6 on compressibilities and Section 2.6 on the deformation of cavities). An equivalent expression is

$$\frac{1}{K_{dry}} = \frac{1}{K_0} + \frac{\phi}{K_\phi}$$

where K_ϕ is the pore space stiffness (see Section 2.6) and ϕ is the porosity, so that

$$\frac{1}{K_\phi} = \frac{1}{v_p}\frac{dv_p}{d\sigma}\bigg|_{P_P=\text{constant}} \quad ; \quad \beta = \frac{dv_p}{dV}\bigg|_{P_P=\text{constant}} = \frac{\phi K_{dry}}{K_\phi}$$

where v_p is the pore volume, V is the bulk volume, σ is confining pressure, and P_P is pore pressure. The parameters β and K_ϕ are two equivalent descriptions of the pore space stiffness. Ascertaining β versus ϕ or K_ϕ versus ϕ determines the rock bulk modulus K_{dry} versus ϕ.

Krief et al. (1990) used the data of Raymer et al. (1980) to find a relation for β versus ϕ empirically as follows:

$$(1 - \beta) = (1 - \phi)^{m(\phi)} \quad \text{where } m(\phi) = 3/(1 - \phi)$$

Next, they used the empirical result shown by Pickett (1963) and others that the dry rock Poisson's ratio is often approximately equal to the mineral Poisson's ratio, or $\mu_{dry}/K_{dry} = \mu_0/K_0$. Combining these two empirical results gives

$$K_{dry} = K_0(1 - \phi)^{m(\phi)} \quad \text{where } m(\phi) = 3/(1 - \phi)$$

$$\mu_{dry} = \mu_0(1 - \phi)^{m(\phi)}$$

Plots of K_{dry} versus ϕ, μ_{dry} versus ϕ, and β versus ϕ are shown in Figure 7.8.10.

It is clear from these plots that the effective moduli K_{dry} and μ_{dry} display the critical porosity behavior, for they approach zero at $\phi \approx 0.4$–0.5 (see previous

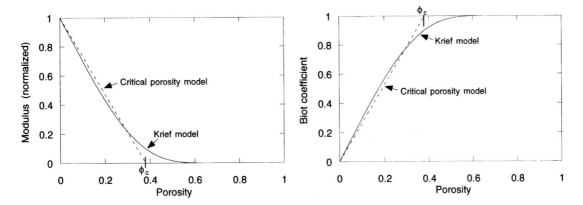

Figure 7.8.10. Left: Bulk and shear moduli (same curves when normalized by their mineral values) as predicted by Krief's model and a linear critical porosity model. Right: Biot coefficient predicted by both models.

discussion). This is no surprise because $\beta(\phi)$ is an empirical fit to shaley sand data, which always exhibit this behavior.

Compare these with the linear moduli–porosity relations suggested by Nur et al. (1995) for the critical porosity model

$$K_{dry} = K_0\left(1 - \frac{\phi}{\phi_c}\right)$$

$$0 \le \phi \le \phi_c$$

$$\mu_{dry} = \mu_0\left(1 - \frac{\phi}{\phi_c}\right)$$

where K_0 and μ_0 are the mineral moduli and ϕ_c is the critical porosity. These imply a Biot coefficient of

$$\beta = \begin{cases} \phi/\phi_c & 0 \le \phi \le \phi_c \\ 1 & \phi \le \phi_c \end{cases}$$

As shown in Figure 7.8.10, these linear forms of K_{dry}, μ_{dry}, and β are essentially the same as Krief's expressions in the range $0 \le \phi \le \phi_c$.

The Reuss average values for the moduli of a suspension, $K_{dry} = \mu_{dry} = 0$; $\beta = 1$ are essentially the same as Krief's expressions for $\phi > \phi_c$. Krief's nonlinear form results from trying to fit a single function $\beta(\phi)$ to the two mechanically distinct domains, $\phi < \phi_c$ and $\phi > \phi_c$. The critical porosity model expresses the result with simpler piecewise functions.

Expressions for any other pore fluid are obtained by combining the expression $K_{dry} = K_0(1 - \beta)$ of Krief et al. with Gassmann's equations. Although these are also nonlinear, they suggest a simple approximation

$$\frac{V_{\text{P-sat}}^2 - V_{\text{fl}}^2}{V_{\text{S-sat}}^2} = \frac{V_{\text{P0}}^2 - V_{\text{fl}}^2}{V_{\text{S0}}^2}$$

where $V_{\text{P-sat}}$, V_{P0}, and V_{fl} are the P-wave velocities of the saturated rock, the mineral, and the pore fluid, respectively, and $V_{\text{S-sat}}$ and V_{S0} are the S-wave velocities in the saturated rock and mineral. Rewriting slightly gives

$$V_{\text{P-sat}}^2 = V_{\text{fl}}^2 + V_{\text{S-sat}}^2 \left(\frac{V_{\text{P0}}^2 - V_{\text{fl}}^2}{V_{\text{S0}}^2} \right)$$

which is a straight line (in velocity-squared) connecting the mineral point $(V_{\text{P0}}^2, V_{\text{S0}}^2)$ and the fluid point $(V_{\text{fl}}^2, 0)$. We suggest that a more accurate (and nearly identical) model is to recognize that velocities tend toward those of a suspension at high porosity rather than toward a fluid, which yields the modified form

$$\frac{V_{\text{P-sat}}^2 - V_{\text{R}}^2}{V_{\text{S-sat}}^2} = \frac{V_{\text{P0}}^2 - V_{\text{R}}^2}{V_{\text{S0}}^2}$$

where V_{R} is the velocity of a suspension of minerals in a fluid given by the Reuss average (Section 4.2) or Wood's relation (Section 4.3) at the critical porosity.

It is easy to show that this modified form of Krief's expression is exactly equivalent to the linear (modified Voigt) K versus ϕ and μ versus ϕ relations in the critical porosity model with the fluid effects given by Gassmann.

Greenberg and Castagna (1992) have given empirical relations for estimating V_S from V_P in multimineralic, brine-saturated rocks based on empirical, polynomial V_P–V_S relations in pure monomineralic lithologies (Castagna et al. 1993). The shear wave velocity in brine-saturated composite lithologies is approximated by a simple average of the arithmetic and harmonic means of the constituent pure lithology shear velocities:

$$V_S = \frac{1}{2} \left\{ \left[\sum_{i=1}^{L} X_i \sum_{j=0}^{N_i} a_{ij} V_P^j \right] + \left[\sum_{i=1}^{L} X_i \left(\sum_{j=0}^{N_i} a_{ij} V_P^j \right)^{-1} \right]^{-1} \right\}$$

$$\sum_{i=1}^{L} X_i = 1$$

where

L = number of pure monomineralic lithologic constituents

X_i = volume fractions of lithological constituents

a_{ij} = empirical regression coefficients

N_i = order of polynomial for constituent i

V_P, V_S = P- and S-wave velocities (km/s) in composite brine-saturated, multimineralic rock

TABLE 7.8.1. **Regression coefficients for pure lithologies.**[a]

Lithology	a_{i2}	a_{i1}	a_{i0}	R^2
Sandstone	0	0.80416	−0.85588	0.98352
Limestone	−0.05508	1.01677	−1.03049	0.99096
Dolomite	0	0.58321	−0.07775	0.87444
Shale	0	0.76969	−0.86735	0.97939

[a] V_P and V_S in km/s: $V_S = a_{i2} V_P^2 + a_{i1} V_P + a_{i0}$ (Castagna et al., 1993).

Castagna et al. (1993) gave representative polynomial regression coefficients for pure monomineralic lithologies as detailed in Table 7.8.1. Note that the preceding relation is for 100 percent brine-saturated rocks. To estimate V_S from measured V_P for other fluid saturations, Gassmann's equation has to be used in an iterative manner. In the following, the subscript b denotes velocities at 100 percent brine saturation, and the subscript f denotes velocities at any other fluid saturation (e.g., oil or a mixture of oil, brine, and gas). The method consists of iteratively finding a (V_P, V_S) point on the brine relation that transforms, with Gassmann's relation, to the measured V_P and the unknown V_S for the new fluid saturation. The steps are as follows:

1) Start with an initial guess for V_{Pb}.
2) Calculate V_{Sb} corresponding to V_{Pb} from the empirical regression.

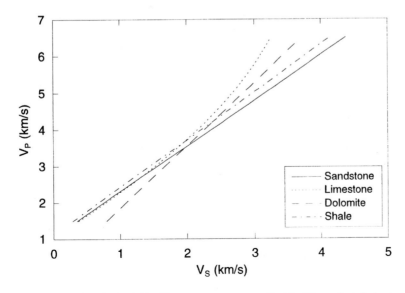

Figure 7.8.11. Typical V_P–V_S curves corresponding to the regression coefficients in Table 7.8.1.

3) Perform fluid substitution using V_{Pb} and V_{Sb} in the Gassmann equation to get V_{Sf}.

4) With the calculated V_{Sf} and the measured V_{Pf}, use the Gassmann relation to get a new estimate of V_{Pb}. Check the result against the previous value of V_{Pb} for convergence. If convergence criterion is met, stop; if not, go back to step 2 and continue.

When the measured P-velocity and desired S-velocity are for 100 percent brine saturation, then of course iterations are not required. The desired V_S is obtained from a single application of the empirical regression. This method requires prior knowledge of the lithology, porosity, saturation, and elastic moduli and densities of the constituent minerals and pore fluids.

EXAMPLE

Estimate, using the Greenberg–Castagna empirical relations, the shear wave velocity in a brine-saturated shaley sandstone (60 percent sandstone, 40 percent shale) with $V_P = 3.0$ km/s.

Here $L = 2$ with X_1(sandstone) $= 0.6$, and X_2 (shale) $= 0.4$.
The regressions for pure lithologic constituents give us

$$V_{S\text{-sand}} = 0.80416V_P - 0.85588 = 1.5566 \text{ km/s}$$

$$V_{S\text{-shale}} = 0.76969V_P - 0.86735 = 1.4417 \text{ km/s}$$

The weighted arithmetic and harmonic means are

$$V_{S\text{-arith}} = 0.6V_{S\text{-sand}} + 0.4V_{S\text{-shale}} = 1.5106 \text{ km/s}$$

$$V_{S\text{-harm}} = (0.6/V_{S\text{-sand}} + 0.4/V_{S\text{-shale}})^{-1} = 1.5085 \text{ km/s}$$

and finally the estimated V_S is given by

$$V_S = \frac{1}{2}(V_{S\text{-arith}} + V_{S\text{-harm}}) = 1.51 \text{ km/s}$$

Williams (1990) used empirical V_P–V_S relations from acoustic logs to differentiate hydrocarbon-bearing sandstones from water-bearing sandstones and shales statistically. His least-squares regressions are

$$V_P/V_S = 1.182 + 0.00422\Delta t_S \quad \text{(water-bearing sands)}$$

$$V_P/V_S = 1.276 + 0.00374\Delta t_S \quad \text{(shales)}$$

where Δt_S is the shear wave slowness in μs/ft. The effect of replacing water with more compressible hydrocarbons is a large decrease in P-wave velocity with little change (slight increase) in S-wave velocity. This causes a large reduction in the V_P/V_S ratio in hydrocarbon sands compared with water-saturated sands having a similar Δt_S. A measured V_P/V_S and Δt_S is classified as either water-bearing or hydrocarbon-bearing by comparing it with the regression and using a statistically determined threshold to make the decision. Williams chose the threshold so that the probability of correctly identifying a water-saturated sandstone is 95 percent. For this threshold a measured V_P/V_S is classified as water-bearing if

$$V_P/V_S \text{ (measured)} \geq \min[V_P/V_S \text{ (sand)}, V_P/V_S \text{ (shale)}] - 0.09$$

and as potentially hydrocarbon-bearing otherwise. Williams found that when $\Delta t_S < 130$ μs/ft (or $\Delta t_P < 75$ μs/ft), the rock is too stiff to give any statistically significant V_P/V_S anomaly upon fluid substitution.

Xu and White (1995) developed a theoretical model for velocities in shaley sandstones. The formulation uses the Kuster–Toksöz and differential effective medium theories to estimate the dry rock P- and S-velocities, and the low-frequency saturated velocities are obtained from Gassmann's equation. The sand–clay mixture is modeled with ellipsoidal inclusions of two different aspect ratios. The sand fraction has stiffer pores with aspect ratio $\alpha \approx 0.1$–0.15, whereas the clay-related pores are more compliant with $\alpha \approx 0.02$–0.05. The velocity model simulates the "V"-shaped velocity–porosity relation of Marion et al. (1992) for sand–clay mixtures. The total porosity $\phi = \phi_{\text{sand}} + \phi_{\text{clay}}$, where ϕ_{sand} and ϕ_{clay} are the porosities associated with the sand and clay fractions, respectively. These are approximated by

$$\phi_{\text{sand}} = (1 - \phi - V_{\text{clay}})\frac{\phi}{1 - \phi} = V_{\text{sand}}\frac{\phi}{1 - \phi}$$

$$\phi_{\text{clay}} = V_{\text{clay}}\frac{\phi}{1 - \phi}$$

where V_{sand} and V_{clay} denote the volumetric sand and clay content, respectively. Shale volume from logs may be used as an estimate of V_{clay}. Though the log-derived shale volume includes silts and overestimates clay content, results obtained by Xu and White justify its use. The properties of the solid mineral mixture are estimated by a Wyllie time average of the quartz and clay mineral velocities and arithmetic average of their densities by

$$\frac{1}{V_{P_0}} = \left(\frac{1 - \phi - V_{\text{clay}}}{1 - \phi}\right)\frac{1}{V_{P_{\text{quartz}}}} + \frac{V_{\text{clay}}}{1 - \phi}\frac{1}{V_{P_{\text{clay}}}}$$

$$\frac{1}{V_{S_0}} = \left(\frac{1 - \phi - V_{\text{clay}}}{1 - \phi}\right)\frac{1}{V_{S_{\text{quartz}}}} + \frac{V_{\text{clay}}}{1 - \phi}\frac{1}{V_{S_{\text{clay}}}}$$

$$\rho_0 = \left(\frac{1 - \phi - V_{\text{clay}}}{1 - \phi}\right)\rho_{\text{quartz}} + \frac{V_{\text{clay}}}{1 - \phi}\rho_{\text{clay}}$$

where subscript 0 denotes the mineral properties. These mineral properties are then used in the Kuster–Toksöz formulation along with the porosity and clay content to calculate dry rock moduli and velocities. The limitation of small pore concentration of the Kuster–Toksöz model is handled by incrementally adding the pores in small steps so that the noninteraction criterion is satisfied in each step. Gassmann's equations are used to obtain low-frequency saturated velocities. High-frequency saturated velocities are calculated by using fluid-filled ellipsoidal inclusions in the Kuster–Toksöz model.

The model can be used to predict shear wave velocities (Xu and White, 1994). Estimates of V_S may be obtained from known mineral matrix properties and measured porosity and clay content or from measured V_P and either porosity or clay content. Xu and White recommend using measurements of P-wave sonic log because it is more reliable than estimates of shale volume and porosity.

USES

The relations discussed in this section can be used to relate P and S velocity and porosity empirically for use in lithology detection and direct fluid identification.

ASSUMPTIONS AND LIMITATIONS

Strictly speaking, the empirical relations discussed in this section apply only to the set of rocks studied.

7.9 VELOCITY–DENSITY RELATIONS

SYNOPSIS

Many seismic modeling and interpretation schemes require, as a minimum, P-wave velocity V_P, S-wave velocity, V_S, and bulk density ρ_b. Laboratory and log measurements can often yield all three together. But there are many applications where only V_P is known, and density or V_S must be estimated empirically from V_P. Section 7.8 summarizes some V_P–V_S relations. We summarize here some popular and useful V_P–density relations. Castagna et al. (1993) give a very good summary of the topic.

Density is a simple volumetric average of the rock constituent densities and is closely related to porosity by

$$\rho_b = (1 - \phi)\rho_0 + \phi\rho_{fl}$$

where

$\rho_0 = $ density of mineral grains
$\rho_{fl} = $ density of pore fluids
$\phi = $ porosity

The problem is that velocity is often not very well related to porosity (and therefore to density). Cracks and crack-like flaws and grain boundaries can substantially decrease V_P and V_S, even though the cracks may have near-zero porosity. Velocity–porosity relations can be improved by fluid saturation and high effective pressures, both of which minimize the effect of these cracks. Consequently, we also expect velocity–density relations to be more reliable under high effective pressures and fluid saturation.

TABLE 7.9.1. Polynomial and power-law forms of the Gardner et al. (1974) velocity–density relationships presented by Castagna et al. (1993). Units are km/s and g/cm³ for velocity and density, respectively.

Coefficients for the equation $\rho_b = aV_P^2 + bV_P + c$

Lithology	a	b	c	V_P range (km/s)
Shale	−0.0261	0.373	1.458	1.5–5.0
Sandstone	−0.0115	0.261	1.515	1.5–6.0
Limestone	−0.0296	0.461	0.963	3.5–6.4
Dolomite	−0.0235	0.390	1.242	4.5–7.1
Anhydrite	−0.0203	0.321	1.732	4.6–7.4

Coefficients for the equation $\rho_b = dV_P^f$

Lithology	d	f	V_P range (km/s)
Shale	1.75	0.265	1.5–5.0
Sandstone	1.66	0.261	1.5–6.0
Limestone	1.50	0.225	3.5–6.4
Dolomite	1.74	0.252	4.5–7.1
Anhydrite	2.19	0.160	4.6–7.4

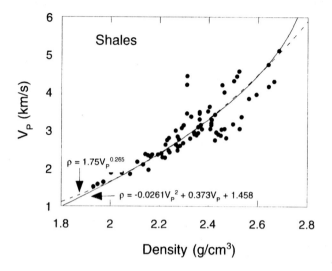

Figure 7.9.1. Both forms of Gardner's relations applied to log and laboratory shale data, as presented by Castagna et al. (1993).

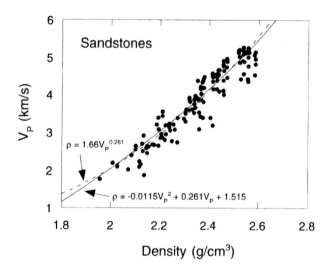

Figure 7.9.2. Both forms of Gardner's relations applied to log and laboratory sandstone data, as presented by Castagna et al. (1993).

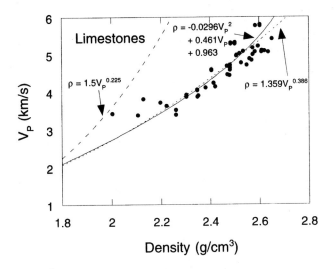

Figure 7.9.3. Both forms of Gardner's relations applied to laboratory limestone data. Note that the published power-law form does not fit as well as the polynomial. We also show a power-law form fit to these data, which agrees very well with the polynomial.

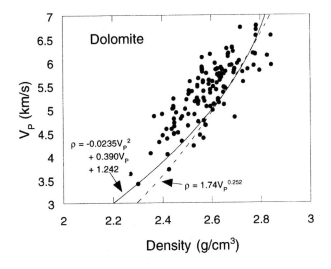

Figure 7.9.4. Both forms of Gardner's relations applied to laboratory dolomite data.

Gardner, Gardner, and Gregory (1974) suggested a useful empirical relation between P-wave velocity and density that represents an average over many rock types:

$$\rho_b \approx 1.741 V_P{}^{0.25}$$

where V_P is in km/s and ρ_b is in g/cm^3, or

$$\rho_b \approx 0.23 V_P{}^{0.25}$$

where V_P is in ft/s.

More useful predictions can be obtained by using the lithology-specific forms given by Gardner et al. Castagna et al. (1993) suggested slight improvements to Gardner's relations and summarized these, as shown in Table 7.9.1, in both polynomial and power-law form.

ASSUMPTION AND LIMITATIONS

Gardner's relations are empirical.

PART 8

FLOW AND DIFFUSION

SYNOPSIS

It was established experimentally by Darcy (1856) that the fluid flow rate is linearly related to the pressure gradient in a fluid-saturated porous medium by the following equation:

$$V_x = -\frac{\kappa}{\eta}\frac{\partial P}{\partial x}$$

where

$$V_x = \text{fluid flow rate in the } x \text{ direction}$$
$$\kappa = \text{permeability of the medium}$$
$$\eta = \text{viscosity of the fluid}$$
$$P = \text{fluid pressure}$$

This can be expressed more generally as

$$V = -\frac{\kappa}{\eta}\text{ grad } P$$

where

$$V = \text{vector fluid velocity field}$$
$$\kappa = \text{permeability tensor}$$

Permeability, κ, has units of area (m^2 in SI units), but the more convenient and traditional unit is the **Darcy**:

$$1 \text{ Darcy} = 0.986923 \times 10^{-12} \text{ m}^2$$

In a water-saturated rock with a permeability of 1 Darcy, a pressure gradient of 1 bar/cm gives a flow velocity of 1 cm/s.

Darcy's law for **multiphase flow** of immiscible fluids in porous media (with porosity ϕ) is often stated as

$$V_i = -\frac{\kappa_{ri} \kappa}{\eta_i} \text{ grad } P_i$$

where the subscript i refers to each phase, and κ_{ri} is the relative permeability of phase i. Simultaneous flow of multiphase immiscible fluids is possible only when the saturation of each phase is greater than the irreducible saturation and each phase is continuous within the porous medium. The relative permeabilities depend on the saturations S_i and show hysteresis, for they depend on the path taken to reach a particular saturation. The pressures P_i in any two phases are related by the capillary pressure P_c, which itself is a function of the saturations. For a two-phase system with fluid 1 as the wetting fluid and fluid 2 as the nonwetting fluid, $P_c = P_2 - P_1$. The presence of a residual nonwetting fluid can interfere considerably with the flow of the wetting phase. Hence the maximum value of κ_{r1} may be substantially less than 1. Extensions of Darcy's law for multiphase flow have been given (Dullien, 1992) that take into account cross-coupling between the fluid velocity in phase i and pressure gradient in phase j. The cross-coupling becomes important only at very high viscosity ratios ($\eta_i/\eta_j \gg 1$) because of an apparent lubricating effect.

In a one-dimensional immiscible displacement of fluid 2 by fluid 1 (e.g., water displacing oil in a water flood) the time history of the saturation $S_1(x, t)$ is governed by the following equation (Marle, 1981):

$$\frac{V}{\phi}\left[\frac{d\xi_1}{dS_1}\frac{\partial S_1}{\partial x} + \frac{\partial}{\partial x}\left(\Psi_1 \frac{\partial S_1}{\partial x}\right)\right] + \frac{\partial S_1}{\partial t} = 0$$

where

$$V = V_1 + V_2$$

$$\xi_1 = \frac{\frac{\eta_2}{\kappa_{r2}} + \frac{\kappa(\rho_1 - \rho_2)g}{V}}{\frac{\eta_1}{\kappa_{r1}} + \frac{\eta_2}{\kappa_{r2}}}$$

$$\Psi_1 = \frac{\kappa}{V}\frac{1}{\frac{\eta_1}{\kappa_{r1}} + \frac{\eta_2}{\kappa_{r2}}}\frac{dP_c}{dS_1}$$

where V_1 and V_2 are the Darcy fluid velocities in phases 1 and 2, respectively, and g is the acceleration due to gravity. The requirement that the two phases completely fill the pore space implies $S_1 + S_2 = 1$. Neglecting the effects of capillary pressure gives the **Buckley–Leverett** equation for immiscible displacement:

$$\left(\frac{V}{\phi} \frac{d\xi_1}{dS_1} \right) \frac{\partial S_1}{\partial x} + \frac{\partial S_1}{\partial t} = 0$$

This represents a saturation wave front traveling with velocity $(V/\phi)\, d\xi_1/dS_1$.

DIFFUSIVITY

If the fluid and the matrix containing it are compressible and elastic, the saturated system can take on the behavior of the diffusion equation. If we combine Darcy's law with the equation of mass conservation given by

$$\nabla \cdot (\rho V) + \frac{\partial(\rho\phi)}{\partial t} = 0$$

plus Hooke's law expressing the compressibility of the fluid, β_{fl}, and of the pore volume, β_{pv}, and drop nonlinear terms in pressure, we obtain the classical diffusion equation

$$\nabla^2 P = \frac{1}{D} \frac{\partial P}{\partial t}$$

where D is the diffusivity:

$$D = \frac{\kappa}{\eta\phi(\beta_{fl} + \beta_{pv})}$$

ONE-DIMENSIONAL DIFFUSION

Consider the one-dimensional diffusion that follows an initial pressure pulse:

$$P = P_0 \delta(x)$$

We get the standard result, illustrated in Figure 8.1.1, that

$$P(x, t) = \frac{P_0}{\sqrt{4\pi Dt}} e^{-x^2/4Dt} = \frac{P_0}{\sqrt{4\pi Dt}} e^{-\tau/t}$$

where the characteristic time depends on the length scale x and the diffusivity:

$$\tau = \frac{x^2}{4D}$$

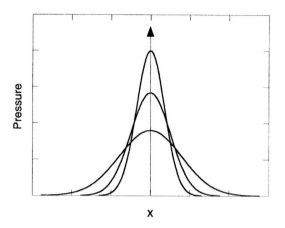

Figure 8.1.1

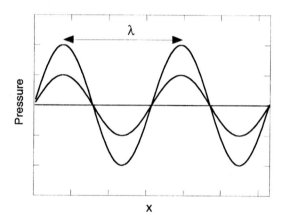

Figure 8.1.2

SINUSOIDAL PRESSURE DISTURBANCE

Consider an instantaneous sinusoidal pore-pressure disturbance in the saturated system as shown in Figure 8.1.2.

The disturbance will decay approximately as $e^{-t/\tau}$, where the diffusion time is again related to the length and diffusivity by

$$\tau_d = \frac{\lambda^2}{4D}$$

It is interesting to ask, When is the diffusion time equal to the period of the seismic wave causing such a disturbance? The seismic period is

$$\tau_s = \frac{\lambda}{V_p}$$

Equating τ_d to τ_s gives

$$\tau_d = \tau_s \Rightarrow \frac{\lambda^2}{4D} = \frac{\lambda}{V_p}$$

which finally gives

$$\tau = \frac{4D}{V_p}$$

For a rock with a permeability of 1 milliDarcy (1 mD), the critical frequency $(1/\tau)$ is 10 MHz.

ASSUMPTIONS AND LIMITATIONS

The following considerations apply to the use of Darcy's law:

- Darcy's law applies to representative elementary volume much larger than the grain or pore scale.
- Darcy's law is applicable when inertial forces are negligible in comparison to pressure gradient and viscous forces, and the Reynolds number Re is small (Re $\approx$ 1 to 10). The Reynolds number for porous media is given by Re $= \frac{\rho v l}{\eta}$, where ρ is the fluid density, η is the fluid viscosity, v is the fluid velocity, and l is a characteristic length of fluid flow determined by pore dimensions. At high Re, inertial forces can no longer be neglected in comparison with viscous forces, and Darcy's law breaks down.
- Some authors mention a minimum threshold pressure gradient below which there is very little flow (Bear, 1972). Non-Darcy behavior below the threshold pressure gradient has been attributed to streaming potentials in fine-grained soils, immobile adsorbed water layers, and clay–water interaction, giving rise to non-Newtonian fluid viscosity.
- When the mean free path of gas molecules is comparable to or larger than the dimensions of the pore space, the continuum description of gas flow becomes invalid. In these cases, measured permeability to gas is larger than the permeability to liquid. This is sometimes thought of as the increase in apparent gas permeability caused by slip at the gas–mineral interface. This is known as the Klinkenberg effect (Bear, 1972). The Klinkenberg correction is given by

$$\kappa_g = \kappa_l \left(1 + \frac{4c\lambda}{r} \right) = \kappa_l \left(1 + \frac{b}{P} \right)$$

where

κ_g = the gas permeability

κ_l = the liquid permeability

λ = the mean free path of gas molecules at the pressure P at which κ_g is measured

$c \approx 1$ = a proportionality factor

b = an empirical parameter that is best determined by measuring κ_g at several pressures.

r = the radius of the capillary

EXTENSIONS

When inertial forces are not negligible (large Reynolds number) Forchheimer suggested a nonlinear relation between fluid flux and pressure gradient (Bear, 1972) as follows:

$$\frac{dP}{dx} = aV + bV^2$$

where a and b are constants.

8.2 KOZENY–CARMAN RELATION FOR FLOW

SYNOPSIS

The Kozeny–Carman (Carman, 1961) relation provides a way to estimate the permeability of a porous medium in terms of generalized parameters such as porosity, surface area, particle size, and so forth. The derivation is based on flow through a pipe having a circular cross section with radius R. The flux in the pipe can be written as

$$Q = -\frac{\pi R^4}{8\eta}\frac{\partial P}{\partial x}$$

where p is the pressure and η is the viscosity. Comparison with Darcy's law,

$$Q = -\frac{\kappa A}{\eta}\frac{\partial P}{\partial x}$$

where κ is the permeability and A is the cross-sectional area, gives an effective permeability for the pipe expressed as

$$\kappa = \frac{\pi R^4}{8A} = \left(\frac{\pi R^2}{A}\right)\frac{R^2}{8}$$

The porosity, ϕ, and the specific surface area, S (defined as the pore surface area divided by the sample volume), can be expressed in terms of the properties of the pipe by the following relations:

$$\phi = \frac{\pi R^2}{A} \qquad S = \frac{2\pi R}{A}$$

Finally, we can express the permeability of the pipe in terms of the more general properties, ϕ and S, to get the Kozeny–Carman relation:

$$\kappa = \frac{B\phi^3}{\tau^2 S^2}$$

where B is a geometric factor and τ is the tortuosity (defined as the ratio of total flow-path length to length of the sample). An alternative form is given by

$$\kappa = \frac{B\phi^3 d^2}{\tau}$$

where d is a characteristic grain or pore dimension. The d^2 dependence of permeability has been experimentally verified numerous times.

A common extension of the Kozeny–Carman relation is to consider a packing of spheres. This allows a direct estimate of the specific surface area in terms of the porosity, $S = (3/2)(1 - \phi)/d$, which leads to the permeability expression

$$\kappa = B\frac{\phi^3}{(1 - \phi)^2}d^2$$

where the factor of 3/2 has been absorbed into the constant B.

Bourbié et al. (1987) discuss a more general form

$$\frac{\kappa}{d^2} \propto \phi^n$$

in which n has been observed experimentally to vary with porosity from $n \geq 7$ ($\phi < 5\%$) to $n \leq 2$ ($\phi > 30\%$). The Kozeny–Carman value of $n = 3$ appears to be appropriate for very clean materials such as Fontainebleau sandstone and sintered glass, whereas $n = 4$ or 5 is probably more appropriate for more general natural materials (Bourbié et al., 1987).

Mavko and Nur (1997) suggest that one explanation for the apparent dependence of the coefficient, n, on porosity is the existence of a percolation porosity, ϕ_c, below which the remaining porosity is disconnected and does not contribute to flow. Experiments suggest that this percolation porosity is on the order of 1–3 percent, although it depends on the mechanism of porosity reduction. The percolation effect can be incorporated into the Kozeny–Carman relations simply by replacing ϕ by $(\phi - \phi_c)$. The idea is that it is only the porosity *in excess* of the threshold porosity that determines the permeability. Substituting this into the Kozeny–Carman equation gives:

$$\kappa = B \frac{(\phi - \phi_c)^3}{(1 + \phi_c - \phi)^2} d^2$$

$$\approx B(\phi - \phi_c)^3 d^2$$

The result is that the derived $n = 3$ behavior can be retained while fitting the permeability behavior over a large range in porosity.

EXAMPLE

Calculate the permeability of a sandstone sample whose porosity is 0.32 and average grain size is 100 microns. Use the Kozeny–Carmen relation modified for a percolation porosity:

$$\kappa = B \frac{(\phi - \phi_c)^3}{(1 + \phi_c - \phi)^2} d^2$$

Assume that $B = 15$ and $\phi_c = 0.035$. The units of B in this equation are such that expressing d in microns gives permeability in millidarcy. Then

$$\kappa = B \frac{(\phi - \phi_c)^3}{(1 + \phi_c - \phi)^2} d^2 = 15 \frac{(0.32 - 0.035)^3}{(1 + 0.035 - 0.32)^2} 100^2$$

$$= 6792.3 \text{ milliDarcy} = 6.79 \text{ Darcy}$$

Compare the permeabilities κ_1 and κ_2 of two sandstones which have the same porosity and pore microstructure, but different average grain sizes, $d_1 = 80$ microns and $d_2 = 240$ microns.

Assuming that B and ϕ_c are the same for both sandstones since they have the same pore microstructure, we can express the ratio of their permeabilities as

$$\frac{\kappa_1}{\kappa_2} = \frac{d_1^2}{d_2^2} = \frac{80^2}{240^2} = \frac{1}{9}$$

The sandstone with larger average grain size has a higher permeability (by a factor of 9) even though both have the same total porosity.

MIXED PARTICLE SIZES

Extensive tests (Rumpf and Gupte, 1971; Dullien, 1991) on laboratory data for granular media with mixed particle sizes (poor sorting) suggest that the Kozeny–

Carman can still be applied if an effective or average particle size $\bar{D}$ is used defined by

$$\frac{1}{\bar{D}} = \frac{\int D^2 n(D) \, dD}{\int D^3 n(D) \, dD}$$

where $n(D)$ is the number distribution of each size particle. This can be written in terms of a discrete size distribution as follows:

$$\frac{1}{\bar{D}} = \frac{\sum_i D_i^2 n_i}{\sum_i D_i^3 n_i}$$

This can be converted to a mass distribution by noting that

$$m_i = n_i \frac{4}{3} \pi r_i^3 \rho_i = n_i \left(\frac{\pi \rho_i}{6} \right) D_i^3$$

where m_i, r_i, and ρ_i are the mass, radius, and density of the ith particle size. Then, one can write

$$\frac{1}{\bar{D}} = \frac{\sum_i \frac{m_i}{\rho_i D_i}}{\sum_i \frac{m_i}{\rho_i}}$$

If the densities of all particles are the same, then

$$\frac{1}{\bar{D}} = \frac{\sum_i \frac{m_i}{D_i}}{\sum_i m_i} = \sum_i \frac{f_i}{D_i}$$

where f_i is either the mass fraction or the volume fraction of each particle size.

An equivalent description of the particle mixture is in terms of specific surface areas. The specific surface area (pore surface area divided by bulk sample volume) for a mixture of spherical particles is

$$\bar{S} = 3(1 - \phi) \sum_i \frac{f_i}{r_i}$$

USES

The Kozeny–Carman relation can be used to estimate the permeability from geometric properties of a rock.

ASSUMPTIONS AND LIMITATIONS

The following assumptions and limitations apply to the Kozeny–Carman relation:

- The derivation is heuristic. Strictly speaking, it should hold only for rocks with porosity in the form of circular pipes. Nevertheless, in practice it often gives reasonable results. When possible it should be tested and calibrated for the rocks of interest.
- The rock is isotropic.
- Fluid-bearing rock is completely saturated.

8.3 VISCOUS FLOW

SYNOPSIS

In a Newtonian, viscous, incompressible fluid, stresses and velocities are related by Stokes law (Segel, 1987):

$$\sigma_{ij} = -P\delta_{ij} + 2\eta D_{ij}$$

$$D_{ij} = \frac{1}{2}\left(\frac{\partial v_i}{\partial x_j} + \frac{\partial v_j}{\partial x_i}\right)$$

where σ_{ij} denotes the elements of the stress tensor, v_i represents the components of the velocity vector, P is pressure, and η is dynamic viscosity.

For a simple shear flow between two walls, this law is called the Newton friction law and is expressed as

$$\tau = \eta\frac{dv}{dy}$$

where τ is the shear stress along the flow, v is velocity along the flow, and y is the coordinate perpendicular to the flow.

The Navier–Stokes equation for a Newtonian viscous incompressible flow is (e.g., Segel, 1987)

$$\rho\left(\frac{\partial v}{\partial t} + v \cdot \text{grad } v\right) = -\text{grad } P + \eta\Delta v$$

where ρ is density, t is time, and Δ is the Laplace operator.

USEFUL EXAMPLES OF VISCOUS FLOW

a) Steady two-dimensional laminar flow between two walls (Lamb, 1945):

$$v(y) = -\frac{1}{2\eta}\frac{dP}{dx}(R^2 - y^2), \qquad Q = -\frac{2}{3\eta}\frac{dP}{dx}R^3$$

where $2R$ is the distance between the walls and Q is the volumetric flow rate per unit width of the slit.

b) Steady two-dimensional laminar flow in a circular pipe:

$$v(y) = -\frac{1}{4\eta}\frac{dP}{dx}(R^2 - y^2), \qquad Q = -\frac{\pi}{8\eta}\frac{dP}{dx}R^4$$

where R is the radius of the pipe.

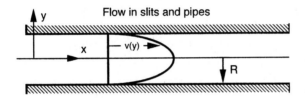

Figure 8.3.1

c) Steady laminar flow in a pipe of elliptical cross section (Lamb, 1945):

$$Q = -\frac{\pi}{4\eta}\frac{dP}{dx}\frac{a^3 b^3}{a^2 + b^2}$$

where a and b are the semiaxes of the cross section.

d) Steady laminar flow in a pipe of rectangular cross section:

$$Q = -\frac{1}{24\eta}\frac{dP}{dx}ab(a^2 + b^2) + \frac{8}{\pi^5\eta}\frac{dP}{dx}\sum_{n=1}^{\infty}\frac{1}{(2n-1)^5}$$

$$\times \left[a^4 \tanh\left(\pi b\frac{2n-1}{2a}\right) + b^4 \tanh\left(\pi a\frac{2n-1}{2b}\right) \right]$$

where a and b are the sides of the rectangle. For a square this equation yields

$$Q = -0.035144\frac{a^4}{\eta}\frac{dP}{dx}$$

e) Steady laminar flow in a pipe of equilateral triangular cross section with the length of a side b:

$$Q = -\frac{\sqrt{3}b^4}{320\eta}\frac{dP}{dx}$$

f) Steady laminar flow past a sphere (pressure on the surface of the sphere depends on the x coordinate only):

$$P = -\frac{3}{2}\eta \frac{x}{R^2} v$$

where v is the undisturbed velocity of the viscous flow (velocity at infinity), and the origin of the x-axis is in the center of the sphere of radius R. The total resistance force is $6\pi \eta v R$.

The combination

$$\text{Re} = \frac{\rho v R}{\eta}$$

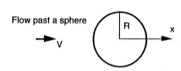

Flow past a sphere

Figure 8.3.2

where R is the characteristic length of a flow (e.g., pipe radius) is called the **Reynolds number**. Flows where $\text{Re} < 1$ are called **creeping flows** (viscous forces are the dominant factors in such flows). A flow becomes turbulent (i.e., nonlaminar) if $\text{Re} > 2000$.

USES

The equations presented in this section are used to describe viscous flow in pores.

ASSUMPTIONS AND LIMITATIONS

The equations presented in this section assume that the fluid is incompressible and Newtonian.

8.4 CAPILLARY FORCES

SYNOPSIS

A **surface tension**, γ, exists at the interface between two fluids or between a fluid and a solid. The surface tension at a fluid interface acts tangentially to an interface surface. If τ is a force acting on length l of the surface, then the surface tension is defined as $\gamma = \tau/l$. The unit of surface tension is force per unit length (N/m). A surface tension may be different at interfaces between different fluids. For example, γ "oil–water" (γ_{ow}) differs from γ "oil–gas" (γ_{og}) and from γ "water–gas" (γ_{wg}).

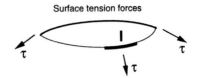

Surface tension forces

Figure 8.4.1

The equilibrium condition at a triple point "liquid–gas–solid" is

$$\gamma_{lg} \cos(\theta) + \gamma_{ls} = \gamma_{gs}, \qquad \cos(\theta) = \frac{\gamma_{gs} - \gamma_{ls}}{\gamma_{lg}}$$

The liquid is wetting if $\theta < 90°$ and nonwetting if $\theta > 90°$. The equilibrium will not exist if

$$|(\gamma_{gs} - \gamma_{ls})/\gamma_{lg}| > 1$$

If a sphere of oil (radius R) is floating inside water, pressure in the sphere P_0 will be greater than pressure in the water P_W resulting from surface tension. The difference between these two pressures is called **capillary pressure**: $P_c = P_0 - P_W$. The Laplace equation for capillary pressure is

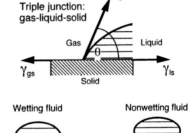

Figure 8.4.2

$$P_c = 2\gamma/R$$

In general, capillary pressure inside a surface of two principal radii of curvature R_1 and R_2 is

$$P_c = \gamma \left(\frac{1}{R_1} \pm \frac{1}{R_2} \right)$$

where the plus sign corresponds to the case in which the centers of curvature are located on the same side of the interface (e.g., a sphere), and the minus sign corresponds to the case in which the centers of the curvature are located on opposite sides (e.g., a torus).

If pore fluid is wetting, a pendular ring may exist at the point of contact of two grains. The capillary pressure in the ring depends on the contact angle θ of the liquid with the grains and on the radii of curvature, R_1 and

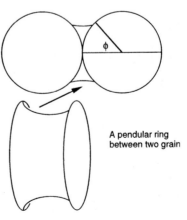

A pendular ring between two grain

Figure 8.4.3

R_2 as follows

$$P_c = \gamma \left(\frac{1}{R_1} - \frac{1}{R_2} \right) \cos(\theta)$$

If $\theta = 0$, we have at the contact point of two identical spherical grains of radius R (Gvirtzman and Roberts, 1991)

$$R_1 = R \frac{\sin(\phi) + \cos(\phi) - 1}{\cos(\phi)}, \qquad R_2 = R \frac{1 - \cos(\phi)}{\cos(\phi)}$$

where ϕ is the angle between the lines connecting the centers of the spheres and the line from the center of the sphere to the edge of the pendular ring. The ring will exist as long as its capillary pressure is smaller than the external pressure. The preceding formulas show that this condition is valid for $\phi < 53°$. The maximum volume of this ring is about 0.09 of the volume of an individual grain.

This maximum volume tends to decrease with increasing angle θ (e.g., it is 0.04 of the grain volume for $\theta = 32°$).

USES

The formulas presented in this section can be used to calculate wetting pore-fluid geometry in granular rocks.

ASSUMPTIONS AND LIMITATIONS

The formulas presented in this section presuppose that the grains are spherical.

8.5 DIFFUSION AND FILTRATION – SPECIAL CASES

SYNOPSIS

NONLINEAR DIFFUSION

Some rocks, such as coals, exhibit strong sensitivity of permeability (κ) and porosity (ϕ) to net pressure changes. During an injection test in a well, apparent permeability may be 10 to 20 times larger than that registered during a production test in the same well. In this situation, the assumption of constant permeability, which leads to the linear diffusion equation, is not valid. The following nonlinear diffusion equation must be used (Walls, Nur, and Dvorkin, 1991):

$$\frac{\eta\phi(\beta_{fl} + \beta_{pv})}{\kappa} \frac{\partial P}{\partial t} = \Delta P + (\beta_{fl} + \gamma)(\nabla P)^2$$

where fluid compressibility β_{fl} and pore-volume compressibility β_{pv} are defined as

$$\beta_{fl} = \frac{1}{\rho} \frac{\partial \rho}{\partial P}, \qquad \beta_{pv} = \frac{1}{v_p} \frac{\partial v_p}{\partial P}$$

where v_p is the pore volume. The permeability–pressure parameter γ is defined as

$$\gamma = \frac{1}{\kappa} \frac{\partial \kappa}{\partial P}$$

For one-dimensional plane filtration, the diffusion equation given above is

$$\frac{\eta\phi(\beta_{fl} + \beta_{pv})}{\kappa} \frac{\partial P}{\partial t} = \frac{\partial^2 P}{\partial x^2} + (\beta_{fl} + \gamma)\left(\frac{\partial P}{\partial x}\right)^2$$

For one-dimensional radial filtration it is

$$\frac{\eta\phi(\beta_{fl} + \beta_{pv})}{\kappa} \frac{\partial P}{\partial t} = \frac{\partial^2 P}{\partial r^2} + \frac{1}{r} \frac{\partial P}{\partial r} + (\beta_{fl} + \gamma)\left(\frac{\partial P}{\partial r}\right)^2$$

HYPERBOLIC EQUATION OF FILTRATION (DIFFUSION)

The diffusion equations presented above imply that changes in pore pressure propagate through a reservoir with infinitely high velocity. This artifact results from using the original Darcy's law, which states that volumetric fluid flow rate and pressure gradient are linearly related. In fact, according to Newton's second law, pressure gradient (or acting force) should also be proportional to acceleration (time derivative of the fluid flow rate). This modified Darcy's law was first used by Biot (1956) in his theory of dynamic poroelasticity:

$$-\frac{\partial P}{\partial x} = \frac{\eta}{\kappa} V_x + \frac{\tau\rho}{\phi} \frac{\partial V_x}{\partial t}$$

where τ is tortuosity. The latter equation, if used instead of the traditional Darcy's law

$$-\frac{\partial P}{\partial x} = \frac{\eta}{\kappa} V_x$$

yields the following hyperbolic equation that governs plane one-dimensional filtration:

$$\frac{\partial^2 P}{\partial x^2} = \tau \rho (\beta_{fl} + \beta_{pv}) \frac{\partial^2 P}{\partial t^2} + \frac{\eta \phi (\beta_{fl} + \beta_{pv})}{\kappa} \frac{\partial P}{\partial t}$$

This equation differs from the classical diffusion equation because of the inertia term

$$\tau \rho (\beta_{fl} + \beta_{pv}) \frac{\partial^2 P}{\partial t^2}$$

Changes in pore pressure propagate through a reservoir with a finite velocity, c:

$$c = \sqrt{\frac{1}{\tau \rho (\beta_{fl} + \beta_{pv})}}$$

This is the velocity of the slow Biot wave in a very rigid rock.

USES

The equations presented in this section can be used to calculate fluid filtration and pore-pressure pulse propagation in rocks.

ASSUMPTIONS AND LIMITATIONS

The equations presented in this section assume that the fluid is Newtonian and the flow is isothermal.

PART 9

ELECTRICAL PROPERTIES

9.1 BOUNDS AND EFFECTIVE MEDIUM MODELS

SYNOPSIS

If we wish to predict the effective dielectric permittivity ε of a mixture of phases theoretically, we generally need to specify (1) the volume fractions of the various phases, (2) the dielectric permittivity of the various phases, and (3) the geometric details of how the phases are arranged relative to each other. If we specify only the volume fractions and the constituent dielectric permittivities, then the best we can do is predict the upper and lower bounds.

BOUNDS: The best bounds, defined as giving the narrowest possible range without specifying anything about the geometries of the constituents, are the Hashin–Shtrikman (Hashin and Shtrikman, 1962) bounds. For a two-phase composite, the Hashin–Shtrikman bounds for dielectric permittivity are given by

$$\varepsilon^{\pm} = \varepsilon_1 + \frac{f_2}{(\varepsilon_2 - \varepsilon_1)^{-1} + \frac{f_1}{3\varepsilon_1}}$$

where

$$\varepsilon_1, \varepsilon_2 = \text{dielectric permittivity of individual phases}$$
$$f_1, f_2 = \text{volume fractions of individual phases}$$

Upper and lower bounds are computed by interchanging which material is termed 1 and which is termed 2. The expressions give the upper bound when the material with higher permittivity is termed 1 and the lower bound when the lower permittivity material is termed 1.

A more general form of the bounds, which can be applied to more than two phases (Berryman, 1995), can be written as

$$\varepsilon^{HS+} = \Sigma(\varepsilon_{max}), \qquad \varepsilon^{HS-} = \Sigma(\varepsilon_{min})$$

$$\Sigma(z) = \left\langle \frac{1}{\varepsilon(r) + 2z} \right\rangle^{-1} - 2z$$

where z is just the argument of the function $\Sigma(\cdot)$, and r is the spatial position. The brackets $\langle \cdot \rangle$ indicate an average over the medium, which is the same as an average over the constituents weighted by their volume fractions.

SPHERICAL INCLUSIONS: Estimates of the effective dielectric permittivity, ε^*, of a composite may be obtained by using various approximations, both self-consistent and non-self-consistent. The Clausius–Mossotti formula for a two-component material with spherical inclusions of material 2 in a host of material 1 is given by

$$\frac{\varepsilon_{CM}^* - \varepsilon_2}{\varepsilon_{CM}^* + 2\varepsilon_2} = f_1 \frac{\varepsilon_1 - \varepsilon_2}{\varepsilon_1 + 2\varepsilon_2}$$

or equivalently

$$\varepsilon_{CM}^* = \Sigma(\varepsilon_2)$$

This non-self-consistent estimate, also known as the Lorentz–Lorenz or Maxwell–Garnett equation, actually coincides with the Hashin–Shtrikman bounds. The two bounds are obtained by interchanging the role of spherical inclusions and host material.

The self-consistent (SC) or **coherent potential approximation** (Bruggeman, 1935; Landauer, 1952; Berryman, 1995) for the effective dielectric permittivity ε_{SC}^* of a composite made up of spherical inclusions of N phases may be written as

$$\sum_{i=1}^{N} f_i \frac{\varepsilon_i - \varepsilon_{SC}^*}{\varepsilon_i + 2\varepsilon_{SC}^*} = 0$$

or

$$\varepsilon_{SC}^* = \Sigma\left(\varepsilon_{SC}^*\right)$$

The solution, which is a fixed point of the function $\Sigma(\varepsilon)$, is obtained by iteration. In this approximation, all N components are treated symmetrically with no preferred host material.

In the differential effective medium (DEM) approach (Bruggeman, 1935; Sen, Scala, and Cohen, 1981), infinitesimal increments of inclusions are added to the host material until the desired volume fractions are reached. For a two-component composite with material 1 as the host containing spherical inclusions of material 2, the effective dielectric permittivity ε_{DEM}^* is obtained by solving the differential equation

$$(1 - y)\frac{d}{dy}\left[\varepsilon_{DEM}^*(y)\right] = \frac{\varepsilon_2 - \varepsilon_{DEM}^*(y)}{\varepsilon_2 + 2\varepsilon_{DEM}^*(y)}\left[3\varepsilon_{DEM}^*(y)\right]$$

where $y = f_2$, the volume fraction of spherical inclusions. The analytic solution with the initial condition $\varepsilon_{DEM}^*(y = 0) = \varepsilon_1$ is (Berryman, 1995)

$$\left[\frac{\varepsilon_2 - \varepsilon_{DEM}^*(y)}{\varepsilon_2 - \varepsilon_1}\right]\left[\frac{\varepsilon_1}{\varepsilon_{DEM}^*(y)}\right]^{\frac{1}{3}} = 1 - y$$

The DEM results are path dependent and depend on which material is chosen as the host. The Hanai–Bruggeman approach (Bruggeman, 1935; Hanai, 1968) starts with the rock as the host into which infinitesimal amounts of spherical inclusions of water are added. This results in a rock with zero dc conductivity because at each stage the fluid inclusions are isolated and there is no conducting path (usually the rock mineral itself does not contribute to the dc electrical conductivity). Sen et al. (1981) in their self-similar model of coated spheres start with water as the initial host and incrementally add spherical inclusions of mineral material. This leads to a composite rock with a finite dc conductivity because a conducting path always exists through the fluid. Both the Hanai–Bruggeman and the Sen et al. formulas are obtained from the DEM result with the appropriate choice of host and inclusion.

Bounds and estimates for electrical conductivity, σ, can be obtained from the preceding equations by replacing ε everywhere with σ. This is because the governing relations for dielectric permittivity and electrical conductivity (and other properties such as magnetic permeability and thermal conductivity) are mathematically equivalent (Berryman, 1995). The relationship between the dielectric constant, the electrical field, E, and the displacement field, D, is $D = \varepsilon E$. In the absence of charges $\nabla \cdot D = 0$, and $\nabla \times E = 0$ because the electric field is the gradient of a potential. Similarly, for

- Electrical conductivity, σ, $J = \sigma E$ from Ohm's law, where J is the current density, $\nabla \cdot J = 0$ in the absence of current source and sinks, and $\nabla \times E = 0$.
- Magnetic permeability, μ, $B = \mu H$, where B is the magnetic induction, H is the magnetic field, $\nabla \cdot B = 0$, and in the absence of currents $\nabla \times H = 0$.
- Thermal conductivity, κ, $q = -\kappa \nabla \theta$ from Fourier's law for heat flux q and temperature θ, $\nabla \cdot q = 0$ when heat is conserved, and $\nabla \times \nabla \theta = 0$.

ELLIPSOIDAL INCLUSIONS: Estimates for the effective dielectric permittivity of composites with nonspherical, ellipsoidal inclusions require the use of depolarizing factors L_a, L_b, L_c along the principal directions a, b, c of the ellipsoid. The generalization of the Clausius–Mossotti relation for randomly arranged ellipsoidal inclusions in an isotropic composite is (Berryman, 1995)

$$\frac{\varepsilon_{CM}^* - \varepsilon_m}{\varepsilon_{CM}^* + 2\varepsilon_m} = \sum_{i=1}^{N} f_i(\varepsilon_i - \varepsilon_m)R^{mi}$$

and the self-consistent estimate for ellipsoidal inclusions in an isotropic composite is (Berryman, 1995)

$$\sum_{i=1}^{N} f_i(\varepsilon_i - \varepsilon_{SC}^*)R^{*i} = 0$$

where R is a function of the depolarizing factors L_a, L_b, L_c:

$$R^{mi} = \frac{1}{9} \sum_{j=a,b,c} \frac{1}{L_j\varepsilon_i + (1 - L_j)\varepsilon_m}$$

The superscripts m and i refer to the host matrix phase and the inclusion phase. In the self-consistent formula, the superscript $*$ on R indicates that ε_m should be replaced by ε_{SC}^* in the expression for R. Depolarizing factors and the coefficient R for some specific shapes are given in the Table 9.1.1. Depolarizing factors for more general ellipsoidal shapes are tabulated by Osborn (1945) and Stoner (1945).

Ellipsoidal inclusion models have been used to model the effects of pore-scale fluid distributions on the effective dielectric properties of partially saturated rocks theoretically (Knight and Nur, 1987; Endres and Knight, 1992).

LAYERED MEDIA: Exact results for the long-wavelength effective dielectric permittivity of a layered medium (layer thicknesses much smaller than the

TABLE 9.1.1. Coefficient R and depolarizing factors L_j for some specific shapes. The subscripts m and i refer to the background and inclusion materials [from Berryman (1995)].

Inclusion shape	L_a, L_b, L_c	R^{mi}
Spheres	$\dfrac{1}{3}, \dfrac{1}{3}, \dfrac{1}{3}$	$\dfrac{1}{\varepsilon_i + 2\varepsilon_m}$
Needles	$0, \dfrac{1}{2}, \dfrac{1}{2}$	$\dfrac{1}{9}\left(\dfrac{1}{\varepsilon_m} + \dfrac{4}{\varepsilon_i + \varepsilon_m}\right)$
Disks	$1, 0, 0$	$\dfrac{1}{9}\left(\dfrac{1}{\varepsilon_i} + \dfrac{2}{\varepsilon_m}\right)$

wavelength) are given by (Sen et al., 1981)

$$\varepsilon_\parallel^* = \langle \varepsilon_i \rangle$$

and

$$\frac{1}{\varepsilon_\perp^*} = \left\langle \frac{1}{\varepsilon_i} \right\rangle$$

for fields parallel to the layer interfaces and perpendicular to the interfaces, respectively, where ε_i is the dielectric permittivity of each constituent layer. The direction of wave propagation is perpendicular to the field direction.

USES

The equations presented in this section can be used for the following purposes:

- To estimate the range of the average mineral dielectric constant for a mixture of mineral grains.
- To compute the upper and lower bounds for a mixture of mineral and pore fluid.

ASSUMPTIONS AND LIMITATIONS

The following assumption and limitation apply to the equations in this section:

- Most inclusion models assume the rock is isotropic.
- Effective medium theories are valid when wavelengths are much longer than the scale of the heterogeneities.

9.2 VELOCITY DISPERSION AND ATTENUATION

SYNOPSIS

The complex wavenumber associated with propagation of electromagnetic waves of angular frequency ω is given by $k = \omega\sqrt{\varepsilon\mu}$, where both ε, the dielectric permittivity, and μ, the magnetic permeability, are in general frequency dependent, complex quantities denoted by

$$\varepsilon = \varepsilon' - i\left(\frac{\sigma}{\omega} + \varepsilon''\right)$$

$$\mu = \mu' - i\mu''$$

where σ is the electrical conductivity. For most nonmagnetic earth materials, the magnetic permeability equals μ_0, the magnetic permeability of free space. The dielectric permittivity normalized by ε_0, the dielectric permittivity of free space, is often termed the relative dielectric permittivity or **dielectric constant** κ, which is a dimensionless measure of the dielectric behavior. The dielectric susceptibility $\chi = \kappa - 1$.

The real part, k_R, of the complex wavenumber describes the propagation of an electromagnetic wave field, whereas the imaginary part, k_I, governs the decay in field amplitude with propagation distance. A plane wave of amplitude E_0 propagating along the z-direction and polarized along the x-direction may be described by

$$E_x = E_0 e^{i(\omega t - kz)} = E_0 e^{-k_I z} e^{i(\omega t - k_R z)}$$

The **skin depth**, the distance over which the field amplitude falls to $1/e$ of its initial value, is equal to $1/k_I$. The dissipation may also be characterized by the loss tangent, the ratio of the imaginary part of the dielectric permittivity to the real part:

$$\tan \delta = \frac{\sigma}{\omega \varepsilon'} + \frac{\varepsilon''}{\varepsilon'}$$

Relations between the various parameters may be easily derived (e.g., Guéguen and Palciauskas, 1994). With $\mu = \mu_0$:

$$k_R = \omega \sqrt{\mu_0 \varepsilon'} \sqrt{\frac{1 + \cos \delta}{2 \cos \delta}}$$

$$k_I = \omega \sqrt{\mu_0 \varepsilon'} \sqrt{\frac{1 - \cos \delta}{2 \cos \delta}}$$

$$\varepsilon' = \frac{k_R^2 - k_I^2}{\mu_0 \omega^2}, \qquad \frac{\sigma}{\omega} + \varepsilon'' = \frac{2 k_R k_I}{\mu_0 \omega^2}$$

$$\tan \delta = \frac{2 k_R k_I}{k_R^2 - k_I^2}$$

$$V = \frac{\omega}{k_R} = \frac{1}{\sqrt{\mu_0 \varepsilon'}} \sqrt{\frac{2 \cos \delta}{1 + \cos \delta}}$$

where V is the phase velocity. When there is no attenuation $\delta = 0$ and

$$V = \frac{1}{\sqrt{\mu_0 \varepsilon'}} = \frac{c}{\sqrt{\kappa'}}$$

where $c = 1/\sqrt{\mu_0 \varepsilon_0}$ is the speed of light in a vacuum and κ' is the real part of the dielectric constant.

In the high-frequency *propagation* regime ($\omega \gg \sigma/\varepsilon'$), displacement currents dominate, whereas conduction currents are negligible. Electromagnetic waves propagate with little attenuation and dispersion. In this high-frequency limit the wavenumber is

$$k_{\text{hi}f} = \omega\sqrt{\mu_0 \varepsilon'}\sqrt{1 - i\frac{\varepsilon''}{\varepsilon'}}$$

In the low-frequency *diffusion* regime ($\omega \ll \sigma/\varepsilon'$), conduction currents dominate, and an electromagnetic pulse tends to spread out with a $\sqrt{t}$ time dependence characteristic of diffusive processes. The wavenumber in this low-frequency limit is

$$k_{\text{low}f} = (1 - i)\sqrt{\frac{\omega\sigma\mu_0}{2}}$$

For typical crustal rocks the diffusion region falls below about 100 kHz.

The Debye (1945) and the Cole–Cole (1941) models are two common phenomenological models that describe the frequency dependence of the complex dielectric constant. In the **Debye** model, which is identical to the standard linear solid model for viscoelasticity (see Section 3.6), the dielectric constant as a function of frequency is given by

$$\kappa(\omega) = \kappa' - i\kappa'' = \kappa_\infty + \frac{\kappa_0 - \kappa_\infty}{1 + i\omega\tau}$$

$$\kappa' = \kappa_\infty + \frac{\kappa_0 - \kappa_\infty}{1 + (\omega\tau)^2}$$

$$\kappa'' = (\kappa_0 - \kappa_\infty)\frac{\omega\tau}{1 + (\omega\tau)^2}$$

where

$$\tau = \text{characteristic relaxation time}$$

$$\kappa_0 = \text{low-frequency limit}$$

$$\kappa_\infty = \text{high-frequency limit}$$

The **Cole–Cole** model is given by

$$\kappa(\omega) = \kappa_\infty + \frac{\kappa_0 - \kappa_\infty}{1 + (i\omega\tau)^{1-\alpha}}, \quad 0 \le \alpha \le 1$$

When the parameter α equals 0, the Cole–Cole model reduces to the Debye model. Sherman (1988) described the frequency-dependent dielectric constant of brine saturated rocks as a sum of two Debye models – one for interfacial polarization (below 1 GHz) and another for dipole polarization in brine (above 1 GHz).

The amplitude reflection and transmission coefficients for uniform, linearly polarized, homogeneous plane waves are given by **Fresnel's equations** (e.g., Jackson, 1975). For a plane wave incident from medium 1 onto an interface between two isotropic, homogeneous half-spaces, medium 1 and 2, the equations are

- Electric field transverse to the plane of incidence (TE mode):

$$\frac{E_r^\perp}{E_i^\perp} = \frac{\sqrt{\mu_1 \varepsilon_1}\cos\theta_i - \frac{\mu_1}{\mu_2}\sqrt{\mu_2 \varepsilon_2 - \mu_1 \varepsilon_1 \sin^2\theta_i}}{\sqrt{\mu_1 \varepsilon_1}\cos\theta_i + \frac{\mu_1}{\mu_2}\sqrt{\mu_2 \varepsilon_2 - \mu_1 \varepsilon_1 \sin^2\theta_i}}$$

$$\frac{E_t^\perp}{E_i^\perp} = \frac{2\sqrt{\mu_1 \varepsilon_1}\cos\theta_i}{\sqrt{\mu_1 \varepsilon_1}\cos\theta_i + \frac{\mu_1}{\mu_2}\sqrt{\mu_2 \varepsilon_2 - \mu_1 \varepsilon_1 \sin^2\theta_i}}$$

- Electric field in the plane of incidence (TM mode):

$$\frac{E_r^\parallel}{E_i^\parallel} = \frac{\sqrt{\mu_1 \varepsilon_1}\sqrt{\mu_2 \varepsilon_2 - \mu_1 \varepsilon_1 \sin^2\theta_i} - \mu_1 \varepsilon_2 \cos\theta_i}{\sqrt{\mu_1 \varepsilon_1}\sqrt{\mu_2 \varepsilon_2 - \mu_1 \varepsilon_1 \sin^2\theta_i} + \mu_1 \varepsilon_2 \cos\theta_i}$$

$$\frac{E_t^\parallel}{E_i^\parallel} = \frac{2\sqrt{\mu_1 \varepsilon_1 \mu_2 \varepsilon_2}\cos\theta_i}{\sqrt{\mu_1 \varepsilon_1}\sqrt{\mu_2 \varepsilon_2 - \mu_1 \varepsilon_1 \sin^2\theta_i} + \mu_1 \varepsilon_2 \cos\theta_i}$$

where

$$\theta_i = \text{angle of incidence}$$
$$E_i^\perp, E_r^\perp, E_t^\perp = \text{incident, reflected, and transmitted transverse electric field amplitudes}$$
$$E_i^\parallel, E_r^\parallel, E_t^\parallel = \text{incident, reflected, and transmitted parallel electric field amplitudes}$$
$$\mu_1, \mu_2 = \text{magnetic permeability of medium 1 and 2}$$
$$\varepsilon_1, \varepsilon_2 = \text{dielectric permittivity of medium 1 and 2}$$

The positive direction of polarization is taken to be the same for incident, reflected, and transmitted electric fields. In terms of the electromagnetic plane-wave impedance $Z = \sqrt{\mu/\varepsilon} = \omega\mu/k$, the amplitude reflection coefficients may be expressed as

$$R_{\text{TE}}^\perp = \frac{E_r^\perp}{E_i^\perp} = \frac{Z_2\cos\theta_i - Z_1\sqrt{1 - \left(\frac{Z_1 \varepsilon_1}{Z_2 \varepsilon_2}\right)^2 \sin^2\theta_i}}{Z_2\cos\theta_i + Z_1\sqrt{1 - \left(\frac{Z_1 \varepsilon_1}{Z_2 \varepsilon_2}\right)^2 \sin^2\theta_i}}$$

$$R_{\text{TM}}^\parallel = \frac{E_r^\parallel}{E_i^\parallel} = \frac{Z_2\sqrt{1 - \left(\frac{Z_1 \varepsilon_1}{Z_2 \varepsilon_2}\right)^2 \sin^2\theta_i} - Z_1\cos\theta_i}{Z_2\sqrt{1 - \left(\frac{Z_1 \varepsilon_1}{Z_2 \varepsilon_2}\right)^2 \sin^2\theta_i} + Z_1\cos\theta_i}$$

For normal incidence, the incident plane is no longer uniquely defined, and the difference between transverse and parallel modes disappears. The reflection coefficient is then

$$R_{\text{TE}}^{\perp} = \frac{Z_2 - Z_1}{Z_2 + Z_1} = R_{\text{TM}}^{\parallel}$$

Electromagnetic wave propagation in layered media can be calculated by the use of propagator matrices (Ward and Hohmann, 1987). The electric and magnetic fields at the top and bottom of the stack of layers are related by a product of propagator matrices, one for each layer. The calculations are done in the frequency domain and include the effects of all multiples. For waves traveling perpendicularly to the layers with layer impedances and thicknesses Z_j and d_j, respectively

$$\begin{bmatrix} E_y \\ H_x \end{bmatrix}_{j-1} = \mathbf{A}_j \begin{bmatrix} E_y \\ H_x \end{bmatrix}_j$$

Each layer matrix $\mathbf{A}_j$ has the form

$$\mathbf{A}_j = \begin{bmatrix} \cosh(ik_jd_j) & -Z_j\sinh(ik_jd_j) \\ -\frac{1}{Z_j}\sinh(ik_jd_j) & \cosh(ik_jd_j) \end{bmatrix}$$

USES

The results described in this section can be used for computing electromagnetic wave propagation, velocity dispersion, and attenuation.

ASSUMPTIONS AND LIMITATIONS

The results described in this section are based on the following assumptions:

- Isotropic homogeneous media, except for layered media.
- Plane-wave propagation.

9.3 EMPIRICAL RELATIONS

SYNOPSIS

The **Lichtnecker–Rother** empirical formula for the effective dielectric constant κ^* of a mixture of N constituents is given by a simple volumetric power-law average of the dielectric constants of the constituents (Sherman, 1986; Guéguen and Palciauskas, 1994):

$$\kappa^* = \left[\sum_{i=1}^{N} f_i(\kappa_i)^\gamma \right]^{1/\gamma}, \quad -1 \le \gamma \le 1$$

where

$$\kappa_i = \text{dielectric constant of individual phases}$$
$$f_i = \text{volume fractions of individual phases}$$

For $\gamma = 1/2$ this is equivalent to the complex refractive index method (**CRIM**) formula:

$$\sqrt{\kappa^*} = \sum_{i=1}^{N} f_i \sqrt{\kappa_i}$$

The CRIM equation (Meador and Cox, 1975; Endres and Knight, 1992) is analogous to the time average equation of Wyllie (see Section 7.3) because the velocity of electromagnetic wave propagation is inversely proportional to $\sqrt{\kappa}$. The CRIM empirical relation has been found to give reasonable results at high frequencies (above ~ 0.5 GHz). The **Odelevskii** formula for two phases is (Shen, 1985)

$$\varepsilon^* = B + \left(B^2 + \frac{1}{2}\varepsilon_1\varepsilon_1 \right)^{1/2}$$

$$B = \frac{1}{4}[(3f_1 - 1)\varepsilon_1 + (3f_2 - 1)\varepsilon_2]$$

Typically, only the real part of the dielectric permittivity is used in this empirical formula.

Topp's relation (Topp et al., 1980), based on measurements on a variety of soil samples at frequencies of 20 MHz to 1 GHz, is used widely in interpretation of time domain reflectometry (**TDR**) measurements for volumetric soil water content. The empirical relations are

$$\theta_v = -5.3 \times 10^{-2} + 2.92 \times 10^{-2}\kappa_a - 5.5 \times 10^{-4}\kappa_a^2 + 4.3 \times 10^{-6}\kappa_a^3$$

$$\kappa_a = 3.03 + 9.3\,\theta_v + 146.0\,\theta_v^2 - 76.7\,\theta_v^3$$

where κ_a is the apparent dielectric constant as measured by pulse transmission (such as in TDR or coaxial transmission lines) with dielectric losses excluded, and θ_v is the volumetric soil water content, the ratio of volume of water to the total volume of the sample. The estimation error for the data of Topp et al. (1980) was about 1.3 percent. The relations do not violate the Hashin–Shtrikman bounds for most materials. They do not give good estimates for soils with high clay content or organic matter and should be recalibrated for such material. Brisco et al. (1992) published empirical relations between volumetric soil water content and the real part of the dielectric constant measured by portable dielectric probes

TABLE 9.3.1. Regression coefficients and coefficient of determination R^2 for empirical relations between volumetric water content and dielectric constant (real part) at different portable dielectric probe frequencies (Brisco et al., 1992).

$$\theta_v = a + b\kappa' + c\kappa'^2 + d\kappa'^3$$

Frequency (GHz)	a	b	c	d	R^2
9.3 (X-band)	-3.58×10^{-2}	4.23×10^{-2}	-0.153×10^{-4}	17.7×10^{-6}	0.86
5.3 (C-band)	-1.01×10^{-2}	2.62×10^{-2}	-4.71×10^{-4}	4.12×10^{-6}	0.91
1.25 (L-band)	-2.78×10^{-2}	2.80×10^{-2}	-5.86×10^{-4}	5.03×10^{-6}	0.95
0.45 (P-band)	-1.88×10^{-2}	2.46×10^{-2}	-4.34×10^{-4}	3.61×10^{-6}	0.95

(PDP) utilizing frequencies from 0.45 to 9.3 GHz. The PDP measures both the real and imaginary components of the dielectric constant. In general TDR can sample soil layers 0–5 cm or deeper, whereas the PDP samples layers of about 1 cm thickness. Empirical relations in different frequency bands from Brisco et al. are summarized in Table 9.3.1.

Olhoeft (1979) obtained the following empirical relation between the measured effective dielectric constant and density for dry rocks:

$$\kappa' = \left(\kappa_0'^{1/\rho_0}\right)^\rho = 1.91^\rho$$

where

κ_0', ρ_0 = mineral dielectric constant and mineral density (gm/cm^3)
κ', ρ = dry rock dielectric constant and dry bulk density (gm/cm^3)

The coefficient 1.91 was obtained from a best fit to data on a variety of terrestrial and lunar rock samples. The relation becomes poor for rocks with water-containing clays and conducting minerals like sulfides and magnetite.

Knight and Nur (1987) measured the complex dielectric constant of eight different sandstones at different saturations and frequencies. They obtained a power-law dependence of κ' (the real part of the complex dielectric constant) on frequency ω expressed by

$$\kappa' = A\omega^{-\alpha}$$

where A and α are empirical parameters determined by fit to data and depend on saturation and rock type. For the different samples measured by Knight and Nur, α ranged from 0.08 to 0.266 at a saturation of 0.36 by deionized water. At the same saturation log A ranged from about 1.1 to 1.8.

Mazáč et al. (1990) correlated aquifer hydraulic conductivity K determined from pumping tests with electrical resistivities ρ interpreted from vertical electrical sounding. Their relation is

$$K(10^{-5}\,\text{m/s}) = \frac{\rho^{1.195}\,(\text{ohm-m})}{97.5}$$

The correlation coefficient was 0.871.

USES

The equations presented in this section can be used to relate rock and soil properties such as porosity, saturation, soil moisture content, and hydraulic conductivity to electrical measurements.

ASSUMPTIONS AND LIMITATIONS

The relations are empirical and therefore strictly valid only for the data set from which they were derived. The relations may need to be recalibrated to specific locations or rock and soil types.

9.4 ELECTRICAL CONDUCTIVITY IN POROUS ROCKS

SYNOPSIS

Most crustal rocks are made up of minerals that are semiconductors or insulators (silicates and oxides). Conducting currents in fluid-saturated rocks caused by an applied dc voltage arise primarily from the flow of ions within the pore fluids. The ratio of the conductivity of the pore fluid to the bulk conductivity of the fully saturated rock is known as the **formation factor**, F (Archie, 1942):

$$F = \frac{\sigma_w}{\sigma} = \frac{R}{R_w}$$

where

$$\sigma_w, R_w = \text{conductivity and resistivity of pore fluid}$$
$$\sigma, R = \text{conductivity and resistivity of fully saturated rock}$$

The Hashin–Shtrikman lower bound on F for a rock with porosity ϕ is (Berryman, 1995)

$$F^{HS-} = 1 + \frac{3}{2}\frac{1-\phi}{\phi}$$

The differential effective medium (DEM) theory of Sen et al. (1981) predicts (as $\omega \to 0$ where ω is the frequency)

$$\sigma = \sigma_w \phi^{3/2}$$

or

$$F_{DEM} = \phi^{-3/2}$$

In this version of the DEM model, spheres of nonconducting mineral grains are embedded in the conducting fluid host so that a conducting path always exists through the fluid for all porosities (see Section 9.1). The exponent depends on the shape of the inclusions and is greater than 1.5 for platey or needle-like inclusions.

Archie's law (1942), which forms the basis for resistivity log interpretation, is an empirical relation relating formation factor to porosity in *brine*-saturated *clean* (no shale) reservoir rocks:

$$F = \phi^{-m}$$

The exponent m (sometimes termed the cementation exponent) varies approximately between 1.3 to 2.5 for most sedimentary rocks and is close to 2 for sandstones. For natural and artificial unconsolidated sands and glass beads, m is close to 1.3 for spherical grains and increases to 1.9 for thin disk-like grains (Wyllie and Gregory, 1953; Jackson et al., 1978). Carbonates show a much wider range of variation and have m values as high as 5 (Focke and Munn, 1987). The minimum value of m is 1 when porosity is 100 percent and the rock is fully saturated with brine. This corresponds to an open fracture.

Archie's law is sometimes written as

$$F = (\phi - \phi_0)^{-m}$$

or

$$F = a\phi^{-m}$$

where ϕ_0 is a percolation porosity below which there are no conducting pathways and the rock conductivity is zero and a is an empirical constant close to 1. A value different from 1 (usually greater than 1) results from trying to fit an Archie-like model to rocks that do not follow the Archie behavior. Clean, well-sorted sands with electrical conduction occurring only by diffusion of ions in the pore fluid are best described by Archie's law. Shaley sands, rocks with moldic secondary porosity, and rocks with isolated microporous grains are examples of non-Archie rocks (Herrick, 1988).

Archie's second law for saturation relates the dc resistivity, R_t, of a partially saturated rock to the brine saturation, S_w, and the porosity by

$$S_w^{-n} = \frac{R_t}{R} = \phi^m \frac{R_t}{R_w}$$

where R is the dc resistivity of the same rock at $S_w = 1$, and the saturation exponent, n, derived empirically, is around 2. The value of n depends on the type of the pore fluid and is different for gas-brine saturation versus oil-brine saturation. Experimentally, saturation exponents for oil-wet porous media have been found to be substantially higher ($n \approx 2.5$ to 9.5) than for water-wet media (Sharma, Garrouch, and Dunlap, 1991). In terms of conductivity Archie's second law may be expressed as

$$\sigma_t = \left(S_w^n \phi^m \right) \sigma_w$$

where $\sigma_t = 1/R_t$ is the conductivity of the partially saturated rock. Archie's empirical relations have been found to be applicable to a remarkably wide range of rocks (Ransom, 1984)

SHALEY SANDS: Electrical conductivity in shaley sands is complicated by the presence of clays. Excess ions in a diffuse double layer around clay particles provide current conduction pathways along the clay surface in addition to the current flow by ions diffusing through the bulk pore fluid. The conductivity of this surface layer depends on the brine conductivity, and hence the overall bulk conductivity of the saturated rock is a nonlinear function of the brine conductivity. A wide variety of formulations have been used to model conductivity in shaley sands, and Worthington (1985) describes over thirty shaley sand models used in well log interpretation. Almost all of the models try to modify Archie's relation and account for the excess conductivity by introducing a shale conductivity term X:

$$\sigma = \frac{1}{F}\sigma_w, \qquad \text{clean sands, Archie}$$

$$\sigma = \frac{1}{F}\sigma_w + X, \quad \text{shaley sands}$$

The various models differ in their choice of X. Some of the earlier models described X in terms of the volume of shale V_{sh}, as determined from logs:

$$\sigma = \frac{1}{F}\sigma_w + V_{sh}\sigma_{sh} \quad \text{Simandoux (1963)}$$

$$\sqrt{\sigma} = \sqrt{\frac{1}{F}\sigma_w + V_{sh}^\alpha \sqrt{\sigma_{sh}}}, \quad \alpha = 1 - \frac{V_{sh}}{2}$$

Poupon and Leveaux (1971) "Indonesia formula"

where σ_{sh} is the conductivity of fully brine-saturated shale. Although these equations are applicable to log interpretation and may be used without calibration with core data, they do not have much physical basis and do not allow a complete representation of conductivity behavior for all ranges of σ_w. More recent models attempt to capture the physics of the diffuse ion double layer surrounding clay particles. Of these, the Waxman–Smits model (Waxman and Smits, 1968) and its various modifications such as the dual-water model (Clavier, Coates, and Dumanoir, 1984) and the Waxman–Smits–Juhász model (Juhász, 1981) are the most widely accepted. The Waxman–Smits formula is

$$\sigma = \frac{1}{F}(\sigma_w + BQ_v)$$

$$B = 4.6\left(1 - 0.6e^{-\sigma_w/1.3}\right)$$

$$Q_v = \frac{CEC(1 - \phi)\rho_0}{\phi}$$

where

$$CEC = \text{cation exchange capacity}$$
$$\rho_0 = \text{mineral grain density}$$

Note that here and in the following shaley sand equations $F = a\phi^{-m}$ and not σ_w/σ. The cation exchange capacity is a measure of the excess charges and Q_v is the charge per unit pore volume. Clays often have an excess negative electrical charge within the sheet-like particles. This is compensated by positive counterions clinging to the outside surface of the dry clay sheets. The resulting positive surface charge is a property of the dry clay mineral and is called the cation exchange capacity (Clavier et al., 1984). In the presence of an electrolytic solution such as brine, the electrical forces holding the positive counterions at the clay surface are reduced. The counterions can move along the surface contributing to the electrical conductivity. The average mobility of the ions is described by B. The parameter B is a source of uncertainty, and several expressions for it have been developed since the original paper. Juhász (1981) gives the following expressions for B:

$$B = \frac{-5.41 + 0.133T - 1.253 \times 10^{-4}T^{-2}}{1 + R_w^{1.23}(0.025T - 1.07)}$$

where T is the temperature in degrees Fahrenheit or

$$B = \frac{-1.28 + 0.225T - 4.059 \times 10^{-4}T^{-2}}{1 + R_w^{1.23}(0.045T - 0.27)}$$

for temperature in degrees Celsius. Application of the Waxman–Smits equation requires calibration with core CEC measurements, which are not always available.

The normalized Waxman–Smits or Waxman–Smits–Juhász model (Juhász, 1981) does not require CEC data because it uses V_{sh} derived from logs to estimate

Q_v by normalizing it to the shale response. In this model

$$BQ_v = Q_{vn}(\sigma_{wsh} - \sigma_w)$$

$$Q_{vn} = \frac{Q_v}{Q_{vsh}} = \frac{V_{sh}\phi_{sh}}{\phi}$$

where ϕ is the total porosity (density porosity), ϕ_{sh} is the total shale porosity, and σ_{wsh} is the shale water conductivity obtained from $\sigma_{wsh} = F\sigma_{sh}$, where σ_{sh} is the conductivity of 100 percent brine-saturated shale. The normalized Q_v ranges from 0 in clean sands to 1 in shales. Brine saturation S_w can be obtained from these models by solving (Bilodeaux, 1997)

$$S_w = \left[\frac{FR_w}{R_t\left(1 + \frac{R_wBQ_v}{S_w}\right)} \right]^{\frac{1}{n}}$$

For $n = 2$ the explicit solution (ignoring the negative root) is

$$S_w = \sqrt{\frac{FR_w}{R_t} + \left(\frac{BQ_vR_w}{2}\right)^2} - \frac{BQ_vR_w}{2}$$

The dual water model divides the total water content into the bound clay water, whose conductivity depends only on the clay counterions, and the far water, away from the clay, whose conductivity corresponds to the ions in the bulk formation water (Clavier et al., 1984). The bound water reduces the water conductivity σ_w by a factor $(1 - \alpha v_Q Q_v)$. The dual water model formula is (Clavier et al., 1984; Sen and Goode, 1988)

$$\sigma = \phi^m[\sigma_w(1 - \alpha v_Q Q_v) + \beta Q_v]$$

where v_Q is the amount of clay water associated with 1 milliequivalent of clay counterions, β is the counterion mobility in the clay double layer, and α is the ratio of the diffuse double-layer thickness to the bound water layer thickness. At high salinities (salt concentration exceeding 0.35 mol/mL) $\alpha = 1$. At low salinities it is a function of σ_w, and is given by

$$\alpha = \sqrt{\frac{\gamma_1\langle n_1\rangle}{\gamma\langle n\rangle}}$$

$\langle n\rangle$ = salt concentration in bulk water at 25°C in mol/mL

γ = NaCl activity coefficient at that concentration

$\langle n_1\rangle = 0.35$ mol/mL

$\gamma_1 = 0.71$, the corresponding NaCl activity coefficient

Although v_Q and β have temperature and salinity dependence, Clavier et al. (1984) recommend the following values for v_Q and β:

$$v_Q = 0.28 \text{ mL/meq}$$

$$\beta = 2.05 \text{ (S/m)/(meq/cm}^3)$$

These values are based on analysis of CEC data for clays and conductivity data on core samples. At low salinities, v_Q varies with $\sqrt{T}$ and increases by about 26 percent from 25 to 200°C.

Generalizing from theoretical solutions for electrolytic conduction past charged spheres in the presence of double layers, Sen and Goode (1988) proposed the following shaley–sand equation:

$$\sigma = \frac{1}{F}\left(\sigma_w + \frac{AQ_v}{1 + CQ_v/\sigma_w}\right) + EQ_v$$

The constants A and C depend on pore geometry and ion mobility, and the term EQ_v accounts for conductivity by surface counterions even when water conductivity is zero. Sen and Goode were able to express the relation in terms of Archie's exponent m by fitting to core data (about 140 cores)

$$\sigma = \phi^m\left(\sigma_w + \frac{m1.93Q_v}{1 + 0.7/\sigma_w}\right) + 1.3\phi^m Q_v$$

where conductivities are in mho/m and Q_v is in meq/mL.

In the limit of no clay ($Q_v = 0$) or no counterion mobility ($A = C = 0$) the expression reduces to Archie's equation. In the limits of high and low brine conductivity, with nonzero Q_v, the expression becomes

$$\sigma = \frac{1}{F}(\sigma_w + AQ_v) + EQ_v \quad \text{high } \sigma_w \text{ limit}$$

$$\sigma = \frac{1}{F}\left(1 + \frac{A}{C}\right)\sigma_w + EQ_v \quad \text{low } \sigma_w \text{ limit}$$

At low σ_w the σ versus σ_w has a higher slope than at the high σ_w limit. At high σ_w the electric current is more concentrated in the pore space bulk fluid than in the clay double layer, whereas for low σ_w the currents are mostly concentrated within the double layer. This gives rise to the curvature in the σ versus σ_w behavior.

USES

The equations presented in this section can be used to interpret resistivity logs.

ASSUMPTIONS AND LIMITATIONS

Models for log interpretation involve much empiricism, and empirical relations should be calibrated to specific locations and formations.

PART 10

APPENDIXES

CHALKS

Low-porosity samples (Brevik, 1995) from cretaceous gas-condensate reservoir.
Each data point represents average values (averaged over an interval of porosity). Saturation (gas–brine) in these samples is highly correlated with porosity: $S_w \approx 0.70$ at $\phi \approx 0.1$ and $S_w \approx 0.15$ at $\phi \approx 0.35$.

High-porosity samples (Urmos and Wilkens, 1993), water-saturated.

Measurement type: sonic logs, log porosity.

Data source: Urmos and Wilkens (1993).

Brevik (1995).

	Minimum	Maximum	Mean	Standard deviation
V_P (km/s)	1.53	4.30	2.16	0.32
V_S (km/s)	1.59	2.51	2.03	0.30
V_P/V_S	1.62	1.79	1.67	0.06
Porosity	0.10	0.75	0.50	0.08
Density (g/cm^3)	1.43	2.57	1.85	0.13
Impedance 10^6 (kg/m^3)(m/s)	2.30	10.99	4.02	0.89

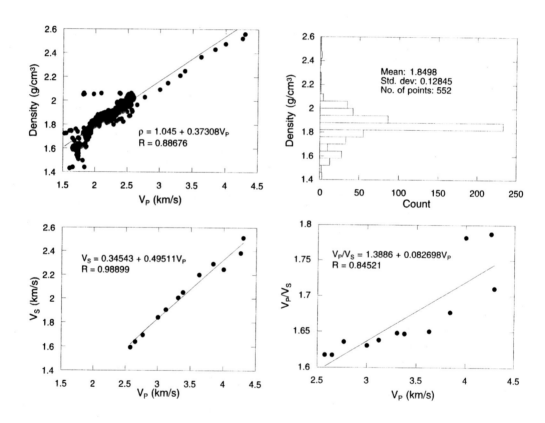

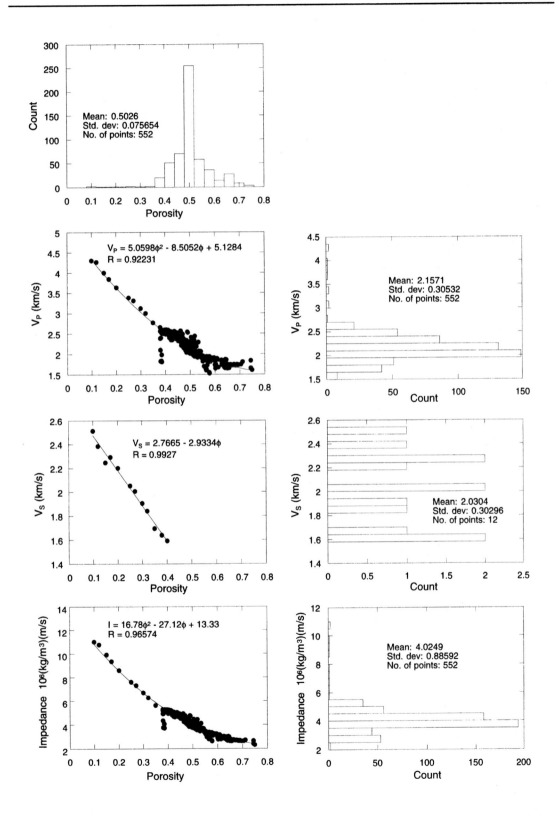

DOLOMITE

Saturation: Water-saturated, calculated from dry data using Gassmann's equations.

Measurement type: Ultrasonic.

Effective pressure: 10, 15, 35 MPa.

Data source: Geertsma (1961).

Yale and Jamieson (1994).

	Minimum	Maximum	Mean	Standard deviation
V_P (km/s)	3.41	7.02	5.39	0.69
V_S (km/s)	2.01	3.64	2.97	0.37
V_P/V_S	1.59	2.09	1.82	0.07
Porosity	0.00	0.32	0.13	0.06
Density (g/cm^3)	2.27	2.84	2.59	0.12
Impedance 10^7 (kg/m^3)(m/s)	0.78	1.93	1.40	0.23

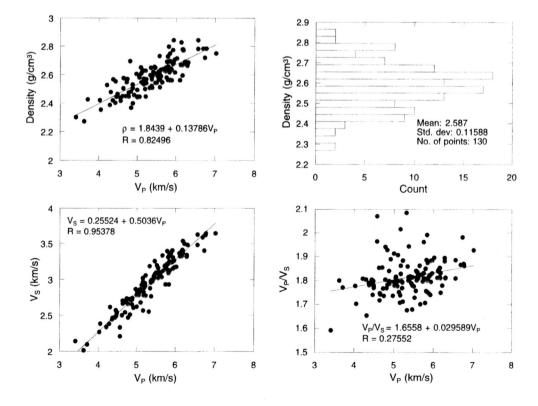

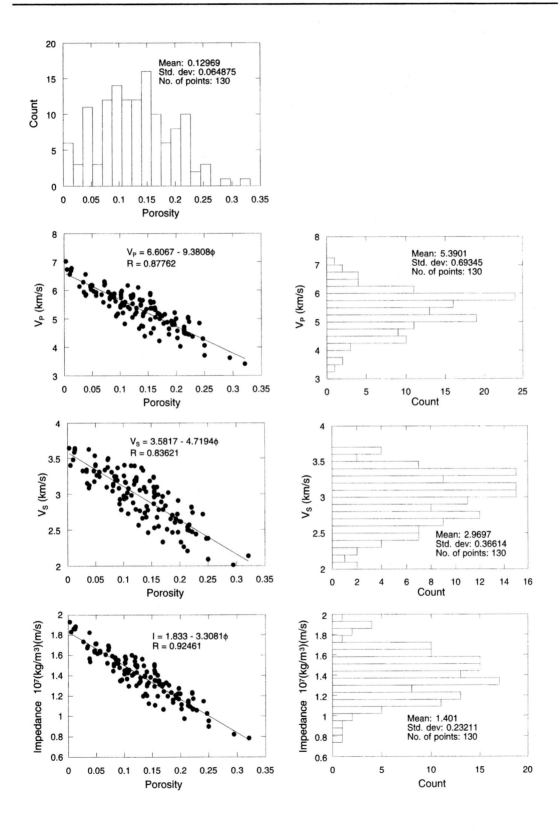

SANDSTONES

Saturation: Water-saturated.
Measurement type: Ultrasonic.
Effective pressure: 30, 40 MPa.
Data source: Han (1986).

	Minimum	Maximum	Mean	Standard deviation
V_P (km/s)	3.13	5.52	4.09	0.51
V_S (km/s)	1.73	3.60	2.41	0.40
V_P/V_S	1.53	1.89	1.71	0.08
Porosity	0.04	0.30	0.16	0.07
Density (g/cm^3)	2.09	2.64	2.37	0.13
Impedance 10^6 (kg/m^3)(m/s)	6.60	13.97	9.73	1.64

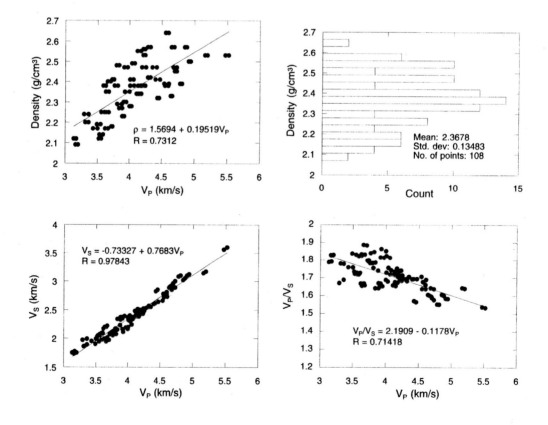

$\rho = 1.5694 + 0.19519 V_P$
$R = 0.7312$

Mean: 2.3678
Std. dev: 0.13483
No. of points: 108

$V_S = -0.73327 + 0.7683 V_P$
$R = 0.97843$

$V_P/V_S = 2.1909 - 0.1178 V_P$
$R = 0.71418$

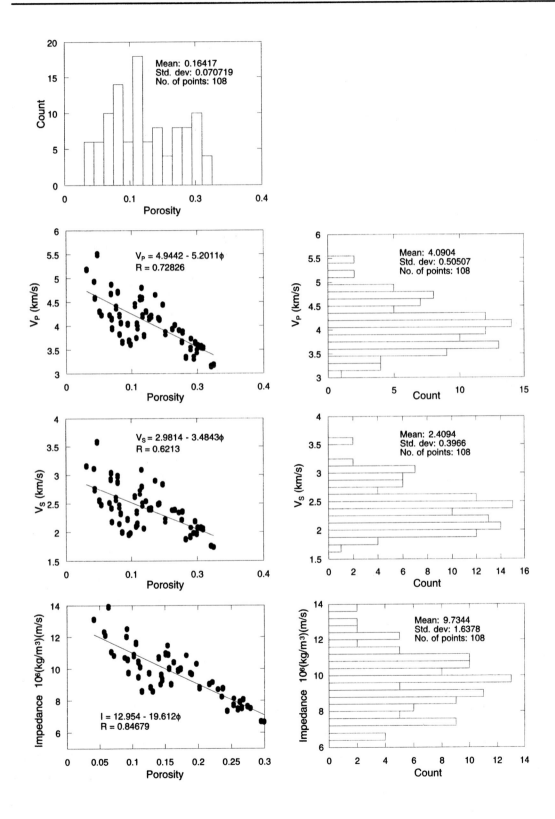

TIGHT-GAS SANDSTONES

Saturation: Dry.
Measurement type: Ultrasonic.
Effective pressure: 40 MPa.
Data source: Jizba (1991).

	Minimum	Maximum	Mean	Standard deviation
V_P (km/s)	3.81	5.57	4.67	0.38
V_S (km/s)	2.59	3.50	3.06	0.23
V_P/V_S	1.42	1.68	1.53	0.05
Porosity	0.01	0.14	0.05	0.04
Density (g/cm^3)	2.26	2.67	2.51	0.11
Impedance 10^7 (kg/m^3)(m/s)	0.89	1.49	1.17	0.13

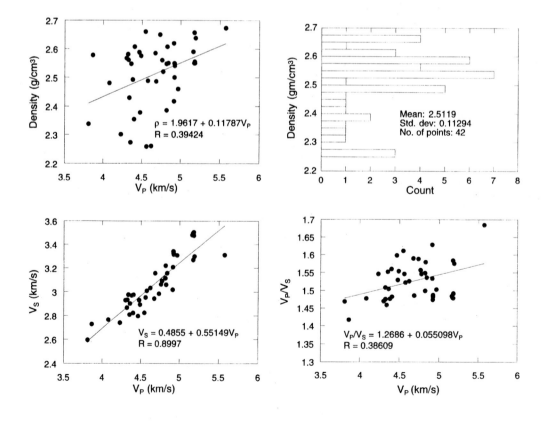

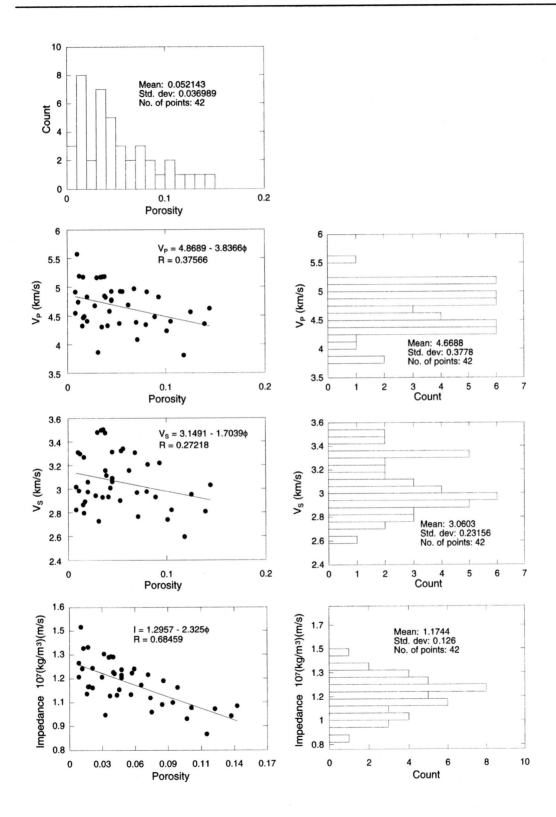

LIMESTONE

Saturation: Some water-saturated; others water-saturated as calculated from dry data using Gassmann's equations.

Measurement type: Ultrasonic, resonant bar.

Effective pressure: 10, 30, 40, 50 MPa.

Data source: Cadoret (1993).

Lucet (1989).

Yale and Jamieson (1994).

	Minimum	Maximum	Mean	Standard deviation
V_P (km/s)	3.39	5.79	4.63	0.66
V_S (km/s)	1.67	3.04	2.44	0.37
V_P/V_S	1.72	2.04	1.88	0.08
Porosity	0.03	0.41	0.15	0.09
Density (g/cm^3)	2.00	2.65	2.43	0.16
Impedance 10^7 (kg/m^3)(m/s)	0.69	1.51	1.43	0.22

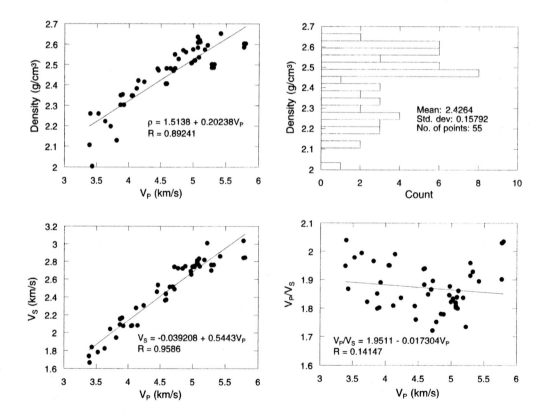

$\rho = 1.5138 + 0.20238V_P$
$R = 0.89241$

Mean: 2.4264
Std. dev: 0.15792
No. of points: 55

$V_S = -0.039208 + 0.5443V_P$
$R = 0.9586$

$V_P/V_S = 1.9511 - 0.017304V_P$
$R = 0.14147$

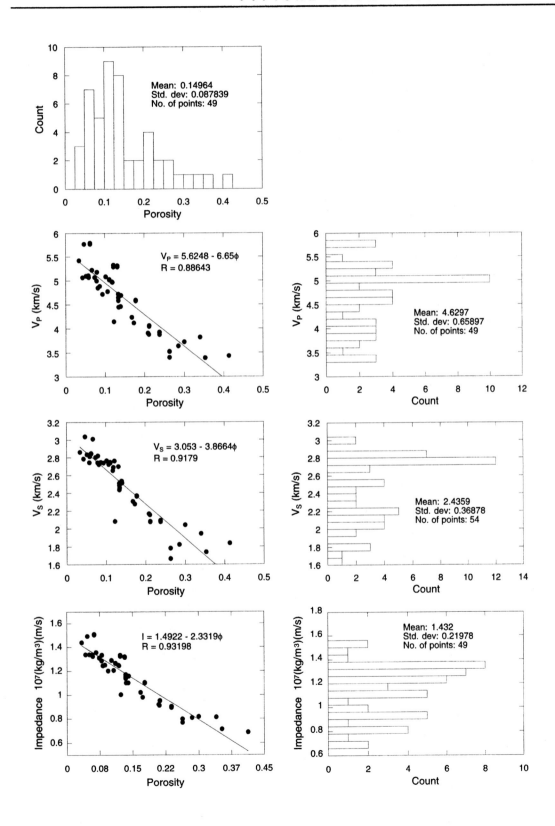

HIGH-POROSITY SANDSTONES

Saturation: Water-saturated.
Measurement type: Ultrasonic.
Effective pressure: 35, 40 MPa.
Data source: Strandenes (1991).

	Minimum	Maximum	Mean	Standard deviation
V_P (km/s)	3.46	4.79	3.80	0.24
V_S (km/s)	1.95	2.66	2.16	0.15
V_P/V_S	1.68	1.88	1.75	0.13
Porosity	0.02	0.32	0.18	0.08
Density (g/cm^3)	2.12	2.69	2.33	0.13
Impedance 10^6 (kg/m^3)(m/s)	7.57	9.98	8.57	0.67

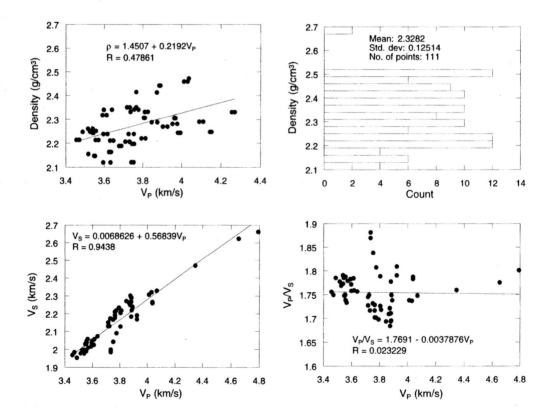

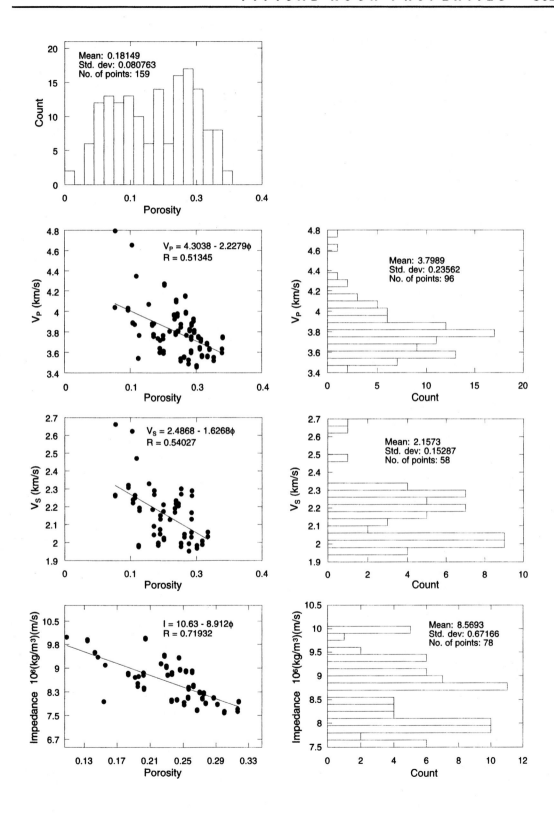

POORLY CONSOLIDATED SANDSTONES

Saturation: Water-saturated.
Measurement type: Ultrasonic.
Effective pressure: 30 MPa.
Data source: Blangy (1992).

	Minimum	Maximum	Mean	Standard deviation
V_P (km/s)	2.43	3.14	2.73	0.18
V_S (km/s)	1.21	1.66	1.37	0.12
V_P/V_S	1.88	2.24	2.02	0.09
Porosity	0.22	0.36	0.31	0.04
Density (g/cm^3)	2.01	2.23	2.11	0.05
Impedance 10^6 (kg/m^3)(m/s)	4.89	7.02	5.77	0.51

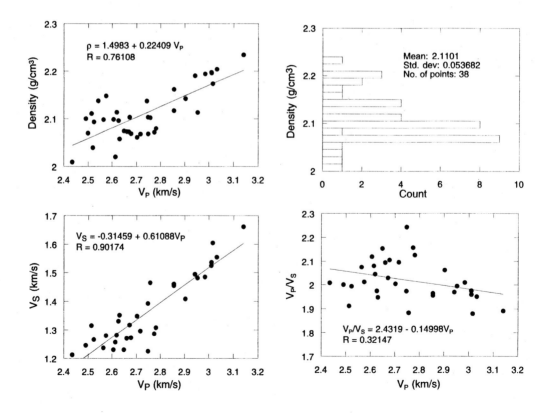

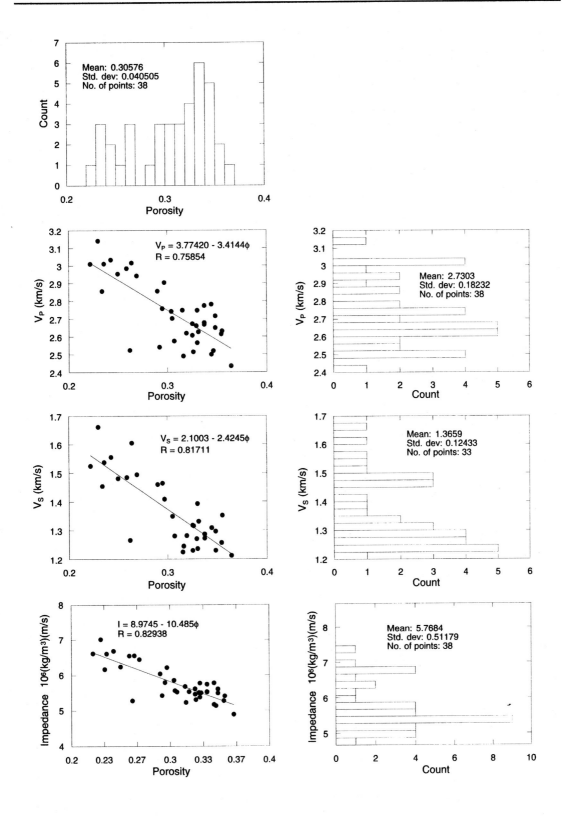

10.2 CONVERSIONS

MASS/WEIGHT

1 g	$= 10^{-3}$ kg
1 kg	$= 2.204623$ lb
1 lb	$= 0.4535924$ kg
1 ton (USA)	$= 2,000$ lb
	$= 907.2$ kg
1 ton (imperial)	$= 2,240$ lb
	$= 1,016$ kg
1 ton (metric)	$= 1,000$ kg
	$= 2,204.622$ lb
1 oz (avdp.)	$= 28.3495$ g
1 oz (troy)	$= 31.10348$ g

LENGTH

1 m	$= 39.37$ in
	$= 3.2808399$ ft
1 cm	$= 0.3937$ in
	$= 0.032808399$ ft
	$= 0.01$ m
1 in	$= 2.540005$ cm
	$= 0.02540005$ m
1 ft	$= 30.48006$ cm
	$= 0.3048006$ m
1 km	$= 0.62137$ mile
1 mile	$= 1.60935$ km
1 nautical mile	$= 1.15077$ miles
	$= 1.852$ km
1 μm	$= 10^{-6}$ m
	$= 10^{-4}$ cm
	$= 3.937 \times 10^{-5}$ in
1 Å	$= 10^{-10}$ m
	$= 10^{-8}$ cm
	$= 3.937 \times 10^{-9}$ in

DENSITY

1 g/cm^3	$= 0.036127$ lb/in^3
	$= 62.42797$ lb/ft^3
	$= 1,000$ kg/m^3

1 lb/in^3 = 27.6799 g/cm^3
             = 27,679.9 kg/m^3
1 lb/ft^3 = 0.016018 g/cm^3

VOLUME

1 cm^3 = 0.0610238 in^3
1 in^3 = 16.38706 cm^3
1 liter   = 0.264172 gallons
          = 0.035315 ft^3
          = 1.056688 qt
          = 1000 cm^3
1 bbl   = 0.158987 m^3
          = 42 gallons
1 m^3 = 6.2898106 bbls

FORCE

1 N   = 1 kg·m/s^2
1 dyn   = 10^{-5} N
1 kg·force = 9.80665 N
            = 9.80665 × 10^5 dyne

PRESSURE

1 atm (76 cm Hg)   = 1.01325 bar
                    = 1.033227 kg·force/cm^2
                    = 14.695949 psi
1 bar   = 10^6 dyne/cm^2
                    = 10^5 N/m^2
                    = 0.1 MPa
1 kg·force/cm^2 = 9.80665 10^5 dyne/cm^2
                    = 0.96784 atm
1 psi   = 0.070307 kg/cm^2
                    = 0.006895 MPa
                    = 0.06895 bar
1 Pa   = 1 N/m^2
                    = 1.4504 × 10^{-4} psi
1 MPa   = 10^6 Pa
                    = 145.0378 psi
                    = 10 bar
1 kbar   = 100 MPa

PRESSURE GRADIENTS (OR MUD WEIGHT TO PRESSURE GRADIENT)

1 psi/ft	$= 144$ lb/ft^3
	$= 19.24$ lb/gallons
	≈ 0.0225 MPa/m
	$= 22.5$ kPa/m
lb/gallon	$= 0.052$ psi/ft

MUD DENSITY TO PRESSURE GRADIENT

1 psi/ft $\Leftrightarrow 2.31$ g/cm^3

VISCOSITY

1 Poise	$= 1$ dyne·s/cm^2
1 cP	$= 0.01$ Poise

PERMEABILITY

1 Darcy	$= 0.986923 \times 10^{-12}$ m^2
	$= 0.986923$ μm^2
	$= 0.986923 \times 10^{-8}$ cm^2
	$= 1.06 \times 10^{-11}$ ft^2

GAS–OIL RATIO

1 liter/liter $= 5.615$ ft^3/bbl

10.3 MODULI AND DENSITY OF COMMON MINERALS

The following table summarizes moduli, densities, and velocities for many common minerals. The data have been taken from a variety of sources, but we drew heavily from extensive compilations by Ellis et al. (1988), Blangy (1992), Castagna et al. (1993), and Carmichael (1989).

Mineral	Bulk modulus (GPa)	Shear modulus (GPa)	Density (g/cc)	V_P (km/s)	V_S (km/s)	Poisson ratio	References
Olivines							
Forsterite	129.8	84.4	3.32	8.54	5.04	0.23	[1–3]
"Olivine"	130	80	3.32	8.45	4.91	0.24	[55]
Garnets							
Almandine	176.3	95.2	4.18	8.51	4.77	0.27	[1]
Zircon	19.8	19.7	4.56	3.18	2.08	0.13	[4, 7]
Epidotes							
Epidote	106.5	61.1	3.40	7.43	4.24	0.26	[9]
Dravite	102.1	78.7	3.05	8.24	5.08	0.19	[4–6]
Pyroxenes							
Diopside	111.2	63.7	3.31	7.70	4.39	0.26	[8, 9]
Augite	94.1	57.0	3.26	7.22	4.18	0.25	[9]
	13.5	24.1	3.26	3.74	2.72	0.06	[10]
Sheet silicates							
Muscovite	61.5	41.1	2.79	6.46	3.84	0.23	[11]
	42.9	22.2	2.79	5.10	2.82	0.28	[56]
	52.0	30.9	2.79	5.78	3.33	0.25	[47]
Phlogopite	58.5	40.1	2.80	6.33	3.79	0.22	[11]
	40.4	13.4	2.80	4.56	2.19	0.35	[56]
Biotite	59.7	42.3	3.05	6.17	3.73	0.21	[11]
	41.1	12.4	3.05	4.35	2.02	0.36	[56]
Clays							
Kaolinite	1.5	1.4	1.58	1.44	0.93	0.14	[10]
"Gulf clays" (Han)[a]	25	9	2.55	3.81	1.88	0.34	[51, 54]
"Gulf clays" (Tosaya)[a]	21	7	2.6	3.41	1.64	0.35	[50, 54]
Mixed clays[a]				3.40	1.60		[50]
				3.41	1.63		[51]
Montmorillonite–illite mixture[a]				3.60	1.85		[52]
Illite[a]				4.32	2.54		[53]
Framework silicates							
Perthite	46.7	23.63	2.54	5.55	3.05	0.28	[55]
Plagioclase Feldspar (Albite)	75.6	25.6	2.63	6.46	3.12	0.35	[10]
"Average" feldspar	37.5	15.0	2.62	4.68	2.39	0.32	
Quartz	37	44.0	2.65	6.05	4.09	0.08	[55]
	36.6	45.0	2.65	6.04	4.12	0.06	[14–16]
	36.5	45.6	2.65	6.06	4.15	0.06	[44]
	37.9	44.3	2.65	6.05	4.09	0.08	[48]
Quartz with clay (Han)	39	33.0	2.65	5.59	3.52	0.17	[51, 54]

Continues

Continued

Mineral	Bulk modulus (GPa)	Shear modulus (GPa)	Density (g/cc)	V_P (km/s)	V_S (km/s)	Poisson ratio	References
Oxides							
Corundum	252.9	162.1	3.99	10.84	6.37	0.24	[17, 18]
Hematite	100.2	95.2	5.24	6.58	3.51	0.14	[19, 20]
	154.1	77.4	5.24	7.01	3.84	0.28	[10, 12]
Rutile	217.1	108.1	4.26	9.21	5.04	0.29	[21, 22]
Spinel	203.1	116.1	3.63	9.93	5.65	0.26	[1]
Magnetite	161.4	91.4	5.20	7.38	4.19	0.26	[4, 23, 24]
	59.2	18.7	4.81	4.18	1.97	0.36	[10]
Hydroxides							
Limonite	60.1	31.3	3.55	5.36	2.97	0.28	[10]
Sulfides							
Pyrite	147.4	132.5	4.93	8.10	5.18	0.15	[25]
	138.6	109.8	4.81	7.70	4.78	0.19	[10]
Pyrrhotite	53.8	34.7	4.55	4.69	2.76	0.23	[10]
Sphalerite	75.2	32.3	4.08	5.38	2.81	0.31	[26, 27]
Sulfates							
Barite	54.5	23.8	4.51	4.37	2.30	0.31	[14]
	58.9	22.8	4.43	4.49	2.27	0.33	[28]
	53.0	22.3	4.50	4.29	2.22	0.32	[7]
Celestite	81.9	21.4	3.96	5.28	2.33	0.38	[4]
	82.5	12.9	3.95	5.02	1.81	0.43	[28]
Anhydrite	56.1	29.1	2.98	5.64	3.13	0.28	[30]
	62.1	33.6	2.96	6.01	3.37	0.27	[49]
Gypsum			2.35	5.80			[29]
Polyhalite			2.78	5.30			[31]
Carbonates							
Calcite	76.8	32.0	2.71	6.64	3.44	0.32	[14]
	63.7	31.7	2.70	6.26	3.42	0.29	[32]
	70.2	29.0	2.71	6.34	3.27	0.32	[33]
	74.8	30.6	2.71	6.53	3.36	0.32	[43]
	68.3	28.4	2.71	6.26	3.24	0.32	[44]
Siderite	123.7	51.0	3.96	6.96	3.59	0.32	[34]
Dolomite	94.9	45.0	2.87	7.34	3.96	0.30	[35]
	69.4	51.6	2.88	6.93	4.23	0.20	[13]
	76.4	49.7	2.87	7.05	4.16	0.23	[45]
Aragonite	44.8	38.8	2.92	5.75	3.64	0.16	[19, 20, 36]
Natronite	52.6	31.6	2.54	6.11	3.53	0.26	[54, 55]

Continues

Continued

Mineral	Bulk modulus (GPa)	Shear modulus (GPa)	Density (g/cc)	V_P (km/s)	V_S (km/s)	Poisson ratio	References
Phosphates							
Hydroxyapatite	83.9	60.7	3.22	7.15	4.34	0.21	[4]
Fluorapatite	86.5	46.6	3.21	6.80	3.81	0.27	[37]
Halides							
Fluorite	86.4	41.8	3.18	6.68	3.62	0.29	[38, 39]
Halite	24.8	14.9	2.16	4.55	2.63	0.25	[14, 40–42]
			2.16	4.50	2.59		[46]
Sylvite	17.4	9.4	1.99	3.88	2.18	0.27	[40]
Organic							
Kerogen	2.9	2.7	1.3	2.25	1.45	0.14	[54, 55]
Zeolites							
Narolite	46.6	28.0	2.25	6.11	3.53	0.25	[54, 55]

[a]Clay velocities were interpreted by extrapolating empirical relations for mixed lithologies to 100-percent clay (Castagna et al., 1993).

REFERENCES CITED IN THE TABLE

1) Verma, R.K., 1960. Elasticity of some high-density crystals. *J. Geophys. Res.*, 65, 757–766.

2) Graham, E.K., Jr., and Barsch, G.R., 1969. Elastic constants of single-crystal forsterite as a function of temperature and pressure, *J. Geophys. Res.*, 74, 5949–5960.

3) Kumazawa, M., and Anderson, O.L., 1969. Elastic moduli, pressure derivatives, and temperature derivative of single-crystal olivine and single-crystal forsterite. *J. Geophys. Res.*, 74, 5961–5972.

4) Hearmon, R.F.S., 1956. The elastic constants of anistropic materials II. *Adv. Phys.*, 5, 323–382.

5) Mason, W.P., 1950. *Piezoelectric Crystals and Their Application to Ultrasonics.* D. Van Nostrand Co., Inc., New York.

6) Voigt, W., 1890. Bestimmung der Elastizitätskonstanten des brasilianischen Turmalines. *Annalen der Physik und Chemie.* 41, 712–729.

7) Huntington, H.B., 1958. The elastic constants of crystals, in *Solid State Physics*, Vol. 7, F. Seitz and D. Turnbull, eds. Academic Press, New York, 213–351.

8) Ryzhova, T.V., Aleksandrov, K.S., and Korobkova, V.M., 1966. The elastic properties of rock-forming minerals V. Additional data on silicates. *Bull. Acad. Sci. USSR, Earth Phys. Ser.* English translation No. 2, 111–113.

9) Alexandrov, K.S., Ryzhova, T.V., and Belikov, B.P., 1964. The elastic properties of pyroxenes. *Sov. Phys. Crystallog.*, 8, 589–591.

10) Woeber, A.F., Katz, S., and Ahrens, T.J., 1963. Elasticity of selected rocks and minerals. *Geophys.*, 28, 658–663.

11) Aleksandrov, K.S., and Ryzhova, T.V., 1961. Elastic properties of rock-forming minerals II. Layered silicates, *Bull. Acad. Sci. USSR, Geophys. Ser.* English translation No. 12, 1165–1168.

12) Wyllie, M.R.J., Gregory, A.R., and Gardner, L.W., 1956. Elastic wave velocities in heterogeneous and porous media. *Geophys.*, 21, 41–70.

13) *Log Interpretation Charts*, 1984. Publication SMP-7006. Schlumberger Ltd., Houston.

14) Simmons, G., 1965. Single crystal elastic constants and calculated aggregate properties. *J. Grad. Res. Center*, Southern Methodist University, 34, 1–269.

15) Mason, W.P., 1943. Quartz crystal applications. *Bell Systems Tech. J.*, 22, 178.

16) Koga, I., Aruga, M., and Yoshinaka, Y., 1958. Theory of plane elastic waves in a piezoelectric crystalline medium and determination of elastic and piezoelectric constants of quartz. *Phys. Rev.*, 109, 1467–1473.

17) Wachtman, J.B., Jr., Tefft, W.E., Lam, D.G., Jr., and Strinchfield, R.P., 1960. Elastic constants of synthetic single crystal corundum at room temperature. *J. Res. Natl. Bur. Stand.*, 64A, 213–228.

18) Bernstein, B.T., 1963. Elastic constants of synthetic sapphire at 27 degrees Celsius. *J. Appl. Phys.*, 34, 169–172.

19) Hearmon, R.F.S., 1946. The elastic constants of anistropic materials. *Rev. Mod. Phys.*, 18, 409–440.

20) Voigt, W., 1907. Bestimmung der Elastizitätskonstanten von Eisenglanz. *Annalen der Physik*, 24, 129–140.

21) Birch, F., 1960. Elastic constants of rutile—a correction to a paper by R.K. Verma, "Elasticity of some high-density crystals." *J. Geophys. Res.*, 65, 3855–3856.

22) Joshi, S.K., and Mitra, S.S., 1960. Debye characteristic temperature of solids. *Proc. Phys. Soc., London*, 76, 295–298.

23) Doraiswami, M.S., 1947. Elastic constants of magnetite, pyrite, and chromite. *Proc. Ind. Acad. Sci.*, A, 25, 414–416.

24) Alexandrov, K.S., and Ryzhova, T.V., 1961. The elastic properties of crystal. *Sov. Phys. Crystallogr.*, 6, 228–252.

25) Simmons, G., and Birch, F., 1963. Elastic constants of pyrite. *J. Appl. Phys.*, 34, 2736–2738.

26) Einspruch, N.G., and Manning, R.J., 1963. Elastic constants of compound semi-conductors ZnS, PbTe, GaSb. *J. Acoust. Soc. Am.*, 35, 215–216.

27) Berlincourt, D., Jaffe, H., and Shiozawa, L.R., 1963. Electroelastic properties of the sulfides, selenides, and tellurides of Zn and Cd. *Phys. Rev.*, 129, 1009–1017.

28) Seshagiri Rao, T., 1951. Elastic constants of barytes and celestite. *Proc. Ind. Acad. Sci.*, A, 33, 251–256.

29) Tixier, M.P., and Alger, R.P., 1967. Log evaluation of non-metallic mineral deposits. *Trans. SPWLA 8th Ann. Logging Symp.*, Denver, June 11–14, Paper R.

30) Schwerdtner, W.M., Tou, J.C.-M, and Hertz, P.B., 1965. Elastic properties of single crystals of anhydrite. *Can. J. Earth Sci.*, 2, 673–683.

31) *Formation Evaluation Data Handbook*, 1982. Gearhart Industries, Inc., Fort Worth, Texas, 237 pp.

32) Bhimasenacher, J., 1945. Elastic constants of calcite and sodium nitrate, *Proc. Ind. Acad. Sci.*, A, 22, 199–207.

33) Peselnick, L., and Robie, R.A., 1963. Elastic constants of calcite. *J. Appl. Phys.*, 34, 2494–2495.

34) Christensen, N.I., 1972. Elastic properties of polycrystalline magnesium, iron, and manganese carbonates to 10 kilobars. *J. Geophys. Res.*, 77, 369–372.

35) Humbert, P., and Plicque, F., 1972. Propriétés élastiques de carbonate rhomboedriques monocristallins: calcite, magnesite, dolomie. *Comptes Rendus de l'Academie des Sciences*, 275, serie B, 391–394.

36) Birch, F., 1960. The velocity of compressional waves in rocks to 10 kilobars. *J. Geophys. Res.*, 65, 1083–1102.

37) Yoon, H.S., and Newnham, R.E., 1969. Elastic properties of fluorapatite. *Am. Mineralog.*, 54, 1193–1197.

38) Bergmann, L., 1954. *Der Ultraschall und seine Anwendung in Wissenschaft und Technik*. S. Hirzel, Zurich.

39) Huffman, D.F., and Norwood, M.H., 1960. Specific heat and elastic constants of calcium fluoride at low temperatures, *Phys. Rev.*, 117, 709–711.

40) Spangenburg, K., and Haussuhl, S., 1957. Die elastischen Konstanten der Alkalihalogenide. *Zeitschrift für Kristallographie*, 109, 422–437.

41) Lazarus, D., 1949. The variation of the adiabatic elastic constants of KCl, NaCl, CuZn, Cu, and Al with pressure to 10,000 bars. *Phys. Rev.* 76, 545–553.

42) Papadakis, E.P., 1963. Attenuation of pure elastic modes in NaCl single crystals. *J. Appl. Phys.*, 34, 1872–1876.

43) Dandekar, D.P., 1968. Pressure dependence of the elastic constants of calcite. *Phys. Rev.*, 172, 873.

44) Anderson, O.L., and Liebermann, R.C., 1966. Sound velocities in rocks and minerals. *VESIAC State-of-the-Art Report No. 7885-4-x*, University of Michigan.

45) Nur, A., and Simmons, G., 1969. The effect of viscosity of a fluid phase on velocity in low-porosity rocks. *Earth and Planetary Sci. Lett.*, 7, 99–108.

46) Birch, F., 1966. Compressibility; Elastic constants. In *Handbook of Physical Constants*, S.P. Clark, ed. Geolog. Soc. Am., Memoir, 97, 97–174.

47) Aleksandrov, K.S., and Ryzhova, T.V., 1961. The elastic properties of crystals. *Sov. Phys. Crystallog.*, 6, 228.

48) McSkimin, H.J., Andreatch, P., Jr., and Thurston, R.N., 1965. Elastic moduli of quartz vs. hydrostatic pressure at 25 and 195.8 degrees Celsius. *J. Appl. Phys.*, 36, 1632.

49) Rafavich, F., Kendal, C.H.St.C., and Todd, T.P., 1984. The relationship between acoustic properties and the peterographic character of carbonate rocks. *Geophys.*, 49, 1622–1636.

50) Tosaya, C.A., 1982. *Acoustical properties of clay-bearing rocks*. Ph.D. dissertation, Stanford University.

51) Han, D.-H., Nur, A., and Morgan, D., 1986. Effects of porosity and clay content on wave velocities in sandstones. *Geophys.*, 51, 2093–2107.

52) Castagna, J.P., Batzle, M.L., and Eastwood, R.L., 1985. Relationships between compressional-wave and shear-wave velocities in elastic silicate rocks. *Geophys.*, 50, 571–581.

53) Eastwood, R.L., and Castagna, J.P., 1986. Interpretation of V_p/V_s ratios from sonic logs, in *Shear Wave Exploration, Geophysical Developments No. 1*, S.H. Danbom and S.N. Domenico, eds., Soc. Expl. Geophys. Tulsa, Oklahoma.

54) Blangy, J.D., 1992. *Integrated Seismic Lithologic Interpretation: The Petrophysical Basis*. Ph.D. dissertation, Stanford University.

55) Carmichael, R.S., 1989. *Practical Handbook of Physical Properties of Rocks and Minerals*. CRC Press, Boca Raton, Florida, 741 pp.

56) Ellis, D., Howard, J., Flaum, C., McKeon, D., Scott, H., Serra, O., and Simmons, G., 1988. Mineral logging parameters: nuclear and acoustic. *Tech. Rev.*, 36(1), 38–55.

REFERENCES

Achenbach, J.D., 1984. *Wave Propagation in Elastic Solids*. Elsevier Science Publication, Amsterdam, 425 pp.

Aki, K., and Richards, P.G., 1980. *Quantitative Seismology: Theory and Methods*, W.H. Freeman and Co., San Francisco, 932 pp.

Archie, G.E., 1942. The electrical resistivity log as an aid in determining some reservoir characteristics. *Trans. Am. Inst. Mech. Eng.*, 146, 54–62.

Auld, B.A., 1990. *Acoustic Fields and Waves in Solids*, Vols. 1, 2. Robert E. Krieger Publication Co., Malabar, Florida, 856 pp.

Backus, G.E., 1962. Long-wave elastic anisotropy produced by horizontal layering. *J. Geophys. Res.*, 67, 4427–4440.

Banik, N.C., 1987. An effective anisotropy parameter in transversely isotropic media. *Geophys.*, 52, 1654.

Banik, N.C., Lerche, I., and Shuey, R.T., 1985. Stratigraphic filtering, Part I: Derivation of the O'Doherty–Anstey formula. *Geophys.*, 50, 2768–2774.

Batzle, M., and Wang, Z., 1992. Seismic properties of pore fluids. *Geophys.*, 57, 1396–1408.

Bear, J., 1972. *Dynamics of Fluids in Porous Media*. Dover Publications, Inc., Mineola, New York.

Beltzer, A.I., 1988. *Acoustics of Solids*, Springer-Verlag, Berlin, 235 pp.

Ben-Menahem, A., and Singh, S., 1981. *Seismic Waves and Sources*, Springer-Verlag, New York.

Beran, M.J., and Molyneux, J., 1966. Use of classical variational principles to determine bounds for the effective bulk modulus in heterogeneous media. *Quart. Appl. Math.*, 24, 107–118.

Berge, P.A., Berryman, J.G., and Bonner, B.P., 1993. Influence of microstructure on rock elastic properties. *Geophys. Res. Lett.*, 20, 2619–2622.

Berge, P.A., Fryer, G.J., and Wilkens, R.H., 1992. Velocity–porosity relationships in the upper oceanic crust: Theoretical considerations. *J. Geophys. Res.*, 97, 15239–15254.

Berryman, J.G., 1980a. Confirmation of Biot's theory. *Appl. Phys Lett.*, 37, 382–384.

Berryman, J.G., 1980b. Long-wavelength propagation in composite elastic media. *J. Acoust. Soc. Am.*, 68, 1809–1831.

Berryman, J.G., 1981. Elastic wave propagation in fluid-saturated porous media. *J. Acoust. Soc. Am.*, 69, 416–424.

Berryman, J.G., 1983. Dispersion of extensional waves in fluid-saturated porous cylinders at ultrasonic frequencies. *J. Acoust. Soc. Am.*, 74, 1805–1812.

Berryman, J.G., 1992. Single-scattering approximations for coefficients in Biot's equations of poroelasticity. *J. Acoust. Soc. Am.*, 91, 551–571.

Berryman, J.G., 1995. Mixture theories for rock properties, in *A Handbook of Physical Constants*, T.J. Ahrens, ed. American Geophysical Union, Washington, D.C., 205–228.

Berryman, J.G., and Milton, G.W., 1991. Exact results for generalized Gassmann's equation in composite porous media with two constituents. *Geophys.*, 56, 1950–1960.

Bilodeaux, B., 1997. Shaley sand evaluation, course notes, Stanford University.

Biot, M.A., 1956. Theory of propagation of elastic waves in a fluid saturated porous solid. I. Low frequency range and II. Higher-frequency range. *J. Acoust. Soc. Am.*, 28, 168–191.

Biot, M.A., 1962. Mechanics of deformation and acoustic propagation in porous media. *J. Appl. Phys.*, 33, 1482–1498.

Birch, F., 1961. The velocity of compressional waves in rocks to 10 kilobars, Part 2. *J. Geophys. Res.*, 66, 2199–2224.

Blair, D.P., 1990. A direct comparison between vibrational resonance and pulse transmission data for assessment of seismic attenuation in rock. *Geophys.*, 55, 51–60.

Blangy, J.P., 1992. *Integrated Seismic Lithologic Interpretation: The Petrophysical Basis.* Ph.D. dissertation, Stanford University.

Born, M., and Wolf, E., 1980. *Principles of Optics*, 6th ed. Pergamon Press, Oxford, U.K., 808 pp.

Bortfeld, R., 1961. Approximation to the reflection and transmission coefficients of plane longitudinal and transverse waves. *Geophys. Prospecting*, 9, 485–503.

Bourbié, T., Coussy, O., and Zinszner, B., 1987. *Acoustics of Porous Media.* Gulf Publishing Co., Houston, 334 pp.

Boyse, W.E., 1986. *Wave Propagation and Inversion in Slightly Inhomogeneous Media.* Ph.D. dissertation, Stanford University.

Bracewell, R., 1965. *The Fourier Transform and Its Application*, McGraw-Hill Book Co., New York, 381 pp.

Brandt, H., 1955. A study of the speed of sound in porous granular media. *J. Appl. Mech.*, 22, 479–486.

Brevik, I., 1995. Chalk data, presented at workshop on effective media, Karlsruhe.

Brisco, B., Pultz, T.J., Brown, R.J., Topp, G.C., Hares, M.A., and Zebchuk, W.D., 1992. Soil moisture measurement using portable dielectric probes and time domain reflectometry. *Water Resources Res.*, 28, 1339–1346.

Brown, R., and Korringa, J., 1975. On the dependence of the elastic properties of a porous rock on the compressibility of the pore fluid. *Geophys.*, 40, 608–616.

Bruggeman, D.A.G., 1935. Berechnung verschiedener physikalischer Konstanten von heterogenen Substanzen. *Annalen der Physik* (Leipzig), 24, 636–679.

Budiansky, B., 1965. On the elastic moduli of some heterogeneous materials. *J. Mech. Phys. Solids*, 13, 223–227.

Budiansky, B., and O'Connell, R.J., 1976. Elastic moduli of a cracked solid. *Int. J. Solids and Structures*, 12, 81–97.

Cadoret, T., 1993. *Effet de la Saturation Eau/Gaz sur les Propriétés Acoustiques des Roches*. Ph.D. dissertation, University of Paris, VII.

Carman, P.C., 1961. *L'écoulement des Gaz á Travers les Milieux Poreux*, Bibliothéque des Sciences et Techniques Nucléaires, Presses Universitaires de France, Paris, 198 pp.

Carmichael, R.S., 1989. *Practical Handbook of Physical Properties of Rocks and Minerals*. CRC Press, Boca Raton, Florida, 741 pp.

Castagna, J.P., 1993. AVO analysis – tutorial and review, in *Offset Dependent Reflectivity – Theory and Practice of AVO Analysis*, J.P. Castagna and M. Backus, eds. Investigations in Geophysics, No. 8, Society of Exploration Geophysicists, Tulsa, Oklahoma, 3–36.

Castagna, J.P., Batzle, M.L., and Eastwood, R.L., 1985. Relationships between compressional-wave and shear-wave velocities in clastic silicate rocks. *Geophys.*, 50, 571–581.

Castagna, J.P., Batzle, M.L., and Kan, T.K., 1993. Rock physics – The link between rock properties and AVO response, in *Offset-Dependent Reflectivity – Theory and Practice of AVO Analysis*, J.P. Castagna and M. Backus, eds. Investigations in Geophysics, No. 8, Society of Exploration Geophysicists, Tulsa, Oklahoma, 135–171.

Chen, W., 1995. *AVO in Azimuthally Anisotropic Media: Fracture Detection Using P-wave Data and a Seismic Study of Naturally Fractured Tight Gas Reservoirs*. Ph.D. dissertation, Stanford University.

Cheng, C.H., 1978. *Seismic Velocities in Porous Rocks: Direct and Inverse Problems*. Sc.D. thesis, MIT, Cambridge, Massachusetts.

Cheng, C.H., 1993. Crack models for a transversely anisotropic medium. *J. Geophys. Res.*, 98, 675–684.

Christensen, R.M., 1991. *Mechanics of Composite Materials*. Robert E. Krieger Publication Co., Malabar, Florida, 348 pp.

Claerbout, J.F., 1985. *Fundamentals of Geophysical Data Processing*. Blackwell Scientific Publications, Palo Alto, 274 pp.

Claerbout, J.F., 1992. *Earth Sounding Analysis: Processing versus Inversion*. Blackwell Scientific Publications, Boston, 304 pp.

Clavier, C., Coates, G., and Dumanoir, J., 1984. Theoretical and experimental bases for the dual-water model for interpretation of shaley sands. *Soc. Pet. Eng. J.*, 24, 153–168.

Cleary, M.P., Chen, I.-W., and Lee, S.-M., 1980. Self-consistent techniques for heterogeneous media. *Am. Soc. Civil Eng. J. Eng. Mech.*, 106, 861–887.

Cole, K.S., and Cole, R.H., 1941. Dispersion and absorption in dielectrics I. Alternating current characteristics. *J. Chem. Phys.*, 9, 341–351.

Corson, P.B., 1974. Correlation functions for predicting properties of heterogeneous materials. *J. Appl. Phys.*, 45, 3159–3179.

Cruts, H.M.A., Groenenboom, J., Duijndam, A.J.W., and Fokkema, J.T., 1995. Experimental verification of stress-induced anisotropy. *Expanded Abstracts, Soc. Expl. Geophys.*, 65th Annual International Meeting, 894–897.

Darcy, H., 1856. *Les Fontaines Publiques de la Ville de Dijon*, Dalmont, Paris.

Debye, P., 1945. *Polar Molecules*. Dover, Mineola, New York, 172 pp.

Dellinger, J., and Vernik, L., 1992. Do core sample measurements record group or phase velocity? *Expanded Abstracts, Soc. Expl. Geophys.*, 62nd Annual International Meeting, 662–665.

Digby, P.J., 1981. The effective elastic moduli of porous granular rocks. *J. Appl. Mech.*, 48, 803–808.

Domenico, S.N., 1976. Effect of brine–gas mixture on velocity in an unconsolidated sand reservoir. *Geophys.*, 41, 882–894.

Dullien, F.A.L., 1991. One and two phase flow in porous media and pore structure, in *Physics of Granular Media*, D. Bideau and J. Dodds, eds. Science Publishers Inc., New York 173–214 pp.

Dullien, F.A.L., 1992. *Porous Media: Fluid Transport and Pore Structure*. Academic Press, San Diego, 574 pp.

Dutta, N.C., and Odé, H., 1979. Attenuation and dispersion of compressional waves in fluid-filled porous rocks with partial gas saturation (White model) – Part 1: Biot theory, Part II: Results. *Geophys.*, 44, 1777–1805.

Dutta, N.C., and Seriff, A.J., 1979. On White's model of attenuation in rocks with partial gas saturation, *Geophysics*, 44, 1806–1812.

Dvorkin, J., Mavko, G., and Nur, A., 1995. Squirt flow in fully saturated rocks. *Geophys.* 60, 97–107.

Dvorkin, J., Nolen-Hoeksema, R., and Nur, A., 1994. The squirt-flow mechanism: macroscopic description. *Geophys.*, 59, 428–438.

Dvorkin, J., and Nur, A., 1993. Dynamic poroelasticity: A unified model with the squirt and the Biot mechanisms. *Geophys.*, 58, 524–533.

Dvorkin, J., and Nur, A., 1996. Elasticity of high-porosity sandstones: Theory for two North Sea datasets. *Geophys.*, 61, 1363–1370.

Eberhart-Phillips, D.M., 1989. *Investigation of Crustal Structure and Active Tectonic Processes in the Coast Ranges, Central California*. Ph.D. dissertation, Stanford University.

Ellis, D., Howard, J., Flaum, C., McKeon, D., Scott, H., Serra, O., and Simmons, G., 1988. Mineral logging parameters: Nuclear and acoustic. *Tech. Rev.*, 36(1), 38–55.

Elmore, W.C., and Heald, M.A., 1985. *Physics of Waves*. Dover Publications, Inc., Mineola, New York, 477 pp.

Endres, A.L., and Knight, R., 1992. A theoretical treatment of the effect of microscopic fluid distribution on the dielectric properties of partially saturated rocks. *Geophys. Prospecting*, 40, 307–324.

Epstein, P.S., 1941. On the absorption of sound waves in suspensions and emulsions. *Theodore Von Karmen Anniversary Volume*, 162–188.

Epstein, P.S., and Carhart, R.R., 1953. The absorption of sound in suspensions and emulsions: I. Water fog in air, *J. Acoust. Soc. Am.*, 25, 553–565.

Eshelby, J.D., 1957. The determination of the elastic field of an ellipsoidal inclusion, and related problems. *Proc. Royal Soc. London*, A241, 376–396.

Focke, J.W., and Munn, D., 1987. Cementation exponents (m) in Middle Eastern carbonate reservoirs. *Soc. Pet. Eng., Form. Eval.*, 2, 155–167.

Frazer, L.N., 1994. A pulse in a binary sediment. *Geophys. J. Int.*, 118, 75–93.

Gardner, G.H.F., 1962. Extensional waves in fluid-saturated porous cylinders. *J. Acoust. Soc. Am.*, 34, 36–40.

Gardner, G.H.F., Gardner, L.W., and Gregory, A.R., 1974. Formation velocity and density – The diagnostic basics for stratigraphic traps. *Geophys.*, 39, 770–780.

Gassmann, F., 1951. Über die Elastizität poröser Medien. *Vier. der Natur. Gesellschaft in Zürich*, 96, 1–23.

Geertsma, J., 1961. Velocity-log interpretation: The effect of rock bulk compressibility. *Soc. Pet. Eng. J.*, 1, 235–248.

Geertsma, J., and Smit, D.C., 1961. Some aspects of elastic wave propagation in fluid-saturated porous solids. *Geophys.*, 26, 169–181.

Gibson, R.L., and Toksöz, M.N., 1990. Permeability estimation from velocity anisotropy in fractured rock. *J. Geophys. Res.*, 95, 15643–15656.

Greenberg, M.L., and Castagna, J.P., 1992. Shear-wave velocity estimation in porous rocks: Theoretical formulation, preliminary verification and applications. *Geophys. Prospecting*, 40, 195–209.

Guéguen, Y., and Palciauskas, V., 1994. *Introduction to the Physics of Rocks*. Princeton University Press, Princeton, New Jersey, 294 pp.

Gvirtzman, H., and Roberts, P., 1991. Pore-scale spatial analysis of two immiscible fluids in porous media. *Water Resources Res.*, 27, 1165–1176.

Han, D.-H., 1986. *Effects of Porosity and Clay Content on Acoustic Properties of Sandstones and Unconsolidated Sediments*. Ph.D. dissertation, Stanford University.

Hanai, T., 1968. Electrical properties of emulsions, in *Emulsion Science*, P. Sherman, ed. Academic Press, New York, 353–478.

Hashin, Z., and Shtrikman, S., 1962. A variational approach to the theory of effective magnetic permeability of multiphase materials. *J. Appl. Phys.*, 33, 3125–3131.

Hashin, Z., and Shtrikman, S., 1963. A variational approach to the elastic behavior of multiphase materials. *J. Mech. Phys. Solids*, 11, 127–140.

Herrick, D.C., 1988. Conductivity models, pore geometry, and conduction mechanisms. *Trans. Soc. Prof. Well Log Analysts, 29th Annual Logging Symposium*, Paper D.

Hill, R., 1952. The elastic behavior of crystalline aggregate. *Proc. Physical Soc., London*, A65, 349–354.

Hill, R., 1963. Elastic properties of reinforced solids: Some theoretical principles. *J. Mech. Phys. Solids*, 11, 357–372.

Hill, R., 1965. A self-consistent mechanics of composite materials. *J. Mech. Phys. Solids*, 13, 213–222.

Hilterman, F., 1989. Is AVO the seismic signature of rock properties? *Expanded Abstracts, Soc. Expl. Geophys., 59th Annual International Meeting*, 559.

Hovem, J.M., and Ingram, G.D., 1979. Viscous attenuation of sound in saturated sand. *J. Acoust. Soc. Am.*, 66, 1807–1812.

Hudson, J.A., 1980. Overall properties of a cracked solid. *Math. Proc. Cambridge Philos. Soc.*, 88, 371–384.

Hudson, J.A., 1981. Wave speeds and attenuation of elastic waves in material containing cracks. *Geophys. J. Royal Astronom. Soc.*, 64, 133–150.

Hudson, J.A., 1990. Overall elastic properties of isotropic materials with arbitrary distribution of circular cracks. *Geophys. J. Int.* 102, 465–469.

Jackson, J.D., 1975. *Classical Electrodynamics*, 2nd ed., John Wiley and Sons, New York, 848 pp.

Jackson, P.D., Taylor-Smith, D., and Stanford, P.N., 1978. Resistivity-porosity-particle shape relationships for marine sands. *Geophys.*, 43, 1250–1262.

Jaeger, J.C., and Cook, N.G.W., 1969. *Fundamentals of Rock Mechanics*. Chapman and Hall Ltd. and Science Paperbacks, London, 515 pp.

Jenkins, G.M., and Watts, D.G., 1968. *Spectral Analysis and Its Applications*, Holden-Day, San Francisco, 525 pp.

Jizba, D.L., 1991. *Mechanical and Acoustical Properties of Sandstones and Shales*. Ph.D. dissertation, Stanford University.

Johnson, D.L., and Plona, T.J., 1982. Acoustic slow waves and the consolidation transition. *J. Acoust. Soc. Am.*, 72, 556–565.

Juhász, I., 1981. Normalised Q_v – The key to shaley sand evaluation using the Waxman–Smits equation in the absence of core data. *Trans. Soc. Prof. Well Log Analysts, 22nd Annual Logging Symposium*, Paper Z.

Kachanov, M., 1992. Effective elastic properties of cracked solids: Critical review of some basic concepts. *Appl. Mech. Rev.*, 45, 304–335.

Keller, J.B., 1964. Stochastic equations and wave propagation in random media. *Proc. Symp. Appl. Math.*, 16, 145–170.

Kennett, B.L.N., 1974. Reflections, rays and reverberations. *Bull. Seismol. Soc. Am.*, 64, 1685–1696.

Kennett, B.L.N., 1983. *Seismic Wave Propagation in Stratified Media*. Cambridge University Press, Cambridge, 342 pp.

Kjartansson, E., 1979. Constant Q wave propagation and attenuation. *J. Geophys. Res.*, 84, 4737–4748.

Knight, R., and Dvorkin, J., 1992. Seismic and electrial properties of sandstones at low saturations, *J. Geophys. Res.*, 97, 17425–17432.

Knight, R., and Nolen-Hoeksema, R., 1990. A laboratory study of the dependence of elastic wave velocities on pore scale fluid distribution. *Geophys. Res. Lett.*, 17, 1529–1532.

Knight, R.J., and Nur, A., 1987. The dielectric constant of sandstones, 60 kHz to 4 MHz. *Geophys.*, 52, 644–654.

Knopoff, L., 1964. Q. *Rev. Geophys.*, 2, 625–660.

Krief, M., Garat, J., Stellingwerff, J., and Ventre, J., 1990. A petrophysical interpretation using the velocities of P and S waves (full-waveform sonic). *The Log Analyst*, 31, November, 355–369.

Kuperman, W.A., 1975. Coherent components of specular reflection and transmission at a randomly rough two-fluid interface. *J. Acoust. Soc. Am.*, 58, 365–370.

Kuster, G.T., and Toksöz, M.N., 1974. Velocity and attenuation of seismic waves in two-phase media. *Geophys.*, 39, 587–618.

Lamb, H., 1945. *Hydrodynamics*. Dover, Mineola, New York, 738 pp.

Landauer, R., 1952. The electrical resistance of binary metallic mixtures. *J. Appl. Phys.*, 23, 779–784.

Lawn, B.R., and Wilshaw, T.R., 1975. *Fracture of Brittle Solids*. Cambridge University Press, Cambridge, 204 pp.

Levin, F.K., 1979. Seismic velocities in transversely isotropic media. *Geophys.*, 44, 918–936.

Liu, H.P., Anderson, D.L., and Kanamori, H., 1976. Velocity dispersion due to anelasticity: Implications for seismology and mantle composition. *Geophys. J. Royal Astronom. Soc.*, 47, 41–58.

Lockner, D.A., Walsh, J.B., and Byerlee, J.D., 1977. Changes in velocity and attenuation during deformation of granite. *J. Geophys. Res.*, 82, 5374–5378.

Lucet, N., 1989. *Vitesse et attenuation des ondes élastiques soniques et ultrasoniques dans les roches sous pression de confinement.* Ph.D. dissertation, University of Paris.

Lucet, N., and Zinszner, B., 1992. Effects of heterogeneities and anisotropy on sonic and ultrasonic attenuation in rocks. *Geophys.,* 57, 1018–1026.

Marion, D., 1990. *Acoustical, Mechanical, and Transport Properties of Sediments and Granular Materials.* Ph.D. dissertation, Stanford University.

Marion, D., and Nur, A., 1991. Pore-filling material and its effect on velocity in rocks. *Geophys.,* 56, 225–230.

Marion, D., Nur, A., Yin, H., and Han, D., 1992. Compressional velocity and porosity in sand–clay mixtures. *Geophys.,* 57, 554–563.

Marle, C.M., 1981. *Multiphase Flow in Porous Media.* Gulf Publishing Company, Houston, 257 pp.

Marple, S.L., 1987. *Digital Spectral Analysis with Applications.* Prentice-Hall, Englwood Cliffs, New Jersey, 492 pp.

Mavko, G., 1980. Velocity and attenuation in partially molten rocks. *J. Geophys. Res.,* 85, 5173–5189.

Mavko, G., Chan, C., and Mukerji, T., 1995. Fluid substitution: Estimating changes in V_P without knowing V_S. *Geophys.,* 60, 1750–1755.

Mavko, G., and Jizba, D., 1991. Estimating grain-scale fluid effects on velocity dispersion in rocks. *Geophys.,* 56, 1940–1949.

Mavko, G.M., Kjartansson, E., and Winkler, K., 1979. Seismic wave attenuation in rocks. *Rev. Geophys.,* 17, 1155–1164.

Mavko, G., and Mukerji, T., 1995. Pore space compressibility and Gassmann's relation. *Geophys.,* 60, 1743–1749.

Mavko, G., Mukerji, T., and Godfrey, N., 1995. Predicting stress-induced velocity anisotropy in rocks. *Geophys.,* 60, 1081–1087.

Mavko, G., and Nolen-Hoeksema, R., 1994. Estimating seismic velocities in partially saturated rocks. *Geophys.,* 59, 252–258.

Mavko, G., and Nur, A., 1975. Melt squirt in the asthenosphere. *J. Geophys. Res.,* 80, 1444–1448.

Mavko, G., and Nur, A., 1978. The effect of nonelliptical cracks on the compressibility of rocks. *J. Geophys. Res.,* 83, 4459–4468.

Mavko, G., and Nur, A., 1997. The effect of a percolation threshold in the Kozeny–Carman relation. *Geophys.,* 62, 1480–1482.

Mazáč, O., Císlerová, M., Kelly, W.E., Landa, I., and Venhodová, D., 1990. Determination of hydraulic conductivities by surface geoelectrical methods, in *Geotechnical and Environmental Geophysics,* vol. II., S.H. Ward, ed. Society of Exploration Geophysicists, Tulsa, Oklahoma, 125–131.

McCoy, J.J., 1970. On the displacement field in an elastic medium with random variations in material properties, in *Recent Advances in Engineering Science,* A.C. Eringen, ed. Gordon and Breach, New York, Vol. 2, 235–254.

Meador, R.A., and Cox, P.T., 1975. Dielectric constant logging: A salinity independent estimation of formation water volume. *Soc. Pet. Eng.,* Paper 5504.

Mehta, C.H., 1983. Scattering theory of wave propagation in a two-phase medium. *Geophys.,* 48, 1359–1372.

Milholland, P., Manghnani, M.H., Schlanger, S.O., and Sutton, G.H., 1980. Geoacoustic modeling of deep-sea carbonate sediments. *J. Acoust. Soc. Am.,* 68, 1351–1360.

Mindlin, R.D., 1949. Compliance of elastic bodies in contact. *J. Appl. Mech.*, 16, 259–268.

Mukerji, T., Berryman, J.G., Mavko, G., and Berge, P.A., 1995. Differential effective medium modeling of rock elastic moduli with critical porosity constraints. *Geophys. Res. Lett.*, 22, 555–558.

Mukerji, T., and Mavko, G., 1994. Pore fluid effects on seismic velocity in anisotropic rocks. *Geophys.*, 59, 233–244.

Mukerji, T., Mavko, G., Mujica, D., and Lucet, N., 1995. Scale-dependent seismic velocity in heterogeneous media. *Geophys.*, 60, 1222–1233.

Müller, G., Roth, M., and Korn, M., 1992. Seismic-wave traveltimes in random media. *Geophys. J. Int.*, 110, 29–41.

Murphy, W.F., III, 1982. *Effects of Microstructure and Pore Fluids on the Acoustic Properties of Granular Sedimentary Materials*. Ph.D. dissertation, Stanford University.

Murphy, W.F., III, 1984. Acoustic measures of partial gas saturation in tight sandstones. *J. Geophys. Res.*, 89, 11549–11559.

Murphy, W.F., Schwartz, L.M., and Hornby, B., 1991. Interpretation physics of V_P and V_S in sedimentary rocks. Transactions *SPWLA 32nd Ann. Logging Symp.*, 1–24.

Murphy, W.F., III, Winkler, K.W., and Kleinberg, R.L., 1984. Contact microphysics and viscous relaxation in sandstones, in *Physics and Chemistry of Porous Media*, D.L. Johnson and P.N. Sen, eds. American Institute of Physics, New York, 176–190.

Nishizawa, O., 1982. Seismic velocity anisotropy in a medium containing oriented cracks – Transversely isotropic case. *J. Phys. Earth*, 30, 331–347.

Nolet, G., 1987. Seismic wave propagation and seismic tomography, in *Seismic Tomography*, G. Nolet, ed. D. Reidel Publication Co, Dordrecht, The Netherlands, 1–23.

Norris, A.N., 1985. A differential scheme for the effective moduli of composites. *Mechanics of Materials*, 4, 1–16.

Norris, A.N., Sheng, P., and Callegari, A.J., 1985. Effective-medium theories for two-phase dielectric media. *J. Appl. Phys.*, 57, 1990–1996.

Nur, A., 1971. Effects of stress on velocity anisotropy in rocks with cracks. *J. Geophys. Res.*, 76, 2022–2034.

Nur, A., Marion, D., and Yin, H., 1991. Wave velocities in sediments, in *Shear Waves in Marine Sediments*, J.M. Hovem, M.D., Richardson, and R.D. Stoll, eds. Kluwer Academic Publishers, Dordrecht, The Netherlands, 131–140.

Nur, A., Mavko, G., Dvorkin, J., and Gal, D., 1995. Critical porosity: The key to relating physical properties to porosity in rocks, in *Proc., 65th Ann. Int. Meeting. Soc. Expl. Geophys.*, 878.

Nur, A., and Simmons, G., 1969. Stress-induced velocity anisotropy in rocks: An experimental study. *J. Geophys. Res.*, 74, 6667.

O'Connell, R.J., and Budiansky, B., 1974. Seismic velocities in dry and saturated cracked solids. *J. Geophys. Res.*, 79, 4626–4627.

O'Connell, R., and Budiansky, B., 1977. Viscoelastic properties of fluid-saturated cracked solids. *J. Geophys. Res.*, 82, 5719–5735.

O'Doherty, R.F., and Anstey, N.A., 1971. Reflections on amplitudes. *Geophys. Prospecting*, 19, 430–458.

Olhoeft, G.R., 1979. Tables of room temperature electrical properties for selected rocks and minerals with dielectric permittivity statistics. *United States Geological Survey Open File Report 79-993*.

Osborn, J.A., 1945. Demagnetizing factors of the general ellipsoid. *Phys. Rev.*, 67, 351–357.

Paterson, M.S., and Weiss, L.E., 1961. Symmetry concepts in the structural analysis of deformed rocks. *Geolog. Soc. Am. Bull.*, 72, 841.

Pickett, G.R., 1963. Acoustic character logs and their applications in formation evaluation. *J. Pet. Technol.*, 15, 650–667.

Poupon, A., and Leveaux, J., 1971. Evaluation of water saturations in shaley formations. *Trans. Soc. Prof. Well Log Analysts, 12th Annual Logging Symposium*, Paper O.

Ransom, R.C., 1984. A contribution towards a better understanding of the modified Archie formation resistivity factor relationship. *The Log Analyst*, 25, 7–15.

Raymer, L.L., Hunt, E.R., and Gardner, J.S., 1980. An improved sonic transit time-to-porosity transform, *Trans. Soc. Prof. Well Log Analysts, 21st Annual Logging Symposium*, Paper P.

Resnick, J.R., Lerche, I., and Shuey, R.T., 1986. Reflection, transmission, and the generalized primary wave. *Geophys. J. Royal Astronom. Soc.*, 87, 349–377.

Reuss, A., 1929. Berechnung der Fliessgrenzen von Mischkristallen auf Grund der Plastizitätsbedingung für Einkristalle, *Zeitschrift für Angewandte Mathematik und Mechanik*, 9, 49–58.

Roth, M., Müller, G., and Sneider, R., 1993. Velocity shift in random media. *Geophys. J. Int.*, 115, 552–563.

Rüger, A., 1995. P-wave reflection coefficients for transversely isotropic media with vertical and horizontal axis of symmetry. *Expanded Abstracts, Soc. Expl. Geophys.*, 65th Annual International Meeting, 278–281.

Rüger, A., 1996. Variation of P-wave reflectivity with offset and azimuth in anisotropic media. *Expanded Abstracts, Soc. Expl. Geophys.*, 66th Annual International Meeting, 1810–1813.

Rumpf, H., and Gupte, A.R., 1971. Einflüsse der Porosität und Korngrössenverteilung im Widerstandsgesetz der Porenströmung. *Chem-Ing.-Tech.*, 43, 367–375.

Sayers, C.M., 1988a. Inversion of ultrasonic wave velocity measurements to obtain the microcrack orientation distribution function in rocks. *Ultrasonics*, 26, 73–77.

Sayers, C.M., 1988b. Stress-induced ultrasonic wave velocity anisotropy in fractured rock. *Ultrasonics*, 26, 311–317.

Sayers, C.M., 1995. Stress-dependent seismic anisotropy of shales, *Expanded Abstracts, Soc. Expl. Geophys.*, 65th Annual International Meeting, 902–905.

Sayers, C.M., Van Munster, J.G., and King, M.S., 1990. Stress-induced ultrasonic anisotropy in Berea sandstone, *Int. J. Rock Mech. Mining Sci. Geomech. Absts.*, 27, 429–436.

Schlichting, H., 1951. *Grenzschicht-Theorie*, G. Braun, Karlsruhe.

Schoenberg, M., and Protázio, J., 1992. 'Zoeppritz' rationalized and generalized to anisotropy. *J. Seismic Explor.*, 1, 125–144.

Segel, L.A., 1987. *Mathematics Applied to Continuum Mechanics*, Dover, Mineola, New York, 590 pp.

Sen, P.N., and Goode, P.A., 1988. Shaley sand conductivities at low and high salinities, *Trans. Soc. Prof. Well Log Analysts, 29th Annual Logging Symposium*, Paper F.

Sen, P.N., Scala, C., and Cohen, M.H., 1981. A self-similar model for sedimentary rocks with application to the dielectric constant of fused glass beads. *Geophys.*, 46, 781–795.

Shapiro, S.A., Hubral, P., and Zien, H., 1994. Frequency-dependent anisotropy of scalar waves in a multilayered medium. *J. Seismic Explor.*, 3, 37–52.

Shapiro, S.A., and Hubral, P., 1995. Frequency-dependent shear-wave splitting and velocity anisotropy due to elastic multilayering. *J. Seismic Explor.*, 4, 151–168.

Shapiro, S.A., and Zien, H., 1993. The O'Doherty–Anstey formula and localization of seismic waves. *Geophys.*, 58, 736–740.

Sharma, M.M., Garrouch, A., and Dunlap, H.F., 1991. Effects of wettability, pore geometry, and stress on electrical conduction in fluid-saturated rocks. *The Log Analyst*, 32, 511–526.

Sharma, M.M., and Tutuncu, A.N., 1994. Grain contact adhesion hysteresis: A mechanism for attenuation of seismic waves. *Geophys. Res. Lett.*, 21, 2323–2326.

Sharma, M.M., Tutuncu, A.N., and Podia, A.L., 1994. Grain contact adhesion hysteresis: A mechanism for attenuation of seismic waves in sedimentary granular media, *Extended Abstracts, Soc. Expl. Geophys.*, 64th Annual International Meeting, Los Angeles, 1077–1080.

Shen, L.C., 1985. Problems in dielectric-constant logging and possible routes to their solution. *The Log Analyst*, 26, 14–25.

Sheriff, R.E., 1991. *Encyclopedic Dictionary of Exploration Geophysics*, 3rd ed. Society of Exploration Geophysics, Tulsa, Oklahoma, 376 pp.

Sherman, M.M., 1986. The calculation of porosity from dielectric constant measurements: A study using laboratory data, *The Log Analyst*, Jan.–Feb., 15–24.

Sherman, M.M., 1988. A model for the frequency dependence of the dielectric permittivity of reservoir rocks, *The Log Analyst*, Sept.–Oct., 358–369.

Shuey, R.T., 1985. A simplification of the Zoeppritz equations. *Geophys.*, 50, 609–614.

Simandoux, P., 1963. Dielectric measurements on porous media: Application to the measurement of water saturations: Study of the behavior of argillaceous formations. *Revue de l'Institut Français du Petrole*, 18, supplementary issue, 193–215.

Smith, G.C., and Gidlow, P.M., 1987. Weighted stacking for rock property estimation and detection of gas. *Geophysic Prospecting*, 35, 993–1014.

Spratt, R.S., Goins, N.R., and Fitch, T.J., 1993. Pseudo-shear – The analysis of AVO, in *Offset Dependent Reflectivity – Theory and Practice of AVO Analysis*, J.P. Castagna and M. Backus, eds. *Invest. Geophys.*, No. 8, Society of Exploration Geophysicists, Tulsa, Oklahoma, 37–56.

Stoll, R.D., 1974. Acoustic waves in saturated sediments, in *Physics of Sound in Marine Sediments*, L.D. Hampton, ed. Plenum, New York, 19–39.

Stoll, R.D., 1977. Acoustic waves in ocean sediments. *Geophys.*, 42, 715–725.

Stoll, R.D., 1989. *Sediment Acoustics*, Springer Verlag, Berlin, 154.

Stoll, R.D., and Bryan, G.M., 1970. Wave attenuation in saturated sediments. *J. Acoust. Soc. Am.*, 47, 1440–1447.

Stoner, E.C., 1945. The demagnetizing factors for ellipsoids. *Philos. Mag.*, 36, 803–821.

Strandenes, S., 1991. *Rock Physics Analysis of the Brent Group Reservoir in the Oseberg Field*. Stanford Rockphysics and Borehole Geophysics Project, special volume 25 pp.

Thomsen, L., 1986. Weak elastic anisotropy. *Geophys.*, 51, 1954–1966.

Thomsen, L., 1993. Weak anisotropic reflections, in *Offset Dependent Reflectivity – Theory and Practice of AVO Analysis*, J.P. Castagna and M. Backus, eds. *Invest. Geophys.*, No. 8, Society of Exploration Geophysicists, Tulsa, Oklahoma, 103–111.

Timoshenko, S.P., and Goodier, J.N., 1934. *Theory of Elasticity*, McGraw-Hill, New York, 567 pp.

Topp, G.C., Davis, J.L., and Annan, A.P., 1980. Electromagnetic determination of soil water content: Measurements in coaxial transmission lines. *Water Resource Res.*, 16, 574–582.

Tosaya, C., and Nur, A., 1982. Effects of diagenesis and clays on compressional velocities in rocks. *Geophys. Res. Lett.*, 9, 5–8.

Tsvankin, I., 1997. Anisotropic parameters and P-wave velocity for orthorhombic media, *Geophys.*, 62, 1292–1309.

Tutuncu, A.N., 1992. *Velocity Dispersion and Attenuation of Acoustic Waves in Granular Sedimentary Media.* Ph.D. dissertation, University of Texas, Austin.

Tutuncu, A.N., and Sharma, M.M., 1992. The influence of grain contact stiffness and frame moduli in sedimentary rocks. *Geophys.*, 57, 1571–1582.

Urmos, J., and Wilkens, R.H., 1993. In situ velocities in pelagic carbonates: New insights from ocean drilling program leg 130, Ontong Java. *J. Geophys. Res.*, 98, No. B5, 7903–7920.

Walls, J., Nur, A., and Dvorkin, J., 1991. A slug test method in reservoirs with pressure sensitive permeability. *Proc. 1991 Coalbed Methane Symp.*, University of Alabama, Tuscaloosa, May 13–16, 97–105.

Walsh, J.B., 1965. The effect of cracks on the compressibility of rock. *J. Geophys. Res.*, 70, 381–389.

Walsh, J.B., 1969. A new analysis of attenuation in partially melted rock. *J. Geophys. Res.*, 74, 4333.

Walsh, J.B., Brace, W.F., and England, A.W., 1965. Effect of porosity on compressibility of glass, *J. Am. Ceramic Soc.*, 48, 605–608.

Walton, K., 1987. The effective elastic moduli of a random packing of spheres. *J. Mech. Phys. Solids*, 35, 213–226.

Wang, Z., and Nur, A., 1992. Seismic and acoustic velocities in reservoir rocks, Vol. 2, Theoretical and model studies. *Soc. Expl. Geophys.*, Geophysics Reprint Series, 457 pp.

Ward, S.H., and Hohmann, G.W., 1987. Electromagnetic theory for geophysical applications, in *Electromagnetic Methods in Applied Geophysics*, Vol. I, Theory, M.N. Nabhigian, ed. Society of Exploration Geophysicists, Tulsa, Oklahoma, 131–311.

Watt, J.P., Davies, G.F., and O'Connell, R.J., 1976. The elastic properties of composite materials. *Rev. Geophys. Space Phys.*, 14, 541–563.

Waxman, M.H., and Smits, L.J.M., 1968. Electrical conductivities in oil-bearing shaley sands. *Soc. Pet. Eng. J.*, 8, 107–122.

White, J.E., 1975. Computed seismic speeds and attenuation in rocks with partial gas saturation. *Geophys.*, 40, 224–232.

White, J.E., 1983. *Underground Sound: Application of Seismic Waves.* Elsevier, New York, 253 pp.

White, J.E., 1986. Biot–Gardner theory of extensional waves in porous rods. *Geophys.*, 51, 742–745.

Wiggins, R., Kenny, G.S., and McClure, C.D., 1983. A method for determining and displaying the shear-velocity reflectivities of a geologic formation. European Patent Application 0113944.

Williams, D.M., 1990. The acoustic log hydrocarbon indicator. *Soc. Prof. Well Log Analysts, 31st Ann. Logging Symp.*, Paper W.

Winkler, K.W., 1983. Contact stiffness in granular porous materials: Comparison between theory and experiment. *Geophys. Res. Lett.*, 10, 1073–1076.

Winkler, K.W., 1985. Dispersion analysis of velocity and attenuation in Berea sandstone. *J. Geophys. Res.*, 90, 6793–6800.

Winkler, K., 1986. Estimates of velocity dispersion between seismic and ultrasonic frequencies, *Geophys.*, 51, 183–189.

Winkler, K.W., and Nur, A., 1979. Pore fluids and seismic attenuation in rocks. *Geophys. Res. Lett.*, 6, 1–4.

Wood, A.W., 1955. *A Textbook of Sound*, The MacMillan Co., New York, 360 pp.

Worthington, P.F., 1985. Evolution of shaley sand concepts in reservoir evaluation. *The Log Analyst*, 26, 23–40.

Wu, T.T., 1966. The effect of inclusion shape on the elastic moduli of a two-phase material. *Int. J. Solids Structures*, 2, 1–8.

Wyllie, M.R.J., Gardner, G.H.F., and Gregory, A.R., 1963. Studies of elastic wave attenuation in porous media. *Geophys.*, 27, 569–589.

Wyllie, M.R.J., and Gregory, A.R., 1953. Formation factors of unconsolidated porous media: Influence of particle shape and effect of cementation. *Trans. Am. Inst. Mech. Eng.*, 198, 103–110.

Wyllie, M.R.J., Gregory, A.R., and Gardner, L.W., 1956. Elastic wave velocities in heterogeneous and porous media. *Geophys.*, 21, 41–70.

Wyllie, M.R.J., Gregory, A.R., and Gardner, G.H.F., 1958. An experimental investigation of factors affecting elastic wave velocities in porous media. *Geophys.*, 23, 459–493.

Xu, S., and White, R.E., 1994. A physical model for shear-wave velocity prediction. *Expanded Abstracts, 56th Eur. Assoc. Expl. Geoscientists Meet. Tech. Exhib.*, Vienna, 117.

Xu, S., and White, R.E., 1995. A new velocity model for clay–sand mixtures. *Geophys. Prospecting*, 43, 91–118.

Yale, D.P., and Jameison, W.H., Jr., 1994. Static and dynamic rock mechanical properties in the Hugoton and Panoma fields. Kansas Society of Petroleum Engineers, Paper 27939. (Presented at the Society of Petroleum Engineers Mid-Continent Gas Symposium, Amarillo, Texas, May 1994.)

Yamakawa, N., 1962. Scattering and attenuation of elastic waves. *Geophys. Mag.* (Tokyo), 31, 63–103.

Yin, H., 1992. *Acoustic Velocity and Attenuation of Rocks: Isotropy, Intrinsic Anisotropy, and Stress-Induced Anisotropy*. Ph.D. dissertation, Stanford University.

Young, H.D., 1962. *Statistical Treatment of Experimental Data*, McGraw-Hill, New York, 172 pp.

Zamora, M., and Poirier, J.P., 1990. Experimental study of acoustic anisotropy and birefringence in dry and saturated Fontainebleau sandstone. *Geophys.*, 55, 1455–1465.

Zener, C., 1948. *Elasticity and Anelasticity of Metals*, University of Chicago Press, Chicago, 170 pp.

Zimmerman, R.W., 1984. The elastic moduli of a solid with spherical pores: New self-consistent method. *Int. J. Rock Mech., Mining Sci. Geomech. Abst.*, 21, 339–343.

Zimmerman, R.W., 1986. Compressibility of two-dimensional cavities of various shapes. *J. Appl. Mech. Trans. Am. Soc. Mech. Engs.*, 53, 500–504.

Zimmerman, R.W., 1991. *Compressibility of Sandstones*, Elsevier, New York, 173 pp.

Zoeppritz, K., 1919. Erdbebenwellen VIIIB, On the reflection and propagation of seismic waves. *Göttinger Nachrichten*, I, 66–84.

INDEX

Printed in the United States
21214LVS00004B/105